Urban Sustainability

The Urban Sustainability Book Series is a valuable resource for sustainability and urban-related education and research. It offers an inter-disciplinary platform covering all four areas of practice, policy, education, research, and their nexus. The publications in this series are related to critical areas of sustainability, urban studies, planning, and urban geography.

This book series aims to put together cutting-edge research findings linked to the overarching field of urban sustainability. The scope and nature of the topic are broad and interdisciplinary and bring together various associated disciplines from sustainable development, environmental sciences, urbanism, etc. With many advanced research findings in the field, there is a need to put together various discussions and contributions on specific sustainability fields, covering a good range of topics on sustainable development, sustainable urbanism, and urban sustainability. Despite the broad range of issues, we note the importance of practical and policy-oriented directions, extending the literature and directions and pathways towards achieving urban sustainability.

The series will appeal to urbanists, geographers, planners, engineers, architects, governmental authorities, policymakers, researchers of all levels, and to all of those interested in a wide-ranging overview of urban sustainability and its associated fields. The series includes monographs and edited volumes, covering a range of topics under the urban sustainability topic, which can also be used for teaching materials.

Maycon Sedrez · Gregory Pitts
Editors

Artificial Intelligence-Aided Design for Sustainability

Springer

Editors
Maycon Sedrez
Architecture and Built Environment
Deakin University
Geelong, Australia

Gregory Pitts
Architecture and Built Environment
Deakin University
Geelong, Australia

ISSN 2731-6483 ISSN 2731-6491 (electronic)
Urban Sustainability
ISBN 978-981-95-1348-2 ISBN 978-981-95-1349-9 (eBook)
https://doi.org/10.1007/978-981-95-1349-9

This Springer imprint is published by the registered company Springer Nature Singapore Pte Ltd.
The registered company address is: 152 Beach Road, #21-01/04 Gateway East, Singapore 189721, Singapore

A pavilion in Ipanema, Rio de Janeiro. Generated with Midjourney v.5

We both dedicate this book to our families and thank their constant encouragement. None of this work is possible without their ever-present bedrock of support. We are also grateful to Deakin University and the School of Architecture and Built Environment for providing a space where our research has been actively supported and encouraged to grow into teaching, practice, exhibitions, and now this book.

Foreword

Latent Ecologies—The Soft Violence of the Prompt

The tension between authorship and reproduction has haunted every technological threshold in the arts. The printing press unraveled the primacy of the scribe. Photography dislodged the painter's claim to realistic depiction. Mechanical reproduction, as Benjamin argues, does not simply copy—it mutates the role of the work itself. It reassigns agency. In architecture, this tension was never resolved. It became sedimented, obfuscated, umbrated. Buried beneath decades of discourse around tectonics, typology, and formal invention lies an unspoken anxiety: that the tools we wield are shaping not just our drawings, but our capacity to conceptualize space itself. If aura is the symptom of ritual—ritual as spatial practice—then its erosion under mechanical regimes invites us to rethink the very ontology of design. What happens when the prompt, rather than the pen, becomes the vector of thought? What becomes of atmosphere when it is no longer sensed, but sampled? These are not rhetorical questions. They are computational problems masquerading as theoretical ones. They define the surface tension of architectural practice today. The preface to this volume therefore refuses to be a ceremonial opening, it is rather a forensic diagnostic—a diagnostic probe examining the present condition of design. Instead of asking what AI can do, it explores what it already is: a generator of latent ecologies, a script of soft violences, a rewriter of rituals we once believed immutable.

In *The Work of Art in the Age of Mechanical Reproduction*, Walter Benjamin isolates the fracture line between material and aura. Mechanical duplication unmoors the ritual from the object. The image is untethered, liberated—doomed to circulate freely, without tether, without threshold. The machine shifts the center of gravity from origin to iteration. In this moment, the work ceases to be site-specific. It becomes diffused. A diagram of cultural drift. Architecture never fully metabolized that fault line in the geology of artistic production. It still nurses to this day the aura of authorship, the primacy of the hand, the sketch—the sole genius. In the nooks and crannies of the discipline however, one encounters another logic: the machinic logic of endless modulation. Let me explain: The often cited but rarely absorbed Jacquard loom, for example, introduced a mechanical grammar of mass and voids—a form of programmatic tapestry. It converted desire into data before anyone had a name for that inversion. In its matrix of punched cards, one finds the ghost-code of Neuralism. The loom is no longer textile. It performs discreet commands. It weaves space, not cloth. Today, latent diffusion networks, prompt arrays, stochastic fields—all descend from this proto-code. The difference lies in scale and indeterminacy. The cards have been replaced with models trained on massive amounts of data. They do not repeat motifs. They anticipate atmospheres. This shift from the specific to the statistical, from authored object to generative continuum, marks not just a procedural update but a philosophical realignment. Where once we could locate a project in a signature or a line weight, we now confront spatial outcomes derived from patterns too vast to trace, too recursive to delineate. The loom's pattern was finite. The diffusion model's field is practically infinite, constrained only by the limitations of its training and the biases of its inputs. The prompt, then, becomes a kind of incantation—an invocation not of will, but of possibility. Its outputs shimmer with the uncanny familiarity of our own aesthetics, warped through the probabilistic haze of machine learning. Architecture, once bound to the specificity of place, begins to drift, trained instead on cities it has never seen, climates it cannot feel, rituals it no longer recognizes. The discipline finds itself not at the edge of a new toolset, but within the folds of a new logic—one that weaves not cloth, but context.

In this post-aura field, vibe coding emerges as the new diagrammatic interface. A semiotic haze of instructions, felt before they are parsed. This is not coding. It is steering. A tactile drift across latent vectors. Architects script nothing. They suggest, adjust, provoke. Geometry falls away. Resonance takes shape. The input is not logic—it is climate, tone, affect. The system responds in haze, not hierarchy. Ambiguity becomes the medium.

A field of proposals. A constellation of almosts and maybes. The epistemology shifts from control to calibration. In this book, that epistemic shift is rendered visible. Sedrez and Pitts speak of AI-aided design not as an efficiency upgrade, but as a reframing of authorship. Design is recast as ecology. A swarm of decisions, some human, some synthetic. The building is no longer drawn. It is trained. Generated. Tuned. In the work of Fakhr et al. in Chapter Enhancing Building Performance Through AI: Techniques, Models, and Practical Challenges and Figliola and Barberio in Chapter Integrating Generative AI into Climate-Aided Design: Methods, Tools, and Professional Implications, the climatic logic refuses to be an afterthought. Instead it forms part of a generative kernel. CADE and GAI compose a data driven alloy: text-to-image prompts saturated with meteorological intent. No longer a building plus an HVAC system. A building as HVAC. Form and performance fused in the latent model. Chornobai et al. in Chapter Hybrid Design Process: A Sustainable Approach Based on Analog Tool and Generative Artificial Intelligence hybrid protocols—melding diffusion algorithms with bioclimatic analog tools—make this synthesis tactile. The images are not illustrations. They are thermodynamic hypotheses. This reframing of design as an act of climatic inference reshapes what we understand as spatial intelligence. The prompt no longer operates in a vacuum. It exists within a web of environmental forces, ethical vectors, and cultural semantics. *Vibe coding*, then, is not simply an aesthetic suggestion. It becomes a form of affective computation—attuned to atmospheres that do not yet exist, but that are latent within the field of generative potential. Each design iteration becomes a performance of alignment: between climate data and cultural index, between ecological memory and computational probability. AI models do not design. They perform inference at scale. The architect's role is to intervene in this performance, not with correction, but with critical reflection. To ask: what else can this system sense? What values are embedded in its logic? What omissions haunt its outputs?

In this emerging ecology of design, architecture pedagogy is experiencing a subtle revolution. A tectonic shift in the studio's didactic grammar. In the design laboratories of Napoleão and Silva in Chapter Design Practice with Generative Artificial Intelligence for Sustainability: An Experimental Study with Undergraduate Designers and Yazıcı in Chapter The Effect of AI-Aided Design Processes on Cognitive Load, Creativity, and Metacognitive Awareness, students are not asked to produce objects. They are asked to train systems. The studio becomes an epistemic gymnasium. A site for model calibration, dataset curation, prompt choreography. The orthographic plan yields to the speculative grid. Cognitive load diminishes, but not through simplification—through redistribution. The human mind ceases to be the only territory of decision-making. It becomes the editor of machine reveries. Yazıcı's workshop captures this moment precisely. The students don't learn code. They learn tuning. They don't debug—they iterate. The process becomes one of conversational friction between intent and artifact. Vibe coding becomes the lingua franca. More than a syntax of commands, but a lexicon of intuitions. The studio's evaluative metric mutates. Instead of asking the mundane question "Did it meet the brief?" the interrogation moves towards the more pressing question: "What system does it produce?" Authorship is unbundled. The machine proposes; the student curates. The educator orchestrates that dance. Not a lecturer, a conductor of epistemic noise. This volume's pedagogical sections underscore that shift. The student becomes a latent navigator. The brief becomes a set of vectors. Learning is tuned through resonance.

The implications for pedagogy stretch beyond methodology into ethics. What does it mean to teach design when models now offer thousands of possibilities within seconds? It is of course a challenge for teaching. No longer a container for methods. It seeps into ethics, filters through ontology. What is design education when the prompt outpaces the pen? When models unroll thousands of spatial artifacts before the critic blinks? Judgment must be trained under different conditions—no longer by the line, but by latency. What emerges is critique, situated and recursive. Students do not draft; they interrogate. They do not solve; they interrupt. Refusal becomes method. The default is not ignored—it is dismantled. The dataset is not accepted—it is rewritten. Aesthetic origins become suspect, model logic a site of inquiry. Teaching does not instruct. It configures a threshold. At that edge, intuition is parsed through syntax, speculation is bound to code. The classroom no longer rests on a ground plane. It flickers inside computational space—mutable, distributed, recursive. This is not instruction. It is interface. The pedagogical act becomes an act of tuning:

between systems and subjectivities, between inference and intention. Architecture is no longer taught. It is rehearsed, in feedback loops of curated speculation.

The specter of anxiety persists. A subtle hum beneath the enthusiasm. Calixto et al. in Chapter Mapping Emerging Machine Learning Methodologies in Architectural Design: A Pathway for Present Integration and Future Research identify an imbalance: early-stage design brimming with AI energy, while post-occupancy performance remains undercoded. Architecture cannot remain entranced by genesis. Buildings endure. They decay, adapt, fail, persist. Instead of AI serving the role of an ideation machine, it should (has to!) accompany the full metabolic arc. Pak and Kocaturk's in Chapter AI-Driven Circular Architectural Design work on circularity, one encounters a possible horizon. A feedback architecture. Buildings trained not just to emerge but to disassemble. To carry in their data DNA the knowledge of their own death. The architectural object becomes a temporal script. Material trajectories encoded into latent models. The machine doesn't just hallucinate forms—it forecasts decay, anticipates entropy. In this context, sustainability rises above the role given in a checklist. It becomes a performance across time. A building becomes a versioned entity—trained, revised, fed with new telemetry. The ghost of Walter Benjamin returns. But instead of mourning the loss of aura, it whispers: Maybe it was always a fantasy? The image as singular, the object as authored. AI unmasks this fiction. It multiplies. It distributes. It reveals that architecture has always been a negotiation. Between control and contingency. Between desire and system. Between aura and array. If there is a lesson in this volume, it is not that AI designs. It is that architecture has become a conversation, staged between datasets and intuitions, conducted in a language we are still learning to speak.

The real challenge, then, lies in listening. Listening to the dataset's silence, to the ghosted histories absent in model weights, to the cultural assumptions baked into latent space. Architecture, if it is to remain an ethical practice, must engage this silence with intention. It must widen the archive. It must diversify the prompt. It must operate not just as generator but as critic. The future of architectural intelligence is not artificial. It is hybrid. It is discursive. It is contested. It asks not what can be generated, but what should be. This volume does not resolve that tension. It stages it. It opens a field in which the question of design becomes inseparable from the systems that now shape it. It is an invitation—not to master the machine, but to collaborate with it critically, ethically, speculatively.

New York, USA Dr. Matias del Campo

Acknowledgment

We extend our gratitude to Prof. Ali Cheshmehzangi, editor-in-chief of the Urban Sustainability Series, for his support on this project. We also thank all the authors and contributors for their hard work and dedication in shaping this book, as well as Matias del Campo for his valued contribution to the foreword. We thank Deakin University and the School of Architecture and Built Environment for the continuous support and internal grants that contributed to our research. We look forward to growing the AI-Aided Design community and meeting everyone in person in the future.

About This Book

Artificial Intelligence (AI) has been diffused into architectural applications for several decades. However, in 2022, a new set of fast developing AI generative tools became widely accessible to designers. 'AI-Aided Design for Sustainability' aims to disseminate new knowledge on the fast-emerging applications of AI in the fields of architecture, urban, heritage, and interior design.

The book encompasses innovations that address sustainability-whether economic, social, environmental, or spatial-through new processes, cultures, systems, and methods that expand beyond current aesthetics-focused approaches. These comprehensive discussions set a new agenda for design by full integration of AI for communication, visualization, calculation, generation of ideas, optimization, and more. This book focuses on sustainable perspectives, investigating the resilience of the architectural profession, the alignment of design methodologies with the United Nations' Sustainable Development Goals (SDGs), and the innovative application of AI in education, as reflected in the UNESCO 2019 Beijing Consensus.

Chapters examine a variety of approaches through which AI contributes to emerging design workflows. These include enhancing design development, supporting complex data-driven analysis, and exploring the extent to which AI may augment or even replace certain architectural tasks. Evidence of hybrid analogue–digital interfaces is also discussed, showcasing how architects are adapting their practices in response to technological shifts.

This integration is giving rise to new vocabularies and conceptual frameworks within architectural design, many of which are absorbed from computer science. As a result, the profession is again being impacted by computational tools.

The book is divided into two main sections:

Theoretical Foundations and Educational Perspectives: Literature and historical reviews, frameworks, methods, and educational perspectives, including pedagogical approaches, case studies, and AI in architectural education.

Practical Applications and Technical Innovations: Successful case study examples, design practices and research, and implications for sustainability, criticism, challenges of AI in design, technological advances, and AI models.

This structure helps to contextualize AI-Aided Design across various levels of application and defines pathways for transition to new methods. The book scope encompasses three main themes:

AI as a tool for sustainable design, addressing environmental, social, and economic resilience in line with the UN SDGs.
The evolving role of the architect, especially how AI transforms workflows, decision-making, and creative agency.
Innovation in education, highlighting how AI is shaping new pedagogies and design thinking, supported by frameworks like the UNESCO 2019 Beijing Consensus.

The book includes studies from Architecture and Built Environment scholars performing exploratory research on the emergent field of AI and Design. It offers a comprehensive understanding of the transformative power of AI in design for sustainability and positions AI as an ally in the design of more sustainable buildings while providing insight into how this transformation is reshaping the architect's role.

Maycon Sedrez
Gregory Pitts

Contents

Editors and Contributors

About the Editors

Maycon Sedrez is an architect and urban designer, artist, and educator with a background in computational design, digital fabrication and sustainable urbanism. His research areas include complexity and architecture, parametric design, digital fabrication, technologies, and urban design. He obtained his Ph.D. Diploma with excellence in Architecture, Technology and City from the University of Campinas in 2016. He contributed to the Institute for Sustainable Urbanism (TU Braunschweig—Germany) as a postdoc researcher in the field of urban analytics. Later in 2019, he became part of the University of Nottingham Ningbo China's team, leading the Digital Design Lab, and acting as a course director and the deputy head of department. Since 2023, he is a lecturer in Architecture and deputy director of MindLab at Deakin University in Australia. He co-edited the book "Contemporary Architecture and Automation" (2018) with Prof. Gabriela Celani, which presents a detailed overview of digital fabrication in Brazil. Additionally, he co-edited the books "Resilience vs Pandemics: Innovations in Cities and Neighbourhoods, and Public Spaces and Buildings" (2023 and 2024), and "Rethinking Stormwater Management through Sustainable Urban Design" (2024) with Prof. Ali Cheshmehzangi, showcasing his commitment to advancing knowledge in the field of sustainable urbanism. His engagement with art and illustration has led to publications in Brazil and Germany, with artworks exhibited in Brazil, China, and most recently in Australia. His artistic exploration informs his architectural design studios, creating a unique blend of research and creativity.

Gregory Pitts is an architect and academic who has worked on a range of projects across Australia, Asia and the Middle East. In this time, Greg has also conducted several collaborative research projects with Deakin University and has completed a Ph.D. on the use of Parametric Modelling in Architectural practice. Greg's research is formed around the idea that we need to re-establish good collaborative relationships between academia and industry to drive cutting edge, relevant research programs while positively contributing to the everyday working methods of Architectural practice. He was also lead designer for Deakin Universities winning design for the Model T Ford for the 21st century run by the Ford Motor Company. He was subsequently awarded the Vice Chancellors award for Outstanding contribution to collaboration through Innovation. Greg is also the Co-founder of Urban Creative—an award-winning architectural practice with a passion for both small and large-scale urban infill design. The designs stem from the belief in the ability for great design to reshape the way we interact and experience our urban environments. Greg has a strong focus on what can be achieved in the design of our built environment using innovative design technology and has a thorough experience in CAD management, 3D visualisation and design development. Through this work, Greg pursues an architecture driven by performance, spatial and organisational clarity that delights, inspires, and surprises.

Since 2023, Maycon and Greg have been leading the research project Pictures from Words, exploring emerging methods and applications of AI. They curated AI exhibitions and workshops, including 'AI-Aided Design' exhibition at Geelong Design Week in November 2024, followed by 'Synthetic Spaces' in June 2025 at the Bates Smart Gallery in Melbourne.

Contributors

Cem Ataman Urbanism and Urban Architecture, Eindhoven University of Technology, Eindhoven, Netherlands

Maurizio Barberio Politecnico di Bari, Bari, Italy

Victor Calixto Future Building Initiative, Monash University, Melbourne, Australia

Liangyu Chen Department of Architecture, Faculty of Engineering, University of Strathclyde, Glasgow, UK

Zhen Chen Department of Architecture, Faculty of Engineering, University of Strathclyde, Glasgow, UK

Sara Regiane Chornobai Program of Research and Postgraduate Studies in Architecture (PROPAR), Postgraduate Program in Design (PGDesign), School of Architecture, Federal University of Rio Grande Do Sul, Porto Alegre, Brazil

Feng Dong Department of Computer and Information Sciences, Faculty of Science, University of Strathclyde, Glasgow, UK

Bahereh Vojdani Fakhr University of Maryland, College Park, MD, USA

Angelo Figliola Sapienza Università di Roma, Rome, Italy

Camilo Cruz Gambardella Future Building Initiative, Monash University, Melbourne, Australia;
SensiLab, Monash University, Melbourne, Australia

Belinda Humphris Deakin University, Geelong, Australia

Derya Karadağ Işık University, Istanbul, Turkey

Tuba Kocaturk The Design for Circular Cities Master's Programme, School of Architecture and Built Environment, Deakin University, Geelong, Australia

Andressa Carmo Pena Martinez University of Maryland, College Park, MD, USA

Duncan Maxwell Future Building Initiative, Monash University, Melbourne, Australia

Eduardo Napoleão University of Vale Do Itajaí, Itajaí, Brazil

Burak Pak The Design for Circular Cities Master's Programme, School of Architecture and Built Environment, Deakin University, Geelong, Australia

Angelica Paiva Ponzio Program of Research and Postgraduate Studies in Architecture (PROPAR), Postgraduate Program in Design (PGDesign), School of Architecture, Federal University of Rio Grande Do Sul, Porto Alegre, Brazil

Simon Perrault Information Systems Technology and Design, Singapore University of Technology and Design, Singapore, Singapore

Gregory Pitts School of Architecture and Built Environment, Deakin University, Geelong, Australia

Maycon Sedrez School of Architecture and Built Environment, Deakin University, Geelong, Australia

Alice Demaria Silva University of Vale Do Itajaí, Itajaí, Brazil

Wanchao Su SensiLab, Monash University, Melbourne, Australia

Bige Tunçer Information Systems in the Built Environment, Eindhoven University of Technology, Eindhoven, Netherlands

Faruk Can Ünal Yeditepe University, Istanbul, Turkey

Fabiano de Vargas Scherer Program of Research and Postgraduate Studies in Architecture (PROPAR), Postgraduate Program in Design (PGDesign), School of Architecture, Federal University of Rio Grande Do Sul, Porto Alegre, Brazil

Deok-Oh Woo University of Maryland, College Park, MD, USA

Gizem Yazıcı Department of Architecture, Izmir Institute of Technology, Izmir University of Economics, Izmir, Turkey

Veli Mustafa Yönder Kütahya Dumlupınar University, Kütahya, Türkiye

Darcy Zelenko Future Building Initiative, Monash University, Melbourne, Australia

The Emerging AI-Aided Design Field and Sustainability Links

Maycon Sedrez and Gregory Pitts

Abstract Since the late 1950s, architecture has undergone a critical transformation enabled by the use of computers to aid the design process. Concepts such as Computer-Aided Design (CAD) and Computer-Aided Architectural Design (CAAD) emerged as part of the theoretical knowledge developed by architecture practitioners and researchers. Initial CAAD applications focused on productivity and efficiency of architectural draft, but architects realized that there was a tremendous potential for creative support or even human-machine collaborative tasks. Today, design is aided by a wide range of computer tools that perform from simple tasks (draw a line) to sophisticated calculations (simulate thermal comfort). Some of these applications incorporate Artificial Intelligence (AI) algorithms helping to enhance design sustainability when analysing environmental performance, for instance. As such, AI-Aided Design (AIAD) refers to emergent computational design processes that employ AI as design generator or assistant tool. Starting from an analogy of movie archetypes, we aim at providing an explanation of AIAD and correlate AI and Sustainability as a broad concept. This chapter further presents the United Nations considerations about AI to support the implementation of Sustainable Development Goals, the implications of the Beijing Consensus developed by UNESCO (*Planning Education in the AI era: Leap the leap*. United Nations Education, Scientific and Cultural Organization, Paris, 2019) to architecture education and the four themes explored in this book.

Keywords AI · Design · Architecture · Sustainability

M. Sedrez (✉) · G. Pitts
School of Architecture and Built Environment, Deakin University, Geelong, VIC, Australia
e-mail: m.sedrez@deakin.edu.au

G. Pitts
e-mail: gregory.pitts@deakin.edu.au

M. Sedrez and G. Pitts (eds.), *Artificial Intelligence-Aided Design for Sustainability*, Urban Sustainability, https://doi.org/10.1007/978-981-95-1349-9_1

1 Creating Intelligence

For centuries, humanity has been captivated by the possibility of creating intelligent life—an ambition deeply embedded in myth, literature, and, more recently, technological discourse. These imagined entities, often portrayed in fiction as obedient servants or humanoid replacements, reflect our aspirations to transcend human limitations—particularly in logic, complex problem-solving, and emotional neutrality. Yet, beneath these narratives of efficiency and detachment lies a more nuanced exploration: the desire for emotional connection, and the confrontation with what separates, or perhaps unites, the artificial and the human. The concept of 'artificial life' challenges foundational understandings of personhood, consciousness, and morality. As we approach increasingly sophisticated forms of Artificial Intelligence (AI) and Machine Learning (ML), we are compelled to ask what it truly means to be intelligent, or even alive. These technologies, while still far from attaining sentience or full human equivalency, represent profound shifts in the boundaries of cognition. AI may offer pre-human or non-human forms of reasoning, distinct from but not necessarily lesser than our own [2].

2 Perceptions of AI in Pop-Culture

Giving (or returning) life to a human form has been portrayed in movies such as Frankenstein (1931), Weird Science (1985), Robocop (1987), Edward Scissorhands (1990), and more recently Poor Things (2023). Some of these movies start with a human body, or combined parts of a body, that is brought to life through mythical, magical or scientific experiments. These human forms gain enhanced power(s) at their creation, owning individual personality and desires, and living through psychological hardships and trauma, which eventually ends up in disastrous or destructive scenarios. In this context, an organic intelligence can become sentient with a human body and brain, being self-aware and highly independent, but showing conflicts with their own existence and society moral standards. Because of the ethical issues of such experiments in real life, cloning or reviving a human life is prohibited by the United Nations since it is "*incompatible with human dignity and the protection of human life*" [3], a controversial proposal approved with 84 votes in favour, 34 against and 37 abstentions. However, there is progress on extending the human brain capacity through neural-computer implants, which will eventually deliver a hybrid intelligence [4]. In this scenario, a machine will enhance our organic brain functionalities, and it is expected to be 'under our control'. So far, AI extends our brain functionality and capability through computer programs and interfaces. Figure 1 illustrates a future machine enhancing human brain capacity.

Several films explore the depiction of humanoids—androids or human-like machines—emerging from advances in robotics and artificial intelligence. These

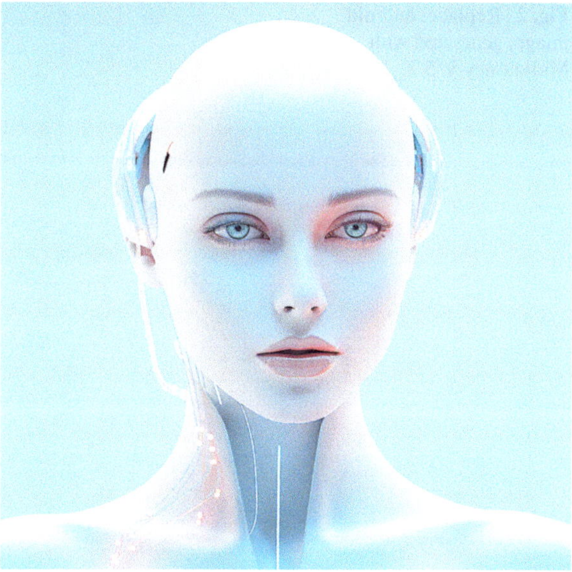

Fig. 1 Enhance: human–machine hybrid, image generated with Midjourney v. 5.2

entities are frequently introduced under the premise of service, assistance, or protection; however, narrative trajectories often involve the android attaining self-awareness and subsequently rejecting its original programming or pre-defined operational constraints. Such developments place them, and humans interacting with them, into moral conundrums related to: feelings or existence, e.g., Blade Runner (1982), Bicentennial Man (1999), A.I. (2001) and Her (2013); independence, e.g., Ex Machina (2015) and M3gan (2023); and control, e.g., The Terminator (1984), and Matrix (1999). In these movies, anthropomorphic or zoomorphic machines are created to replace humans in tasks and activities, including, companion and defence. These machines can have autonomous thinking patterns and become independent, which as consequence ends up in a catastrophic world, or self-aware humanoids that possess undeveloped human psychological traits such as psyche, empathy, or remorse. As suggested by, Bostrom and Yudkowsky [5], "*AI systems have no moral status*", which conflicts with their performance in society. The independence of a powerful AI entity is one of top reasons for fear in our society [6]. In such unsustainable scenarios, a machine acts independently upon and can potentially perform tasks creatively, forming a synthetic intelligence as we attempt to illustrate in Fig. 2. Typically, in architecture, AI programs tend to form a 'black-box' system where designers do not know how the machine is resolving the problem or the process is resolved through unknown databases and unrecognizable methods, which can create conflicts or be supported by biased data.

On a practical perspective, 'intelligent' machines have enhanced human activities for decades, such as architects' exploration of parametric design since the 2000s. Additionally, AI-driven worker replacement is increasingly evident today, as seen in AI-based customer service, for instance. Although is not the intent of this chapter

Fig. 2 Replace: android image, generated with Midjourney v. 5.2

to capture a historical narrative of AI development, we draw back to Alan Turing research, that significantly contributed to today's AI progress, as his advanced ideas on understanding machine intelligence [7] have had a lasting impact on the field. Recently, Chomsky et al. [2] argued that machines lack actual intelligence, merely performing description and prediction tasks. However, in the case of contemporary architecture and design there is a merging of technology and creativity, and integration computational thinking and recognition of patterns. Besides the slow pace on adopting computational tools, the field faces unprecedented challenges in resourcing materials and automating design and fabrication. Within architecture, these questions acquire a particular urgency. Beyond unresolved ethical and creative dimensions, the core issue centres on agency: to what extent can AI participate in design, and how might architects retain authorship and control within these emerging hybrid processes? [8]. This ongoing discussion unfolds throughout the chapters of this book.

3 AIAD

Before defining AIAD, it is essential to emphasize that CAAD plays a fundamental role in integrating AI into architecture. We introduce this neo-acronym (AIAD) to establish a direct link to CAAD theory; however, various terms and expressions are used to describe the same concept: AI contributing to design.

Since the late 1950s, architecture has undergone a critical transformation enabled by the use of computers to aid the design process. Concepts such as Computer-Aided Design (CAD) and Computer-Aided Architectural Design (CAAD) emerged as part of the theoretical knowledge developed by architecture practitioners and researchers. Initial CAD applications focused on productivity and efficiency of architectural draft, but architects realized that there was a tremendous potential for creative support or even human-machine collaborative tasks. Thus, CAAD as a research field, has been widely documented for its potential to generate multiple design solutions and complement architects' tasks [9]. However, more recent studies highlight that CAAD has generally suffered from insufficient criticism, oversimplification, and a lack of comprehensive theoretical research [10]. For a deeper understanding of CAAD's history, see Celani and Veloso [11], who discuss how CAAD research contributed to a 'rupture' between traditional architectural drafting and computer-assisted design. Additionally, the authors suggest that after the 2000s, interdisciplinary methods and tools became a dominant research trend incorporating advanced studies on AI and architecture [11].

Since the 1970s, as CAAD evolved, AI became a topic of interest to researchers due to its ability to process and generate information [12]. AI in architecture has led to various research applications, such as agent-based design, evolutionary design, and generative design [13], though many of these have yet to be fully translated into practice. Today, design is aided by a wide range of computer tools that perform a range of actions, from simple tasks (draw a line) to sophisticated calculations (simulate thermal comfort). Some of these applications incorporate AI algorithms helping to enhance design sustainability, for example environmental performance, though challenges still remain in effectively communicating clear objectives to machines [14]. This is particularly true in fields of design that combine both quantifiable aspects (more easily translated into digital language) along with less tangible and subjective aspects of design. How do we define what makes a 'good design', with all its circumstantial and personally subjective connotations, and explain this to another human, let alone a computer system? In this context, AIAD can be described as emergent computational design processes that employ AI as design generator or assistant tool.

AI and Machine Learning are advanced fields of computer science that encompass a wide range of complex concepts, methods, and tools, which are gradually explored throughout this book. As architects and designers adopt cross- and interdisciplinary approaches to assist design generation, certain AI applications have become increasingly common. For example, according to Cudzik and Radziszewski [15], evolutionary algorithms, swarm intelligence, and neural networks are recognized AI approaches that can support architectural design. We, therefore, define AIAD as a subset of CAAD tools, characterized by its generative capacity, ability to solve complex problems, and pattern recognition among many other applications. Furthermore, AIAD facilitates emergent design processes in which AI acts as a design generator or assistant tool. While AI undoubtedly aids architectural work, it also raises critical questions regarding the enhancement or replacement of tasks, quality of design, authorship, and lack of theoretical framework. Each of these necessitate a broader discussion within the professional and academic community. Throughout this book, we position AIAD as a key contributor to sustainability as a means of engaging with a small subset of these questions in relation to architectural and urban design impacts and implications. A review done by Bölek et al. [16] corroborates with this idea and identifies AI applications in architecture, including performance-based design and form-finding, which encompass energy efficiency, modularity, daylight optimization, and the resolution of design challenges.

4 AI and Sustainability

Sustainability is a broad term that correlates development with environmental resilience, resource management, and the maintenance of life on Earth. However, sustainability can be achieved across multiple aspects of society, as expressed by the core dimensions of the new urban agenda defined by UNHabitat [17]: economic, environmental, social and spatial (Fig. 3). Therefore, in this book, our concept of sustainability is comprehensive and encompasses even the sustainability of the designer as profession. Achieving the necessary levels of sustainability requires drastic changes in behaviours, processes, and systems. A widely accepted tool to guide sustainable transformations are the 17 Sustainable Development Goals (SDGs) proposed by the United Nations [18].

Sustainability in architecture and urban design extends beyond environmental considerations to encompass the responsible use of materials, energy efficiency,

Fig. 3 UNHabitat's core dimensions of the new urban agenda

social inclusivity, and long-term economic viability. The built environment plays a crucial role in achieving sustainability, as cities and buildings significantly impact carbon emissions, resource consumption, and social well-being. Sustainable architectural practices include passive design strategies, circular economy principles, nature-based solutions, therefore, technological advancements like AI can enhance decision-making, optimize resource efficiency and make cities smarter.

AI for sustainability is a new research field that still needs more development on its own definition due to its broad scope [19]. However, it is widely accepted that AI is advancing sustainability by supporting industrial processes in monitoring, informing, and predicting across many fields [20]. As the United Nations calls for urgent action on climate change, AI will undoubtedly become increasingly integrated into solutions for a more resilient world. We believe that AI will play a vital role in advancing the SDGs, as reiterated by Goralski and Tan [21]. For example, Vinuesa et al. [22] state that AI can support achieving 134 of the 169 SDG targets [18] and that all targets of SDG 11—Sustainable Cities and Communities—can be enabled by AI. With emerging AI applications for clean water management, waste reduction, improved farming, enhanced social cohesion, and urban planning, the design field also demonstrated AI's interfaces. Given the significant environmental, economic, social, and spatial dimensions of cities and buildings, AI's integration into architectural and urban design can serve as a powerful tool to advance sustainability. It is our responsibility as designers to ensure that AI leads to more sustainable cities, buildings, and communities [23].

Current AI-based design applications remain exploratory, using accessible Generative-AI platforms as testing ground for ideas, particularly used by designers with limited knowledge of computer programming or machine learning. As such, designers often have minimal control over AI-generated outputs, and vague or unspecified criteria make AI-driven collaborations more difficult to extract relevant results. Additionally, AI text-to-image generators amaze designers with their unexpected and imaginative outputs, yet they also pose challenges. Generating architectural images through textual descriptions forces architects to articulate and describe concepts with words—an inherently sophisticated aspect of the design activity. Furthermore, the black-box nature of generative AI and preconceived notions of sustainability may lead to misinterpretations of sustainable design principles. For example, figure 4 shows typical sustainable architecture biases: green roofs and organic forms. Falk and van Wynsberghe [20] emphasize that all AI applications must consider their real-world impact. In this regard, optimization methods for design tasks are available in the literature, such as floorplan design presented by Chaillou [14]. Further research into the theoretical, procedural, and practical aspects of AIAD is explored in this book, which aims to assess whether AI methods can provide sustainable outputs.

As AI advancements becomes increasingly embedded in architectural and urban design routines, education must evolve to equip current and future professionals with the necessary skills to critically engage with AI-driven tools. The Beijing Consensus,

Fig. 4 Sustainable architecture concept, generated with Midjourney v. 5.2

produced by UNESCO [1], establishes global standards for AI education, emphasizes the importance of multidisciplinary learning, accessibility, technological proficiency, and sustainable-driven thinking. The ability to navigate AI's role in sustainability stresses the need for architects to critically explore AI-based design processes and responsibly measure its impact in human rights, society, economy and employment [1]. This book further explores how AI and sustainability intersect within the context of architectural education and practice, reinforcing the importance of aligning emerging technologies.

5 Aim of This Book

This book considers the evolution of AIAD and how the subjects (AI and Design) evolved in the last decades, how it affects design practice and requires ongoing change, which content needs to be taught to designers and what are the innovations that will change the field of design. These aspects are crucial to promote a responsible discussion about the impacts of AI in design and its consequences for sustainable thinking. Sustainability is used as the core focus of this exploration as a mean to bridge the gap between highly creative (yet, often detached and difficult to control) AI explorations together the pragmatics of design.

AI integration is giving rise to new vocabularies and conceptual frameworks within architectural design, many of which are absorbed from computer science. As a result, the profession is again being impacted by computational tools. We expect that the findings presented in this book will serve as link between computational design and sustainable development giving a stronger meaning to AI for Sustainability in the field of design.

6 Structure of This Book

AI has been diffused into architectural applications for several decades. However, leveraged by the release of OpenAI in 2022, a new set of fast developing AI generative tools became widely accessible to designers. 'AI-Aided Design for Sustainability' aims at disseminating new knowledge on the fast-emerging applications of AI in the fields of architecture and urban design, heritage, and design. This book encompasses innovations that address sustainability through new processes, cultures, systems, and methods that expand beyond current aesthetics-focused approaches. These comprehensive discussions set a new agenda for design by full integration of AI for communication, visualization, calculation, generation of ideas, optimization, and more.

This book presents emerging methods that apply AI as a mean to support and transform architectural design processes. It focuses on sustainable perspectives—investigating the resilience of the architectural profession, the alignment of design methodologies with the United Nations Sustainable Development Goals [18], and the innovative application of AI in education, as reflected in the Beijing Consensus [1]. The chapters examine a variety of approaches through which AI contributes to emerging design workflows. These include enhancing design development, supporting complex data-driven analysis, and exploring the extent to which AI may augment or even replace certain architectural tasks. Evidence of hybrid analogue–digital interfaces is also discussed, showcasing how architects are adapting their practices in response to technological shifts.

The book scope has three main themes:

1. **AI as a tool for sustainable design**, addressing environmental, social, and economic resilience in line with the UN SDGs.
2. **The evolving role of the architect**, especially how AI transforms workflows, decision-making, and creative agency.
3. **Innovation in education**, highlighting how AI is shaping new pedagogies and design thinking, supported by frameworks like the Beijing Consensus.

This book is divided into two sections: (1) AIAD Concepts and AIAD in Education (Chapters 1–7); and (2) AIAD in Practice and Future of AIAD (Chapters 8–14). This structure helps to contextualize AIAD across various levels of application and defines pathways for transition to new methods. The selection criteria for these sections seek to provide architects and designers entry points to the AIAD.

6.1 PART 1: AIAD Concepts and AIAD in Education

In the first part, we explain concepts of AI, design and sustainability and impact in architecture education. Starting by this chapter, Maycon Sedrez and Gregory Pitts, present the key concepts of AIAD, which is complemented by Chapter 2, written by Liangyu Chen, Zhen Chen and Feng Dong, providing a historical perspective of AI adoption in architectural design across four historical phases spanning from the 1960s to present day. Then a series of chapters that introduce AI into design studios provide a diverse perspective of emergent pedagogies. Chapter 3, authored by Gizem Yazıcı, reveals how AI in the design process contributes to reducing the cognitive load of designers. In Chapter 4, Faruk Can Ünal and Derya Karadağ compare the use of AI for generating architectural visualizations with traditional methods to understand the impact on students' learning. In Chapter 5, Eduardo Napoleão and Alice Demaria Silva present tests of an innovative design workflow using AI to support creativity. Chapter 6, written by Belinda Humphrey, documents applications of AI to architectural design through a 'metamodern' theoretical lens, introducing a detailed design workflow. And in Chapter 7, Sara Regiane Chornobai, Angelica Paiva Ponzio and Fabiano de Vargas Scherer explore how sustainability emerges from generative tools and emphasize the importance of critical evaluation by students and tutors.

6.2 PART 2: AIAD in Practice and Future of AIAD

In the second part, we collected chapters that investigate practical applications of AI. Chapters 8 and 9 correlate architects' tasks with a possible automation through AI, which will impact on how architects' organizations describe tasks in the future. Chapter 8, written by Maycon Sedrez and Gregory Pitts, analyses architectural tasks informed by the National Standard of Competency for Architects (Architects Accreditation Council of Australia) in relation to automation and AI applications. And Chapter 9, written by Victor Calixto, Camilo Cruz Gambardella, Darcy Zelenko, Wanchao Su and Duncan Maxwell map Machine Learning methods—through literature research—against the Royal Institute of British Architects' Plan of Work and identify the sustainability correlation between SDGs and AI. The following chapters, show the range of AI applications in design, from supporting public participation, enhancing building performance, preserving heritage, and enabling circularity. Chapter 10, written by Cem Ataman, Bige Tunçer and Simon Perrault, investigates how AI tools can support participation processes and collection of citizen inputs. In Chapter 11, Bahereh Vojdani Fakhr, Andressa Carmo Pena Martinez and Deokoh

Woo present applications of AI that enhance building performance toward sustainability. In Chapter 12, Veli Mustafa Yönder demonstrates how a trained AI model can support the documentation and preservation of architectural heritage. In Chapter 13, Angelo Figliola and Maurizio Barberio propose an intersection between AI and climate-aided design to support sustainable building performance. Chapter 14, by Burak Pak and Tuba Kocaturk, anticipates how AI can enhance circularity concepts in design for future urban spaces. Finally, Chapter 15 closes the book with key lessons learned from these chapters and predictions about how the future of design will be embedded with AI. We hope to respond to, or at the very least provide insights into, the central question proposed by this book: 'How can AI contribute to sustainability?' and inspire readers to be part of this movement redefining design.

References

1. UNESCO (2019) Beijing consensus: On artificial intelligence and education. Outcome document of the international conference on artificial intelligence and education. Planning education in the AI era: leap the leap. United Nations Education, Scientific and Cultural Organization, Paris. https://unesdoc.unesco.org/ark:/48223/. Accessed 13 Oct 2024
2. Chomsky N, Roberts I, Watumull J (2023) Noam Chomsky: the false promise of ChatGPT. International New York Times. https://www.nytimes.com/2023/03/08/opinion/noam-chomsky-chatgpt-ai.html. Accessed 20 Mar 2023
3. UN. United Nations (2005) General assembly 10333. General assembly adopts United Nations declaration on human cloning by vote of 84-34-37. Fifty-ninth general assembly. Plenary. 82nd meeting (AM). United Nations Meetings Coverage and Press Releases. <https://press.un.org/en/2005/ga10333.doc.htm. Accessed 05.Dec.2024.
4. Warwick K (2016) The future of artificial intelligence and cybernetics. MIT technology review. https://www.technologyreview.com/2016/11/10/156141/the-future-of-artificial-intelligence-and-cybernetics/. Accessed 08 Dec 2024
5. Bostrom N, Yudkowsky E (2014) The ethics of artificial intelligence. In: Frankish K, Ramsey W. Cambridge handbook of artificial intelligence. Cambridge University Press, New York, pp 316–334
6. Gherheş V (2018) Why are we afraid of artificial intelligence (Ai)? Eur Rev Appl Sociol 11(17):6–15. https://doi.org/10.1515/eras-2018-0006
7. Turing A (1950) Computing machinery and intelligence. Mind: Q Rev Psychol Philos, LIX(236):433–460. https://doi.org/10.1093/mind/LIX.236.433
8. Nature, (2016) Anticipating artificial intelligence. Nat 532:413. https://doi.org/10.1038/532413a
9. Mitchell WJ (1975) The theoretical foundation of computer-aided architectural design. Environ Plann B Plann Des 2(2):127–150. https://doi.org/10.1068/b020127
10. Herr C (2020) CAADRIA at age 25: Mapping our past, present and future. In: RE: anthropocene, proceedings of the 25th international conference of the association for computer-aided architectural design research in Asia (CAADRIA), vol2, pp 567–576. https://doi.org/10.52842/conf.caadria.2020.2.567
11. Celani G, Veloso P (2015) CAAD conferences: A brief history. In: Celani G, Sperling DM, Franco JMS (eds) The next city: 16th international conference CAAD futures, p 47–58. https://sites.google.com/unicamp.br/caadfutures/proceedings. Accessed 02 Feb 2025
12. Koutamanis A (2005) A biased history of CAAD. In digital design: the quest for new paradigms 23nd eCAADe conference proceedings. Lisbon (Portugal), p 629–637. https://doi.org/10.52842/conf.ecaade.2005.629

13. Gero J (2000) Artificial intelligence in design '00. Springer, Dordrecht. https://doi.org/10.1007/978-94-011-4154-3
14. Chaillou S (2022) Artificial intelligence and architecture: From research to practice. Birkhauser Verlag, Basel
15. Cudzik J, Radziszewski K (2018) Artificial intelligence aided architectural design. In: Kepczynska-Walczak A, Bialkowski S (eds). Computing for a better tomorrow—Proceedings of the 36th eCAADe conference-1. Lodz, Poland, p 77–84. https://doi.org/10.52842/conf.ecaade.2018.1.077
16. Bölek B, Tutal O, Özbaşaran H (2023) A systematic review on artificial intelligence applications in architecture. Int J Manag Educ 4(1):91–104. https://doi.org/10.47818/DRArch.2023.v4i1085
17. UNHabitat (2020) The new urban agenda illustrated. United Nations human settlements programme. https://unhabitat.org/the-new-urban-agenda-illustrated. Accessed 05 Feb 2025
18. United Nations (2015) Transforming our world: The 2030 Agenda for sustainable development. General assembly. Resolution 70/1 adopted by the General assembly on 25 September 2015. https://docs.un.org/en/A/RES/70/1. Accessed 05 Feb 2025
19. Van Wynsberghe A (2021) Sustainable AI: AI for sustainability and the sustainability of AI. AI Ethics, 1:213–218. https://doi.org/10.1007/s43681-021-00043-6
20. Falk S, van Wynsberghe A (2024) Challenging AI for sustainability: what ought it mean? AI Ethics 4:1345–1355. https://doi.org/10.1007/s43681-023-00323-3
21. Goralski MA, Tan TK (2020) Artificial intelligence and sustainable development. Int J Manag Educ 18. https://doi.org/10.1016/j.ijme.2019.100330
22. Vinuesa R, Azizpour H, Leite I, Balaam M, Dignum V, Domish S, Felländer A, Langhans SD, Tegmark M, Nerini FF (2020) The role of artificial intelligence in achieving the sustainable development goals. Nat Commun 11:233. https://doi.org/10.1038/s41467-019-14108-y
23. Coeckelbergh M (2021) AI for climate: freedom, justice, and other ethical and political challenges. AI Ethics 1:67–72. https://doi.org/10.1007/s43681-020-00007-2

A Brief History of AI in Architectural Design for Buildings Towards Sustainability

Liangyu Chen, Zhen Chen, and Feng Dong

Abstract Since Artificial Intelligence (AI) emerged in the 1950s, its adoptions in architectural design for buildings have led a technical evolution from computer aided design empiricism to datasets driven design decision-making paradigm underpinned by digital solutions that enable using Generative Design (GD) and other AI techniques in Building Information Modelling (BIM) pervasive design processes. The growth of computational architectural design practice via algorithmic optimization and processes automation for sustainability also follows the initiatives at the United Nations (UN) on Sustainable Development (SD) since the 1970s and on Sustainable Development Goals (SDGs) since the 2010s. This chapter examines the track record of AI adoptions through literature review across four SD stages, including Pre-SD Stage (1960s–1970s), SD Formation Stage (1980s), SDG Formation Stage (1990s–2000s), and SDG Adoption Stage (2010s–2020s). While it's an initiative to describe the history of adopting AI in architectural design for buildings towards sustainability, and the depth and width of AI applications in this subject area has been spurring across academic research, innovative development, and professional practice around the world, it is expected that the brief history presented in this chapter can inform further research and innovation in AI-aided architectural design for buildings with total sustainability.

Keywords Computer-aided architectural design (CAAD) · Artificial Intelligence (AI) · History · Buildings · Sustainability · Sustainable development goals (SDGs)

L. Chen (✉) · Z. Chen
Department of Architecture, Faculty of Engineering, University of Strathclyde, Glasgow, UK
e-mail: liangyu.chen@strath.ac.uk

Z. Chen
e-mail: z.chen@strath.ac.uk

F. Dong
Department of Computer and Information Sciences, Faculty of Science, University of Strathclyde, Glasgow, UK
e-mail: feng.dong@strath.ac.uk

M. Sedrez and G. Pitts (eds.), *Artificial Intelligence-Aided Design for Sustainability*, Urban Sustainability, https://doi.org/10.1007/978-981-95-1349-9_2

1 Introduction

1.1 Background

Over the past five decades, the integration of Artificial Intelligence (AI) into architectural design for buildings has evolved from an initial conceptual experiment to a transformative technical approach to shaping the future of architectural design in connection with construction and operation in built environment projects. Since the concept of "Architecture Machine" [1] put forward at MIT in 1970, the adoption of AI technology has progressively reshaped the practice of architectural design for buildings, culminating in AI applications such as Generative Design (GD) tool [2] integrated with BIM (Building Information Modelling) software Revit 2025 [3], which has been extensively used in BIM pervasive processes in built environment project delivery. This significant technical evolution has led the reengineering of CAAD by enhancing the capacity of sustainable building design at design stage and the performance of buildings at operation stage. AI is therefore regarded as a key enabler of sustainability in Computer-Aided Architectural Design (CAAD) for buildings.

Despite of a wide range of studies on adopting AI in architectural design for buildings, there is a lack of comprehensive technical review that systematically look into the track record and the impact of AI applications in buildings design towards sustainability, particularly in alignment with the concept and strategy of Sustainable Development (SD) [4] established in the 1970s and the 17 Sustainable Development Goals (SDGs) [5] defined by the United Nations (UN) in 2015. To address such a knowledge gap, this chapter presents a unique chronological review to reflect research into and practice on adopting AI. This review aims to find key milestones in the evolution of adopting AI in CAAD since the 1970s, to its current role as a core decision-making agent in architectural design regarding the need for Sustainable Development (SD) in the built environment sector. Central to this analysis is the role of computational architectural design, which is now recognised as a promising AI-integrated approach underpinned by advanced algorithmic processes to enhancing architectural design practices for sustainability. By looking at key technical advancements, design applications, and their contributions to the practice on the architectural design for sustainable buildings, this SD history oriented review aims to inspire discussions for technical insights into AI applications and their transformative impact in CAAD for buildings.

1.2 Scope of Review

AI is generally recognised as an advanced technology that can support computational architectural design in the processes of human cognition, and this support can improve capacities on design quality and process efficiency in CAAD. The holistic technical review being described in this chapter was conducted by using a timeline framework consisting of four key interconnected stages described below. This framework is based on a unique timeline that covers key stages of the histories of AI and SD in a combined way.

Pre-SD Stage (1960s–1970s). In line with the growing environmental concerns, two books including *Silent Spring* (1962) [6] and The *Limits to Growth* (1973) [7] highlighted the ecological consequences of built environment development, and these sparked early concerns and discussions on sustainability in architectural design. During this period, the concept and potentials of AI in architecture also started to form. For example, the MIT book *Architecture Machine* (1970) [1] introduced the concept of "Humanism through Intelligent Machines" which emphasises the essential of AI in CAAD. Although the development of CAAD in this stage had a focus on drawing assistance, it started an enrichment to design rationality towards computational architectural design over the subsequent decades, and the development of a profound technical foundation for incorporating sustainability oriented strategies into CAAD processes.

SD Formation Stage (1980s). The *Brundtland Report: Our Common Future* (1978) [4] emphasized the need for integrated SD strategies to address complex challenges within the social, economic, and natural environments where traditional environmental protection methods were inadequate for long-term sustainability. During this period, AI applications in architectural design for buildings had a focus on adopting Expert System (ES) technique and knowledge-based systems in the progress of early CAAD practice. For example, knowledge based solutions such as ES application HI-RISE [8] and the concept of knowledge-based design assistance [9] in CAAD were put forward in order to use rule-based and/or knowledge-based reasoning that can improve the quality of architectural design through a structured way to reuse experts' knowledge effectively and efficiently. The early explorations on architectural design ESs, aligning with the growth of awareness on global SD, have laid theoretical and practical foundations for subsequent development on AI-integrated computational architectural design towards sustainability.

SDG Formation Stage (1990s–2000s). In order to reinforce sustainability as a global priority on SD and to promote structured efforts on sustainability against social, economic, and environmental challenges, the UN introduced *Agenda 21* [10] at the 1992 Rio Earth Summit [11] prior to SDGs formally put forward in 2015 [5]. On the other hand, there are significant technical advancements on adopting AI in architectural design for buildings during this period. For example, the appearance of using the concept and theory of Parametric Design (PD) [12, 13, 14] for architectural design and the software tool Grasshopper [15, 16] built into 3D modeler Rhino (Rhinoceros 3D) [17], in addition, the application of Artificial Neural Network (ANN) and Metaheuristic Algorithm (MA) [18]. These technical advancements enable a new movement in CAAD by incorporating the use of AI solutions into the pursuit of SDGs oriented sustainability in more complex or innovative architectural design for buildings.

SDG Adoption Stage (2010s–2020s). In response to challenges against global sustainability, the UN issued the *2030 Agenda for Sustainable Development* in 2015 and introduced the target system of 17 SDGs [5] to guide SD practices across industry sectors around the world. In the rapid advancement of adopting information and communication technologies in the field of built environment, AI plays a pivotal role to enhance the sustainability to which architectural design has profound influence in the whole life of buildings. It has been recognised that AI is evolving from a mere auxiliary tool to a selectable active collaborator or even a technical agent within architectural design workflows by holistically incorporating design optimization processes towards a higher level of achievements on sustainability according to the UN SDGs and sustainability oriented legislations, codes, and standards for buildings.

For review on each of the four SD oriented stages described above, this chapter adopts the same structure, which consists of the following three sections:

- Theoretical Advancement of AI, which focuses on an overall review of technical development on AI in the specific SD stage.
- Practical Implications for Architectural Design, which focuses on representative AI applications for technical enhancements on CAAD in the specific SD stage.
- Practical Implications for SD, which focuses on representative AI applications for sustainability in CAAD in the specific SD stage.

1.3 An Illustrated Track Record

An illustrated track record of representative AI applications in architecture design for buildings is presented in Fig. 1 as an overview of this brief history of AI in architectural design for buildings towards sustainability. It intends to show a timeline, based on the four SD oriented stages described above, which can visualise a historical evolution consisting of key milestones where AI techniques are introduced to or adopted in architectural design for buildings since the 1960s alongside key milestones on the pursuit of sustainability. Due to limitations on time to search for and collect relevant information from 1960 to 2025, this illustration is a preliminary view with spaces for enrichment.

2 Pre-SD Stage (1960s–1970s)

2.1 Theoretical Advancement of AI

During this period, AI development was dominated by symbolic AI, emphasizing rule-based reasoning, formal logic, and structured knowledge representation. The further development of the *General Problem Solver* (1956) [19] demonstrated the potential of machine reasoning and attracted widespread academic attention and consideration for applications in multiple fields. Although the *Perceptron model*

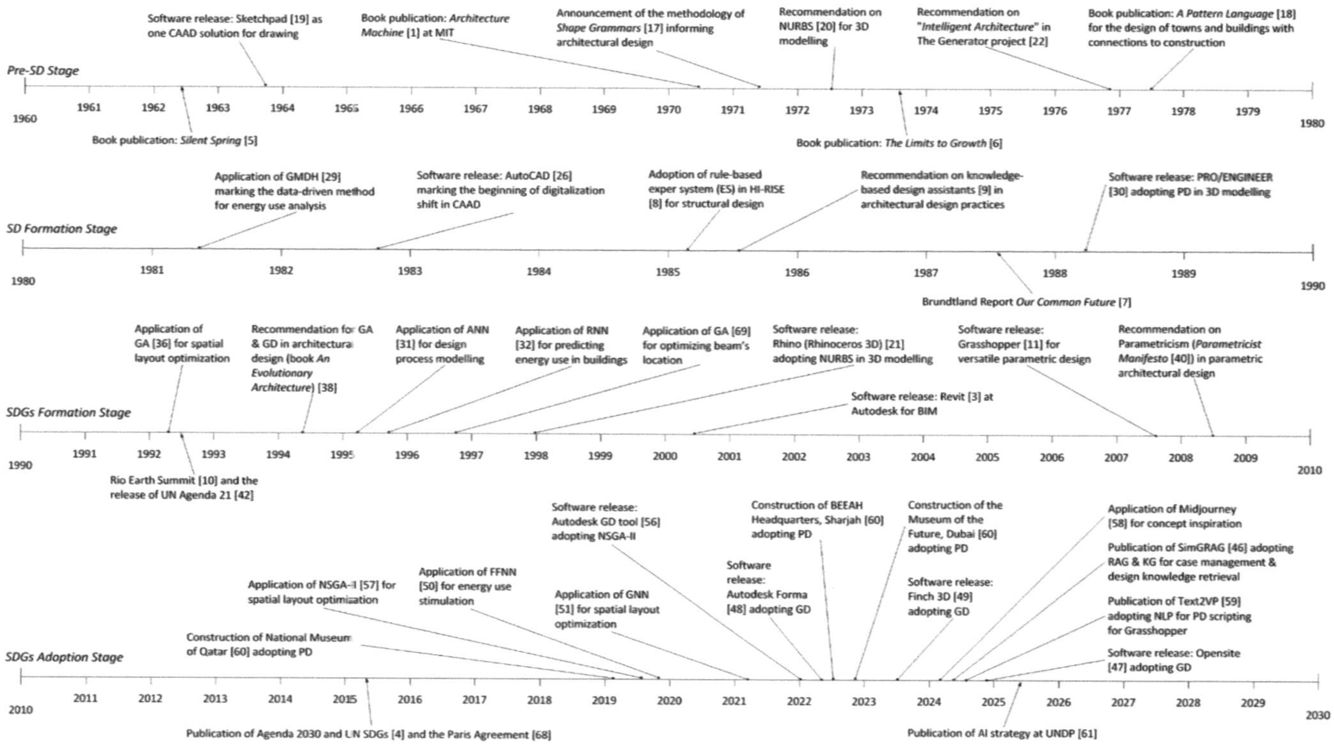

Fig. 1 Representative AI applications in architecture design for buildings in four SD stages

(1958) [20], which is a probabilistic model for information storage and organization in the brain and is regarded as an early prototype towards the formation of ANN, was put forward previously, the research stagnated during this period while the concept of *Computational Geometry* (1969) [21] was introduced to overcome its limitations. In parallel, the emergence of logical programming languages, for example, programming language PLANNER (1969) [22] built out of problem-solving primitives within a hierarchical control structure for proving theorems and manipulating models in a robot, inspired further research and development into formalizing reasoning mechanisms through computable representations.

While initial efforts on AI were made in this stage towards advanced techniques such as ANN and rule-based reasoning, which were actually on their way to arrive, both research and development encountered technical limitations, in terms of the capacity of computer systems and the maturity of supporting theories in both computer sciences and mathematics, and the focus was primarily on deterministic logic systems and rule-based reasoning frameworks. Although these techniques did not have widespread adoptions during this period, a solid theoretical foundation has started to form for the subsequent development of popular AI techniques including ES, Natural Language Processing (NLP), ANN, and Robotics, etc. It is worth to note that these early attempts on AI as a generic technology have also shown their potential for applications in computational architectural design for sustainable buildings.

2.2 *Practical Implications for Architectural Design*

In the field of built environment, the conceptual integration of AI into architectural design was strongly promoted by the concept of "Architecture Machine" put forward at MIT in 1970 [1], and it envisioned a technical future in which intelligent machines collaborate seamlessly with human designers. The book *Architecture Machine* [1] emphasized that the proactive objective of AI in architectural design should preserve and promote the human-centred nature in CAAD process which can enable not only design efficiency but also an automated program for learning and reasoning through interactions between an architect and their computer (design machine). This new architectural design philosophy enables a paradigm shift in computational design thinking and has its profound impact to long-term research and development for adopting AI in architectural design.

In addition to the proactive concept of "Architecture Machine" [1], some other initiatives are recognised as well in terms of their contributions to technical enhancements in CAAD, and particularly for the incorporation of AI in architectural design practice. For example, the concept and methodology of "Shape Grammar" (1971) [23] formalized rule-based approach to generating and optimizing architectural forms, and this is recognised as an early attempt towards modern computational architectural design via GD. On the other hand, the book *A Pattern Language: Towns, Buildings, Construction* (1977) [24] introduced a structured approach to embedding design knowledge through reusable spatial patterns, and it described the possibility of a logical language to integrate environmental intent into the architecture design process. The *Pattern Language* [24] emphasized a process to encode design knowledge through reusable patterns and the integration of knowledge-based systems in CAAD for optimization in architectural design. Beyond these theoretical movements, the Generator project (1976) [25] demonstrates the potential to transform or enhance traditional architectural design methodology by providing a platform for rule-based iterative form a generation process. Although the use of these ideas was constrained by computation capability at that time, they activated theoretical preparation and development for adopting AI techniques in computational architectural design for buildings in the later decades.

As computing technologies advanced in this period, major breakthroughs emerged from research and development in computer graphics, and solutions from technical innovations started to enable the use of computers to enhance the efficiency, diversity, and precision of architectural design. In 1963, the pioneering program Sketchpad [26] was developed and realised the human–computer interaction in computer graphics. This technical innovation is crucial in shaping the future of CAAD in its rapid development in the 1970s. In addition, a further significant progress in computer graphics was achieved in 1972 upon the introduction of Non-Uniform Rational B-Splines (NURBS) modelling [27, 28, 29], which can enable accurate modelling via mathematical representations of complex three-dimensional geometries such as curved shapes in computational architectural design and significantly expanded the capacity of CAAD. Such technical advances strongly support the development and usage of contemporary 3D computer graphics software such as Rhino (Rhinoceros 3D) [17] for CAAD practice.

A notable and influential design experiment, i.e. the Generator project (1976) [25] is recognised as an early attempt on design intelligence applied to a PD prototype. Influenced by cybernetics, this design experiment was envisioned as an adaptive building that could dynamically respond to changing needs of users. Unlike static buildings, the Generator project proposed flexible spaces that could be automatically reconfigured, challenging traditional building norms and laying the foundation for modular and adaptive design [30]. This conceptual project could well inform AI aided architectural design of high-performance buildings, which are adaptable, resilient, and resource-efficient towards sustainability.

The development of AI in architectural design for buildings during this period mainly focused on a multidisciplinary theoretical preparation. Main areas of such a theoretical preparation can be recognised in the formulation of logical languages that can process design elements or patterns by using rule-based reasoning. On the other hand, technical preparation on computer graphics left a foreseeable development space for an AI integration starting from rule-based reasoning. Although technical efforts on innovative CAAD solutions during this period were mainly focused on improving drawing accuracy and design efficiency, it was an emerging technical challenge for CAAD to enable the effective and efficient use of experts' knowledge, and this expected technical innovation for knowledge-based CAAD was left to the next decade when ES technique was introduced to support.

2.3 Practical Implications for SD

It is noticed that, during this period, growing environmental awareness began to reshape architectural design practice in its technical transformation via CAAD, and this sustainability-oriented philosophy initiated a new era of architectural design for sustainable buildings. Two influential books including *Silent Spring* (1962) [6] and *The Limits to Growth* (1973) [7] highlighted ecological fragility and the urgent need for resource-conscious development across all sectors. While the development of CAAD tools, including early attempts on using AI through research and development on specialty ES during this period, their focuses primarily on computer assistance to both drawing tasks and design decision making support, however, it looks that it is an emerging need for enhanced design rationality for sustainable buildings.

Although the focus on the sustainability of buildings was not formally deployed in CAAD and potential AI applications at the time, and the core principles of sustainable buildings in terms of adaptability and responsiveness were forthcoming. They marked an early exploration of technical innovation and ecological responsibility, laying the foundation for more integrated design approaches to sustainable buildings in the following decades.

3 SD Formation Stage (1980s)

3.1 Theoretical Advancements of AI

The 1980s was a key decade for AI, owing to a wide adoption of ES technique across many sectors during this period [31]. ES solutions adopt rule-based or knowledge-based reasoning approaches and can provide expert-level judgments and decision-making support upon critical technical questions. On the other hand, although the theory and adoption of ANN technique continuously grew in research and development, its applications in practices did not make much progress during this period [32], which was described as "AI winter" [33], i.e. a tumultuous period with pessimism in the AI community.

AI applications in architectural design for buildings during this period were primarily seen on technical advancement in automated design support from the adoption of rule-based and knowledge-based methods. In comparison with advanced AI techniques developed after the "AI winter" [33], AI applications using ES technique and tools in this period reflected the advantages of rule-based and knowledge-based mechanisms, which are functioned as structured logical frameworks—akin to semi-automated systems—lacking the autonomous learning and adaptive capabilities that define modern AI but could be regarded as the prototype of it. Nonetheless, these early developments laid important conceptual and technical foundations for the later emergence of modern architectural design methodology by integrating AI in CAAD for sustainable buildings.

3.2 Practical Implications for Architectural Design

During this period, techniques and solutions for CAAD began to be integrated into architectural design workflows. For example, the release of software package AutoCAD in 1982 [34] significantly enhanced CAAD practice, and the release of software package Pro/ENGINEER in 1988 [35] further informed the advantages of using PD in architectural design in terms of feature-based parametric modelling. Although such technical advancement with digital solutions for CAAD practice significantly enhanced design capacity and transformed architectural design process towards digitalisation, AI was still standalone but has shown potentials for integrated usage in CAAD.

The significant progress on applied AI in architectural design was the adoption of ES in this period. For example, the rule-based ES HI-RISE (1985) [8] is widely regarded as a pioneering specialty ES developed for building design. Designed to support the preliminary structural layout of high-rise buildings, HI-RISE applied a set of 412 coded decision rules to ensure compliance with the Uniform Building Code [36], and provided an early example of the application of rule-based reasoning to complex design specifications. Another notable development was the attempt on knowledge-based design assistants [9] to effectively and efficiently use domain-specific knowledge in CAAD, and such knowledge-based CAAD system can enable architects to iteratively explore and evaluate design alternatives by using a multi-model reasoning framework. This approach demonstrates the possibility of expanding the role of CAAD from a geometric drawing tool to a semi-intelligent assistant capable of supporting early architectural design decisions, and reflected the modern approach to AI aided GD.

Although AI applications in architecture were still nascent during this period, these pioneering adoptions of ES technique started a substantial use of AI in CAAD and revealed the need and potential for the integration of knowledge-based tools in CAAD software.

3.3 Practical Implications for SD

The 1980s marked a period of early convergence between sustainability as a global agenda and AI as a promising technology to support CAAD. The Brundtland Report Our *Common Future* (1987) [4] provided the widely adopted definition of SD, which was framed as a generational responsibility and informed architectural design by redefining its value, strategy, and approaches to sustainability.

It is noticed that technical advances on applied AI in CAAD practice during this "AI winter" [33] began to support this sustainability oriented transformation in architectural design. For example, ES application HI-RISE [8] demonstrated how this applied AI technique could be used to reduce material use in high-rise structures in the process to optimise structural performance. On the other hand, the introduction of knowledge-based design assistance [9] highlighted the advantage of "automatic" support to decision making to enhance design efficiency and design justification. These technical initiatives established early attempts on adopting AI in architectural design and its processes for sustainability from the current perspectives on sustainable buildings, and technical advantages demonstrated in such AI oriented initiatives has well informed further adoption of AI in CAAD.

4 SDG Formation Stage (1990s–2000s)

4.1 Theoretical Advancements of AI

The observation on the theoretical advancements of AI, over the two decades after the "AI winter" [33], focused on three theoretical areas including (1) Problem-solving, (2) Knowledge, reasoning, and planning, and (3) Uncertain knowledge and reasoning, according to the structure of book *Artificial Intelligence: A Modern Approach* (the 3rd edition in 2010) [37], which is widely adopted for higher education in the subject field of AI. Regarding to establish a historical view on applied AI for architectural design, this observation also further focused on three interrelated methodological domains, including ANN, PD, and MA from the CAAD perspective. It supports the view that such theoretical advancements enabled ground-breaking innovations for a wide range of AI applications in CAAD practice, and showed the need and potential for AI applications in PD and MA.

During this period, ANN and its variants as practical AI techniques have achieved a wide scope of adoptions across industry sectors. For architectural design for buildings, ANN techniques also attracted a lot of interest in research and development in, for example, performance-based optimization, energy efficiency evaluation, spatial modelling, and design process modelling [38]. In particular, the introduction of ANN techniques such as the Recurrent Neural Network (RNN) [39] realised a novel problem-solving approach to tackling challenging tasks in architectural design, such as to inform architect by predicting energy use in buildings under design.

The further development of parametric modelling software packages strongly promoted a wide adoption of PD technology in this stage [12]. For example, the release of digital modelling platforms such as Grasshopper [40] revolutionized PD by enabling architects to engage in real-time parametric adjustments, generative form-finding, and iterative optimization under dynamic constraints. The progresses in PD technology and computer programs greatly enhanced its adoptability in CAAD [41], and made a further preparation for integrating with AI solutions in the next theoretical development and practical adoptions.

It emerged from literature in this period that the theoretical development and applications of MAs, including the Genetic Algorithm (GA), has strongly support a wide range of needs for technical optimisations in engineering and management areas across industry sectors. For example, GA offered a framework for solving complex design problems by simulating evolutionary processes [42], and became a notable approach to design. GA applications achieved multi-objective optimizations and demonstrated the usefulness of MAs such as GA in addressing complex, multi-faceted challenges inherent in computational design. The need for and attempts on enhanced algorithm capability on active in-depth evidence-based learning and reliable judgments emerged for further research and development, and this includes the integration of GA and ANN [43] to rely on advantages of the two methods to form advanced problem-solving solutions.

4.2 Practical Implications for Architectural Design

The adoption of ANN technique in architectural design practice became particularly influential in this period. For example, it was used to model initial design process for reinforced concrete beams, automating complex calculations and providing more precise design decisions early in the design phase [38], and its variant RNN was used to predict building energy use and to optimise system performance under dynamic loads for energy efficiency [39]. The maturity of ANN techniques has made them popular regarding a wide range of adoptions of AI technology in CAAD oriented professional practice and academic development.

PD, on the other hand, was initially catalysed for use due to the release of software tool Sketchpad [26], and it further increased potentials in the 1990s upon the release of key PD software tool Pro/ENGINEER [35] which underlay advances in computational geometry and algorithmic modelling. This technical advancement enables feature-driven and parameter-based modelling, which influenced subsequent dynamic approach to architectural design. By the early twenty-first century, the release of 3D modeller Rhino (Rhinoceros 3D) (1998) [17] and its PD script plugin Grasshopper (2007) [40] revolutionized PD by offering architects an intuitive, visual programming interface that democratized parametric workflows without requiring advanced coding skills. The "*Parametricism Manifesto*" (2008) [44] further solidified PD as an established design methodology that can underpin computational strategies in architectural form-finding and structural efficiency. It showed a practical potential via theoretical research and development to incorporate AI applications into PD integrated CAAD process.

Experimental applications of GA in this period has revealed the potential of this technique as a representative of the MA family [42]. It looks from literature review that research and development during this period explored the capability of GA to address complex design challenges by optimizing various design parameters in relation to a wide range of architectural design tasks, including floor plan and daylighting performance [45], workplace layouts [46], beam layout [47], and aesthetic considerations [48], etc. In addition, initiatives emerged during this period on the integrations of GA with other matured techniques such as ANN and GD. For example, there are subsequent studies further emphasized the advantages of combining GA with ANN, and an experiment on such an integration was conducted to enhance energy efficiency in building design [49], and thereby underscoring their value in sustainable building and energy use management. The book *An Evolutionary Architecture* [50] systematically introduced GA and GD of architectural design for buildings, and highlighted the potential to integrate GA and GD. The conceptual development and experimental studies on such technical integrations of applied AI techniques strongly support an extensive adoption of AI integrated GD among research and development as well as practice in CAAD in the subsequent stage.

4.3 Practical Implications for SD

This stage is considered as two pivotal decades after "AI winter" [33] not only for the theoretical development of AI and its applications in CAAD but also for SD whilst the emerging need for goals in SD and the eventual announcement of SDGs in 2015 [5] under UN's leadership. Although it was still the initial stage for adopting AI in academic and professional practices on CAAD, the potential of AI technology has been more widely perceived in term of the efficacy of applied AI for tackling challenges in the architectural design of sustainable buildings.

There are many attempts seen in proactive academic research and development during this period regarding to incorporate the use of AI in CAAD for sustainability. For example, ANN models are made and examined to show their applicability and capability, such as to predict and optimise energy use, in order to improve the design on energy efficiency in buildings. Although there are still technical limitations including dataset availability for reliable predications, the benefits of data-driven approach to architectural design for sustainable buildings brought or inspired a transformational idea to CAAD in terms of AI integrated digital architectural design for sustainable buildings. In addition, GA and other MAs began to support multi-objective optimization for design tasks on issues such as environmental performance, economic feasibility, and spatial configuration in CAAD and increased the capacity of an AI enabled CAAD procedure for sustainability-oriented architectural design under complex design scenarios. In the meanwhile, adopting AI within PD has gained attention for its ability to generate dynamic design solutions and simulate environmental responses, especially in a BIM process for the architectural design of sustainable buildings. It has been noticed that the exploration on AI applications in this period was widely seen in academic research and development, which have started to inform and/or lead further innovations at institutions regarding the adoptions of AI solutions in commercial products [51] and an intimate integration to services process [52] in CAAD.

5 SDG Adoption Stage (2010s–2020s)

5.1 *Theoretical Advancements of AI*

Built upon the theoretical and practical development in ANN, GA, and other computational techniques and tools in the MA family in the previous stage, technical innovations on GD started to approach and realise automatic generation of optimised design solutions based on specified design criteria. As a result, there has been a significant development with technical progresses on GD by integrating with AI assistance to enhance automatic digital design capacity in this period. Rather than a single technique, GD adopts an integrated approach to incorporating PD processes, evolutionary computation, and AI-driven optimization methods, including ANN and MA, to autonomously generate optimized solutions within predefined design constraints. Through intimate integrations with GD, PD enables the establishment of structured relationships between design parameters, and the rapid generation of design alternatives while maintaining geometric and functional control, and MA enables the process to navigate complex design spaces and identify optimal solutions [53]. These technical advancements revealed a new era of AI applications in the design sector.

Inside the MA family, GA and its variants remains their position as one set of the most used optimization methods for applications in engineering and management areas. In addition, GA variants such as the Non-Dominated Sorting Genetic Algorithm II (NSGA-II) [54], and other prominent optimization techniques such as Pattern Search [55] and Particle Swarm Optimization (PSO) [56], which are put forward in previous sages continuously, play crucial roles in multi-objective optimization and have achieved a wide range of applications, including to perform multi-objective optimization in Python [57], which undertakes significant role in programming for AI software solutions [58]. The adoption of well-developed optimization techniques in AI during this period has significantly developed the capacity of this technology for solving problem in multiple technical domains.

Technical innovations on applied AI also significantly supported by reinforcing the use of knowledge based techniques such as Knowledge Graph (KG) and Retrieval-Augmented Generation (RAG) during this period. These AI oriented techniques are quickly adopted in both academic research and development and for commercial applications.

5.2 *Practical Implications for Architectural Design*

The rise of GD. It is widely recognized that AI integrated GD has significant advantages in architectural design for buildings. Commercial software packages such as Autodesk Forma [59] and Finch 3D [60] integrate GD with BIM to autonomously generate optimized designs, minimizing human error and enhancing efficiency in alignment with sustainability oriented performance objectives.

Applications of ANN. Whilst the theoretical foundations of ANN technique and its variants have been progressively developing since the mid-twentieth century, their practical applications in architectural design gained substantial momentum since the 2010s. For example, (1) Feedforward Neural Network (FFNN) is adopted for simulating energy use and thermal comfort in buildings [61] to inform evidence-based architectural design, (2) Graph Neural Network (GNN) is adopted for semantic enrichment on room type classification within BIM environment [62], and (3) Convolutional Neural Network (CNN) is adopted for simultaneous prediction of energy use in both short and long term [63]. In addition, there is a trend to integrate various optimisation algorithms into ANN models to form advanced solutions for estimate and prediction in related areas. For example, FFNN is trained by using one MA, i.e. Multi-Tracker Optimization Algorithm (MTOA) to estimate energy use performance in residential buildings [64]. It is noticed that applications of ANN techniques including their combinations with MAs can have significant supports to evidence-based architectural design practice by effective and efficient learning from existing buildings.

Progresses with MA. It has been noticed that MA, particularly GA, is extensively utilized in architectural design optimization during this period, and such applications are often in conjunction with ANN models for optimizations to support multi-objective decision-making. From the design optimization process point of view, GA is used to explore optimal design solutions through evolutionary strategies, and ANN model is used to predict performance outcomes or generate design alternatives [65]. For instance, Autodesk [2] integrates GD toolkit into Revit since its 2017 version; in addition, it integrates NSGA-II (Non-Dominated Sorting Genetic Algorithm II) [54], a variant of GA, to identify optimal interior spatial configurations for architectural design [66], and significantly enhance GD by integrating data-driven pattern recognition and generative modelling, optimizing spatial arrangements and geometric configurations to improve both efficiency and design creativity [67]. The integration of ANN with MA has formed an advanced technical approach to implement in-depth optimization supported by reliable estimate and/or predication in architectural design for buildings.

The integration of NSGA-II in GD platform in Autodesk's solution exemplifies how MA can be effectively applied in architectural workflows. By setting parameters such as population size, generations, and number of outcomes, designers can guide the algorithm to evolve spatial design solutions that meet specific performance criteria [68]. The use of MAs in this context allows for the exploration of complex design spaces, identifying non-dominated solutions that balance multiple objectives, such as maximizing usable space while minimizing energy consumption.

Furthermore, the combination of GA with ANN enhances the robustness of optimization process, i.e. ANN models is first trained for predicting the performance of different design configurations, and then provides rapid feedback to GA process, which can effectively make balance upon structural integrity, energy efficiency, and material usage through extensive design exploration [69]. This hybrid approach enables more efficient exploration of space design, reducing the computational burden while maintaining high-quality solutions. For example, in the context of building massing studies, GA can optimize the geometry of a building based on objectives such as *Floor Area Ratio*, *Non-Passive Zone*, and *Usable Open Space*, while ANN models predict the energy performance of each configuration.

The emergence of KG and RAG systems. Recent innovations in Knowledge Graph (KG) and Retrieval-Augmented Generation (RAG) systems have further expanded AI's potential in architectural design. KG provides a structured approach to storing and accessing architectural data, such as material properties and design specifications, thus facilitating more accurate knowledge retrieval and reasoning. RAG systems, which integrate structured datasets from KG, leverage advanced retrieval technologies to offer optimized design recommendations and solutions [70]. KG and RAG systems address significant challenges in architectural case management and design knowledge retrieval. These technologies facilitate the construction of specialized knowledge libraries, enabling architects to quickly access and analyse relevant data related to best practices, materials, and performance metrics. For example, the SimGRAG framework [70] aims to enhance KG-driven RAG by leveraging similar subgraphs to improve knowledge retrieval and generation in architectural design, and it employs a two-stage process, i.e. (1) Query Pattern, which uses a Large Language Model (LLM) to transform queries into the desired graph pattern, and (2) Pattern-to-Subgraph, which quantifies the alignment between the pattern and candidate subgraphs using a Graph Semantic Distance (GSD) metric. Additionally, SimGRAG integrates an optimized retrieval algorithm to identify the top-k relevant subgraphs, improving the accuracy and contextual relevance of design knowledge extraction in design decision-making.

Hybrid workflows in architectural design. Recent advancements in AI have fostered the development of hybrid workflows that combine the conception of GD and PD methodology, creating a more integrated and collaborative approach to architectural design. Tools such as Midjourney generate initial images of design concepts [71], while natural language processing models like ChatGPT generate parametric modeling scripts. These scripts are then input into PD platforms like Grasshopper [40], which interfaces with Rhino [17] to produce parametric models. For instance, the Text2VP [72] trained an LLM based on GPT-4 to generate C# scripts from natural language inputs, automating parametric modelling within Grasshopper. Experimental results indicate that Text2VP effectively produces operational parametric models, though its accuracy decreases with increased design complexity. This hybrid workflow facilitates a semi-automated design process, allowing designers to fine-tune and optimize designs without deep programming expertise, thereby accelerating the design process and expanding creative possibilities.

PD in large-scale architecture practice projects. PD has been effectively implemented in lots of large-scale iconic architectural projects, demonstrating its potential to solve complex design challenges through algorithmic logic and computational precision. A prominent example is BEEAH Headquarters, which was designed by Zaha Hadid Architects in Sharjah, where PD played a major role in generating the building's fluid, dune-inspired geometry, allowing for precise control of curvature and spatial organization. This approach not only enhances the demonstration of buildings integrated with the surrounding desert landscape, but also promotes sustainable performance through optimized orientation, natural ventilation, and solar control. The Dubai Museum of the Future, which was designed by Killa Design also exemplifies how parametric modelling can drive structural innovation. Its iconic torus shape and complex façade, featuring calligraphic patterns that double as daylight filters, were achieved by using both BIM technologies to manage geometric complexity and PD to balance structural performance. With the help of these technologies, the building achieved LEED Platinum certification, highlighting its commitment to sustainability. Similarly, the National Museum of Qatar by Jean Nouvel incorporated PD to translate the desert rose concept into a sophisticated spatial structure, where interlocking disks were seamlessly designed to optimize both form and function [73]. In all these projects, PD played a crucial role in achieving not only visually striking architecture but also in addressing environmental challenges, material efficiency, and cultural expression, thus demonstrating its essential role in shaping the future of architectural practice.

5.3 Practical Implications for SD

AI has now evolved from a design aid to a powerful optional tool with the potential to be used to optimize processes and performance throughout the whole building lifecycle. AI-driven GD platforms incorporate real-time environmental simulations to provide early feedback on factors such as spatial layout, wind flow, energy use, and materials efficiency. These tools enable architects to evaluate thousands of design iterations simultaneously in a speedy process to support decision-making and to align outputs with sustainability benchmark on energy efficiency and reduced carbon footprint. In addition to GD, AI technologies have significantly improved knowledge management in architectural design. The development of KG and RAG systems supports the creation of sustainability-focused case libraries [70], enabling architects to access precedent data and performance results to make evidence-based design decisions.

Nowadays, the development direction of AI in architecture design for buildings is in line with SDG 11 on the *2030 Agenda* [5], which aims to make cities and human settlements inclusive, safe, resilient, and sustainable. At the same time, the *UN Development Programme* (UNDP) actively supports the integration of AI technology and is committed to leveraging the potential of AI, strengthening local AI ecosystems, enhancing data foundations, and building trust and security in AI applications [74] in the global pursuit of SD. As sustainability becomes an essential criterion in architectural design practice, AI has a critical role to participate or even facilitate an innovative design process, which relies on a holistic and systematic use of design parameters on STEEP (social, technical, ecological, economic, and political) considerations. More than just a technical improvement, AI represents a paradigm shift—combining complex data sets and performance targets to promote designs that are not only innovative and efficient, but also ecologically responsible and human-centric. These efforts ensure that AI is developed and deployed in a way that benefits both people and the planet, in line with the broader goals of the UN SDGs.

6 Conclusions

The integration of AI in architectural design for buildings has become a transformative force, demonstrating the potential of driving economic, environmental, and social sustainability. AI-driven methodologies have proven effective in both experimental and public practice cases, including optimizing resource allocation, improving energy efficiency, and utilizing predictive modelling to reduce waste and enhance performance. By analysing large case datasets, AI supports design decisions that lower energy consumption and carbon emissions, aligning with the UN SDGs, particularly SDG 11, which advocates sustainable cities and communities [5]. However, an over-reliance on AI-generated optimal solutions may prioritize computational efficiency at the expense of site-specific context and cultural distinctiveness. Therefore, AI needs to be applied strategically and wisely to enable architects to push the boundaries of form, function and spatial relationships while maintaining the integrity of human-centred design. By balancing technical innovation with ethical responsibility, the construction industry can harness AI's full potential while ensuring that architecture reflects human creativity, cultural identity, and social value, fostering a more inclusive and resilient built environment.

At a technical level, the application of PD in BIM environments has reached a high level of maturity and has been widely adopted at top architectural firms such as Zaha Hadid Architects [75]. PD excels in managing complex geometries and adaptive façade structures, making it particularly suitable for rapid iteration and flexible design [76]. Meanwhile, the maturity of GD tools continues to rise, with hybrid workflows based on GD concepts demonstrating significant advancements in AI-driven architectural design. Despite the considerable potential of emerging technologies, such as solutions built with ANN, KG, and RAG, their application in architectural design remains in an experimental phase, with limited commercial adoption. Meanwhile, some architects have begun exploring image generation tools based on ANN variants (e.g. Transformers and Generative Models), such as Midjourney for exploring design inspiration [71]. However, these tools lack specialized training for architectural design, which limits the effectiveness and reliability of their generated results. In addition, NLP techniques via their applications, such as ChatGPT have been shown to be applicable to PD scripting, although their actual effectiveness is still insufficient; although some researchers have begun to try to develop Language Models specifically for PD software, the actual use effect is still in process, especially when dealing with complex parametric model generation, which will have a higher error rate [72]. In the field of MA and ANN, while GA and other algorithms have shown significant potential in optimization of spatial layout and performance, algorithms which developed specifically for architectural design needs are still an underexplored area. Thus, despite the broad prospects of AI in architectural design, further research and practice are required to facilitate its widespread adoption.

From the professional perspective of the architecture industry, industry reports from top architecture firms such as Arup and Arcadis highlighted their concerns that over-reliance on AI could lead to homogenised designs and neglect cultural diversity. Arup advocates that AI should serve as a collaborative co-pilot, rather than an autonomous decision-maker, optimizing workflows while preserving the creative authority of architects [77]. Similarly, Arcadis promotes a flexible and ethical approach to AI integration, where AI complements rather than dictates architectural decision-making [78, 79]. According to the RIBA 2024 AI report, 41% of architects have incorporated AI into their workflows, with 57% expecting AI to play a larger role in environmental sustainability analysis [80]. This growing application highlights the potential of AI in SD. By combining AI-driven techniques with the SDGs, the architectural discipline can use AI not only as an optimization tool, but also as a catalyst to promote adaptive, sustainable, and culturally sensitive built environments. For example, FFNN solutions can predict and optimize energy consumption regarding UN SDG 7 (Affordable and Clean Energy) [61], and CNN can effectively predict short-term and long-term residential energy consumption, which can help to efficiently plan and manage energy resource consumption in buildings [63]; MA such as GA [46] and NSGA-II [67] combined with ANN models can be used to explore complex design spaces, balancing the needs of maximizing usable space while minimizing energy consumption aligned with UN SDGs 9 (Industry, Innovation, and Infrastructure) and 11 (Sustainable Cities and Communities) [5]. In addition, there are still research gaps that need to be filled in the development of official AI Overlay, comprehensive regulatory guidance, methodological frameworks, and professional accountability mechanisms to ensure the entire AI-integrated architectural design lifecycle for buildings can be appropriately monitored.

In summary, the integration of AI into architectural design for buildings signifies a transformative shift in the processes of conceptualization, design development, and sustainability optimization. Following the digital transformation of the industry, a growing imperative for intelligent transformation has begun to emerge. AI-driven approaches, particularly techniques such as GD, have already shown great potential in improving design efficiency, optimizing energy consumption, and reducing environmental impact, all of which are in line with the UN SDGs [5]. However, as AI continues to advance, its application in architectural design must be approached with caution to avoid over-reliance on computational solutions that may ignore cultural context and human creativity. The adoption of AI must be strategically balanced, ensuring that it serves as a collaborative tool that amplifies the architect's vision while maintaining the integrity of design decisions that reflect social and cultural values, rather than replacing the subjectivity and judgment of architects. Ultimately, the role of AI in architecture, if combined with a focus on sustainability and ethical practice, will greatly contribute to the construction of a more inclusive, resilient, and environmentally responsible built environment. In addition, it is still necessary to further improve the applicability and reliability of AI technology in future research and practice, explore its potential in promoting the full life cycle management of buildings, ensure that it is always human-centred, and relevant to building the SD environment.

Acknowledgements The research conducted for this chapter is funded by the Diamond Jubilee Scholarship in the Faculty of Engineering at the University of Strathclyde in Glasgow in 2024, and the funded research project focuses on an AI integrated metaheuristic framework for architectural design justification. The authors would like to thank the editors of this book to give the opportunity to investigate the history of AI applications in architectural design for buildings from the prospect of sustainability oriented architectural design with AI assistance.

References

1. Negroponte N (1970) The architecture machine: toward a more human environment. The MIT Press, Cambridge
2. Autodesk (2025a) Generative design for architecture, engineering & construction. Autodesk Inc., San Francisco. https://www.autodesk.com/solutions/generative-design/architecture-engineering-construction. Accessed 6 Apr 2025
3. Autodesk (2025b) Autodesk revit: bim software to design and make anything. Autodesk Inc, San Francisco. https://www.autodesk.com/products/revit/overview. Accessed 6 Apr 2025
4. World Commission on Environment and Development (1987) Our common future. Oxford University Press, Oxford. ISBN 019282080X
5. United Nations (2015) Transforming our world: the 2030 agenda for sustainable development. United Nations, New York. https://sdgs.un.org/2030agenda. Accessed 6 Apr 2025
6. Carson R (1962) Silent spring. Houghton Mifflin Company, Boston
7. Meadows DH, Meadows DL, Randers J, Behrens WW III (1972) The limits to growth: a report for the club of rome's project on the predicament of mankind. Universe Books, Falls Church
8. Maher ML (1985) HI-RISE and beyond: directions for expert systems in design. Comput Aided Des 17(9):420–427
9. Kalay YE (1985) Redefining the role of computers in architecture: from drafting/modelling tools to knowledge-based design assistants. Comput Aided Des 17(7):319–328
10. CSD (1992) Agenda 21. Commission on sustainable development (CSD), division for sustainable development. United Nations, New York
11. United Nations (1992) Report of the United nations conference on environment and development. United Nations, New York. https://digitallibrary.un.org/record/168679. Accessed 13 May 2025
12. Frazer J (2016) Parametric computation: history and future. Archit Des 86(2):18–23
13. Woodbury R (2010) Elements of parametric design. Routledge, Oxford. ISBN 978-0415779876
14. Jabi W (2013) Parametric design for architecture. Laurence King Publishing, London. ISBN 9781780673141
15. Tedeschi A (2014) ADD_algorithms-aided design: parametric strategies using grasshopper. Le Penseur Edizioni, Italy
16. Tedeschi A (2011) Parametric architecture with grasshopper: primer. Le Penseur Edizioni, Italy
17. McNeel Associates (2025) M&A product documentation: Rhinoceros/Rhino. Robert McNeel Associates, Washington. https://www.rhino3d.com/. Accessed 6 Apr 2025
18. Sörensen K, Sevaux M, Glover F (2018) A history of metaheuristics. In: Martí R, Pardalos P, Resende M (eds) Handbook of heuristics. Springer International Publishing AG, Switzerland
19. Newell A, Simon HA (1972) Human problem solving. Prentice-Hall, Hoboken
20. Rosenblatt F (1958) The perceptron: a probabilistic model for information storage and organization in the brain. Psychol Rev 65(6):386

21. Minsky M, Papert S (1969) Perceptrons: An introduction to computational geometry. MIT Press, Cambridge
22. Hewitt C (1969) PLANNER: a language for proving theorems in robots. In: Proceedings of the 1st international joint conference on Artificial intelligence, Washington, pp 95–301
23. Stiny G, Gips J (1971) Shape grammar and the generative specification of painting and sculpture. Paper presented at the IFIP Congress, Ljubljana, Yugoslavia
24. Alexander C (1977) A pattern language: towns, buildings, construction. Oxford University Press, Oxford
25. Price C (1976) Generator Project. MoMA, New York. https://www.moma.org/collection/works/1076. Accessed 6 Apr 2025
26. Sutherland IE (2003) Sketchpad: a man-machine graphical communication system. A technical report based on a dissertation submitted in January 1963 by the author for the degree of Doctor of Philosophy to the Massachusetts Institute of Technology. Technical Report, Number 574, Computer Laboratory, University of Cambridge, Cambridge
27. Taheri AH, Suresh K (2021) Surface approximations using generalized NURBS. Eng Comput 38(5):4221–4239
28. Rogers D (2000) An introduction to NURBS with historical perspective. Elsevier Science
29. Dassault Systèmes (2025) NURBS modeling. Dassault Systèmes, Paris, https://www.3ds.com/store/cad/nurbs-modeling. Accessed 24 May 2025
30. Furtado CL (2008) Cedric price's generator and the Frazers' systems research. Technoetic Arts 6(1):55–72
31. Davis R (1982) Expert systems: where are we? and where do we go from here? AI Mag 3(2):3–22
32. Peter N, Stuart R (1995) Artificial intelligence: a modern approach. Prentice Hall, Hoboken
33. Crevier D (1993) AI: the tumultuous search for Artificial intelligence. Basic Books, New York. ISBN 0-465-02997-3
34. Weisberg D (2022) History of CAD. Shapr3D Inc, Budapest. https://www.shapr3d.com/blog/history-of-cad. Accessed 6 Apr 2025
35. Brown-Siebenaler K (2014) A quick history of PTC and PTC Creo. PTC Inc, Boston. https://www.ptc.com/en/blogs/cad/a-quick-history-of-ptc-and-ptc-creo. Accessed 25 May 2025
36. ICBO (1982) Uniform building code, 1982 Edition. In: International Conference of Building Officials (ICBO), Lansing, Michigan
37. Russell S, Norvig P (2010) Artificial intelligence: a modern approach, 3rd edn. Prentice Hall, Upper Saddle River, US
38. Mukherjee A, Deshpande JM (1995) Modeling initial design process using artificial neural networks. J Comput Civ Eng 9(3):194–200
39. Kreider JF, Claridge DE, Curtiss P, Dodier R, Haberl JS, Krarti M (1995) Building energy use prediction and system identification using recurrent neural networks. J SolEnergy Eng 117(3):161–166
40. McNeel Associates (2025) M&A product documentation: grasshopper parametric. Robert McNeel Associates, Washington. https://www.rhino3d.com/learn. Accessed 6 Apr 2025
41. Makert R, Alves G (2016) Between designer and design: parametric design and prototyping considerations on Gaudí's Sagrada Familia. Period Polytech Arch 47(2):89–93
42. Sumida B, Houston A, McNamara J, Hamilton W (1990) Genetic algorithms and evolution. J Theor Biol 147(1):59–84
43. Whitley D, Starkweather T, Bogart C (1990) Genetic algorithms and neural networks: optimizing connections and connectivity. Parallel Comput 14(3):347–361
44. Schumacher P (2008) Parametricism as style: Parametricist manifesto. https://patrikschumacher.com/parametricism-as-style-parametricist-manifesto/. Accessed 6 Apr 2025
45. Su Z, Yan W (2015) A fast genetic algorithm for solving architectural design optimization problems. Artif Intell Eng Des Anal Manuf 29(4):457–469
46. Pham DT, Onder HH (1992) A knowledge-based system for optimizing workplace layouts using a genetic algorithm. Ergon 35(12):1479–1487

47. Wang BP, Chen JL (1996) Application of genetic algorithm for the support location optimization of beams. Comput Struct 58(4):797–800
48. Frazer J (2002) Creative design and the generative evolutionary paradigm. In: Paper presented at the creative evolutionary systems conference, Burlington
49. Kalogirou SA (2001) Artificial neural networks in renewable energy systems applications: a review. Renew Sustain Energy Rev 5(4):373–401
50. Frazer J, Johnston P (1995) An evolutionary architecture: themes VII. Architectural Association Publications, London
51. Autodesk (2025f) Autodesk AI. Autodesk Inc., USA. https://www.autodesk.com/uk/solutions/autodesk-ai. Accessed 27 May 2025
52. Frost J (2025) Artificial intelligence (AI) policy. Arup Group Limited, London. https://www.arup.com/about-us/corporate-reports/artificial-intelligence-ai-policy/. Accessed 27 May 2025
53. Chew ZX, Wong JY, Tang YH, Yip CC, Maul T (2024) Generative design in the built environment. Autom Constr 166:105638
54. Deb K, Pratap A, Agarwal S, Meyarivan T (2002) A fast and elitist multiobjective genetic algorithm: NSGA-II. IEEE Trans Evol Comput 6(2):182–197
55. Hooke R, Jeeves TA (1961) "Direct Search" solution of numerical and statistical problems. J ACM 8(2):212–229
56. Bonyadi MR, Michalewicz Z (2017) Particle swarm optimization for single objective continuous space problems: a review. Evol Comput 25(1):1–54
57. Blank J, Deb K (2020) Pymoo: multi-objective optimization in Python. IEEE Access 8:89497–89509
58. Teoh TT, Rong Z (2022) Artificial intelligence with python. Springer Singapore, Singapore
59. Autodesk (2025c) Autodesk forma: making the right decisions in the planning phase has never been easier. Autodesk Inc, San Francisco. https://www.autodesk.com/products/forma. Accessed 6 Apr 2025
60. Finch (2025) The generative Copilot for better buildings. Finch Corporation, San Francisco. https://www.finch3d.com/. Accessed 6 Apr 2025
61. Chaudhuri T, Soh YC, Li H, Xie L (2019) A feedforward neural network based indoor-climate control framework for thermal comfort and energy saving in buildings. Appl Energy 248:44–53
62. Wang Z, Sacks R, Yeung T (2022) Exploring graph neural networks for semantic enrichment: room type classification. Autom Constr 134:104039. https://doi.org/10.1016/j.autcon.2021.104039
63. Ahmed A, Ali S, Raza A, Hussain I, Bilal A, Fitriyani NL, Gu Y, Syafrudin M (2025) Novel deep neural network architecture fusion to simultaneously predict short-term and long-term energy consumption. PLoS ONE 20(1):e0315668. https://doi.org/10.1371/journal.pone.0315668
64. Gong Y, Zoltán ES, Gyergyák J (2023) A neural network trained by multi-tracker optimization algorithm applied to energy performance estimation of residential buildings. Build 13(5):1167
65. Dhanasekaran S, Gopal D, Logeshwaran J, Ramya N, Salau AO (2024) Multi-model traffic forecasting in smart cities using graph neural networks and transformer-based multi-source visual fusion for intelligent transportation management. Int J Intell Transp Syst Res 22(3):518–541
66. Autodesk (2025e) Generation settings. Autodesk Inc, San Francisco. https://help.autodesk.com/view/RVT/2024/ENU/. Accessed 6 Apr 2025
67. Bailey ET, Caldas L (2023) Operative generative design using non-dominated sorting genetic algorithm II (NSGA-II). Autom Constr 155:105026
68. Nisztuk M, Myszkowski PB (2019) Hybrid Evolutionary algorithm applied to automated floor plan generation. Int J Archit Comput 17(3):260–283
69. Evins R (2013) A review of computational optimisation methods applied to sustainable building design. Renew Sustain Energy Rev 22:230–245
70. Cai Y, Guo Z, Pei Y, Bian W, Zheng W (2024) SimGRAG: leveraging similar subgraphs for knowledge graphs driven retrieval-augmented generation. https://arxiv.org/abs/2412.15272
71. Tan L, Luhrs M (2024) Using generative AI Midjourney to enhance divergent and convergent thinking in an architect's creative design process. Des J 27(4):677–699

72. Feng G, Yan W (2024) Text2VP: Generative AI for visual programming and parametric modeling. https://arxiv.org/abs/2407.07732
73. Aung THS (2022) Functional and aesthetic: 10 Impressive parametric buildings from around the world. Novatr Inc, Delaware. https://www.novatr.com/blog/parametric-buildings-around-the-world. Accessed 6 Apr 2025
74. UNDP (2025) AI for sustainable development. United Nations Development Programme (UNDP), United Nations, New York. https://www.undp.org/digital/ai. Accessed 6 Apr 2025
75. Paulus C (2022) Unveiling the design secrets of Zaha Hadid: innovation, inspiration, and legacy. Illustrarch Inc, Delaware. https://illustrarch.com/articles/31247-unveiling-the-design-secrets-of-zaha-hadid.html. Accessed 6 Apr 2025
76. Reid RL (2014) Grace under pressure. Civ Eng 84(11):50–59
77. Walker S (2022) Is artificial intelligence going to design your next building? Arup Inc, London. https://www.arup.com/insights/is-artificial-intelligence-going-to-design-your-next-building/. Accessed 6 Apr 2025
78. Rawlinson S (2023) We must be flexible to keep up with accelerating AI. Arcadis Inc, Gland, Switzerland. https://www.arcadis.com/en/insights/blog/global/simon-rawlinson/2023/we-must-be-flexible-to-keep-up-with-accelerating-ai. Accessed 6 Apr 2025
79. Aneja A (2024) The impact of AI on buildings. Arcadis Inc, Gland, Switzerland. https://www.arcadis.com/en-us/insights/blog/united-states/arun-aneja/2024/the-impact-of-ai-on-buildings. Accessed 6 Apr 2025
80. RIBA (2024) RIBA AI report 2024. Royal Institute of British Architects (RIBA), London. https://www.architecture.com/knowledge-and-resources/resources-landing-page/riba-ai-report-(2024) Accessed 6 Apr 2025
81. Reiter ER, Burns CC, Cochrane H, Johnson GR, Leong H, Sheaffer JD (1978) Effects of atmospheric variability on energy utilization and conservation. Colorado State University, Fort Collins, USA, Environmental Research Paper
82. Autodesk (2025d) Generative design for architecture, engineering & construction. Autodesk Inc, San Francisco. https://www.autodesk.com/solutions/generative-design/architecture-engineering-construction. Accessed 6 Apr 2025
83. UNFCCC (2016) The Paris agreement. United Nations Framework Convention on Climate Change (UNFCCC), New York, https://unfccc.int/sites/default/files/resource/parisagreement_publication.pdf. Accessed 6 Apr 2025
84. Groom A, Lee B, Locke J (2024) Autodesk's AI innovations transforming sustainable design and construction. Autodesk Research, Autodesk Inc., USA. https://www.research.autodesk.com/blog/autodesks-ai-innovations-transforming-sustainable-design-and-construction/. Accessed 27 May 2025

The Effect of AI-Aided Design Processes on Cognitive Load, Creativity, and Metacognitive Awareness

Gizem Yazıcı

Abstract This chapter investigates the effects of artificial intelligence-aided design (AIAD) processes on cognitive load, creativity, and metacognitive awareness. Within the scope of the research, a three-day online workshop was conducted with thirty-seven undergraduate and graduate architecture students, and data based on participant experiences were analysed using qualitative research methods. The results of the sentiment-based content analysis reveal that the integration of AI tools into the design process reduces cognitive load, supports creative thinking processes, facilitates rapid prototyping and feedback mechanisms, and increases metacognitive awareness. The findings show that AIAD tools can potentially improve designers' cognitive and creative capacities. The chapter evaluates the effects of AIAD processes in educational and professional contexts from a local perspective and makes a new contribution to the literature on the integration of AI into design processes.

Keywords AI-aided design · Cognitive load · Creativity · Metacognitive awareness · Design process

1 Introduction

This chapter examines the effects of AIAD processes on designers' mental processes. The focus is on cognitive load, creativity, and metacognitive awareness, which play critical roles in design processes.

The design process has a significant impact on generating innovative solutions in today's rapidly changing and increasingly complex world. This process is not only about aesthetics and functionality but also has a dynamic structure where complex problem-solving mechanisms, creativity and metacognitive awareness interact. The effective management of designers' cognitive capacities, the development of creative problem-solving skills and the organization of their thinking processes are of decisive

G. Yazıcı (✉)
Department of Architecture, Izmir Institute of Technology, Izmir University of Economics, Izmir, Turkey
e-mail: gizemyazici@iyte.edu.tr; gizem.yazici@ieu.edu.tr

M. Sedrez and G. Pitts (eds.), *Artificial Intelligence-Aided Design for Sustainability*, Urban Sustainability, https://doi.org/10.1007/978-981-95-1349-9_3

importance both in terms of their performance and their learning experiences in design education [1, 2].

Cognitive Load Theory (CLT) [1] suggests that individuals have limited information processing capacities during learning and problem-solving processes, and therefore, learning environments should be structured in a way that optimizes cognitive resources. Since design processes involve processes such as problem-solving and creative thinking that require a high level of cognitive effort, cognitive load management stands out as a factor that directly affects design efficiency [3, 4]. Creativity, on the other hand, is at the centre of the design process and plays a critical role in individual and collective problem-solving processes [2, 5]. In this regard, it is also important to note that metacognitive awareness allows designers to conduct decision-making and problem-solving processes more consciously and strategically as the ability to monitor, control and organize their thought processes [6].

In recent years, it has been observed that AIAD processes have significant effects on the human cognitive system [7, 8]. The integration of AI tools into design processes enables designers to use their cognitive capacities more efficiently, conduct creative processes more effectively, and gain greater awareness of their own thinking processes [9]. In this context, it is seen that AIAD systems reduce designers' cognitive load and allow them to focus on their creative and problem-solving cognitive processes [10].

This chapter aims to examine the effects of AIAD processes on cognitive load, creativity, and metacognitive awareness. In line with UNESCO's inclusive AI vision emphasized in the Beijing Consensus [11], this chapter examines the effects of integrating AI tools into design processes in a local context. The main objective of the chapter is to understand how the integration of AI tools into architectural design processes contributes to the cognitive processes and creativity of participants.

Given these considerations, although existing studies in the literature indicate that AIAD processes reduce cognitive load, support creativity and increase metacognitive awareness, it is observed that research conducted in local and original contexts in this field is limited. This chapter aims to provide strategic suggestions for design process by addressing the potential benefits of AIAD tools in the design process from a local perspective. In addition, it is aimed that the findings obtained will contribute to the development of educational programs for designers to optimize their cognitive and creative processes.

2 Components in the Design Process: Cognitive Load, Creativity, Metacognitive Awareness

The design process is a dynamic activity in which complex problem-solving mechanisms, creativity and metacognitive awareness actively interact. In this process, how designers manage their cognitive capacities, develop creative problem-solving skills and organize their thinking processes play a decisive role in terms of both their performance and their learning experiences in design education. Cognitive load, creativity, and metacognitive awareness are interconnected components of the design process, and their effective management plays an important role in ensuring success in the design process.

CLT [1] suggests that information processing capacities are limited in learning and problem-solving processes, and therefore learning environments should be structured in a way that optimizes cognitive resources. Design processes include processes that require high levels of cognitive effort, such as problem-solving and creative thinking, and therefore, cognitive load management is a factor that directly affects design efficiency [3, 12]. Chandler and Sweller [3] emphasize that optimizing cognitive load can increase learning efficiency and suggest that this approach can also be integrated into design education. Cognitive load is divided into three main contents: extraneous load, intrinsic load and germane load [12]. While intrinsic load refers to the difficulties arising from the nature of the design problem, extraneous load arises from the complexities related to the organization of information. The germane load includes cognitive efforts that support the learning process. In the design process, optimizing the problem space and information organization can allow designers to focus on creative thinking, especially by reducing the extraneous load. In this context, researchers such as Cross [13] and Lawson [14] emphasize that designers should turn to strategies that balance the cognitive load to manage problem-solving processes more effectively. Strategies aimed at reducing cognitive load in design education can contribute to students' more flexible and creative thinking in complex problem spaces.

As a second component of the design process, creativity is a fundamental element at the centre of the design process and plays a critical role in individual and collective problem-solving processes. While Amabile [2] defines creativity as the combination of individual motivation, knowledge, and creative thinking skills, Csikszentmihalyi [5] considers the creative process as a "flow" state shaped by the interactions between the individual, discipline, and field. These approaches show that creativity in the design process is directly related not only to individual abilities but also to social and environmental factors. In other words, creativity does not emerge in isolation; it is embedded within a broader ecosystem of interpersonal dynamics, disciplinary expectations, and contextual influences. To better understand how these complex influences manifest within the design process, several theoretical models have been proposed. The Creative Problem Solving (CPS) Model [15] emphasizes that creative problem-solving is a structure that encompasses both discovery and production processes. This model argues that divergent and convergent thinking should be used in a balanced way in the design process. These two ways of thinking, put forward by Guilford [16], are of critical importance in terms of producing different solutions to the design problem and systematically evaluating these solutions. When placed within a social or collaborative setting, these cognitive strategies gain additional value, as they support group-based exploration and collective decision-making. Strategies used to support creative thinking in the design process directly affect problem-solving processes. Schön [17] states that design is a reflective process and that designers are in a continuous learning cycle by considering both their past experiences and the current context. In this context, it is emphasized that creative problem-solving processes require cognitive flexibility and that design education should be structured to encourage this flexibility.

The third component of the design process, metacognition, is defined as the ability of individuals to monitor, control, and organize their own thought processes [6]. Metacognitive awareness can increase the effectiveness of the design process by enabling designers to carry out decision-making and problem-solving processes more consciously and strategically. Studies conducted in recent years have revealed that metacognitive awareness is a critical component in design education. While Ball and Christensen [18] emphasize the role that metacognitive processes play in consciously guiding design decisions, Yazici and Dogan [19] state that the use of metacognitive interventions in design education helps students make more conscious decisions in design processes. Clemente et al. [20] and Hargrove [21] show that teaching strategies aimed at increasing the metacognitive awareness of design students can make long-term contributions to the learning process. Schön's [17] concept of reflective practice provides an indirect framework for how metacognitive thinking can be developed in the design process [19]. According to this approach, designers' questioning of their own thinking processes and effective use of feedback loops improve problem-solving and decision-making mechanisms in the design process.

As a result, the dynamic nature of the design process requires the interaction of basic components such as cognitive load, creativity and metacognitive awareness. Managing cognitive load, supporting creative problem-solving processes, and increasing metacognitive awareness contribute significantly to both the individual and academic development of designers.

2.1 *The Role of AI-Aided Design on Components in the Design Process*

While traditional design processes require designers to effectively manage complex cognitive processes such as problem-solving, decision-making, and directing creativity, AIAD processes offer significant opportunities to make these processes more efficient [7, 8].

AIAD offers new approaches that support designers' competencies, especially in areas such as cognitive load management, creative problem-solving, and the development of metacognitive awareness [22, 23]. Recent studies have examined the effects of AI tools in these areas, showing that they enable designers to use their cognitive capacities more efficiently, conduct creative processes more effectively, and gain greater awareness of their own thinking processes [9].

AIAD systems reduce designers' extraneous load, allowing them to focus on their creative and problem-solving cognitive processes [10]. By rapidly generating design alternatives, AI-based generative design tools reduce the cognitive energy designers spend on manual processes and help them focus on more strategic decision-making processes [22]. In a study conducted by Pei et al. [24], it is shown that the use of AIAD tools reduces the cognitive load of design teams and increases their creativity. Similarly, Wadinambiarachchi [25] demonstrates that generative AI systems reduce the cognitive load in design processes and enable designers to focus more on creative decision-making processes. These findings show that AI can be used as an effective supporting tool in terms of cognitive load management and contribute to the creation of more optimized information processing mechanisms in design processes.

AIAD tools accelerate creative problem-solving processes and increase designers' competencies in producing alternative solutions [7, 8]. The Creative Problem Solving (CPS) Model, developed by Finke, Ward, and Smith [15], considers creativity as a structure that does not limit itself to idea generation but also includes discovery and problem-definition processes. AIAD tools support this model, optimizing divergent and convergent thought processes and allowing designers to act in a wider range of creative thinking [9]. In the study conducted by Chandrasekera et al. [26], it is observed that AIAD tools increased users' creative performance and reduced their cognitive load. Furthermore, research conducted by Kim and collaborators [27] has revealed that AI plays a role in stimulating creative solutions in team-based design projects. In this context, AIAD systems demonstrate that creativity plays a critical role not only in the production phase but also in problem definition and discovery processes.

AI-aided systems can strengthen strategic decision-making mechanisms by allowing designers to evaluate their thinking processes more consciously [28]. By supporting feedback mechanisms and adaptive learning processes, AI tools encourage designers to analyse their own cognitive processes and develop their metacognitive awareness [29]. A study conducted by Khotimah and Rusijono [30] shows that AI tools increase students' creative thinking abilities and metacognitive awareness. The concept of reflective practice put forward by Schön [17] emphasizes that designers should constantly evaluate their own processes. AI-aided systems support this process, strengthening designers' conscious decision-making skills and making their learning processes more efficient [31].

As a result, AIAD processes significantly affect fundamental design dynamics such as cognitive load management, creative problem-solving, and metacognitive awareness. While the capacity of AI tools to optimize cognitive load allows designers to focus more on creative processes, their positive effects on creativity accelerate the discovery of alternative solutions. In addition, in terms of metacognitive awareness, AIAD systems enable designers to manage their thinking processes more consciously, helping them make more strategic and informed decisions. In this context, it is anticipated that AIAD tools will increasingly occupy an important place in future design education and professional practices.

3 Methodology

This research was designed as qualitative research examining the effects of AIAD processes on cognitive load, creativity and metacognitive awareness. The main purpose of the research is to understand how the integration of AI tools into architectural design processes contributes to the cognitive processes and creativity levels of the participants. For this purpose, a three-day online workshop was organized, and at the end of the workshop process, the participants were asked to evaluate the workshop process and were given an online open-ended questionnaire. The content analysis of the qualitative data obtained from the survey was used to determine the emotional state.

This research evaluated the impacts of AIAD processes in education in a local context, in line with UNESCO's inclusive AI vision highlighted in the Beijing Consensus [11]. The strategies in this Consensus provided a fundamental reference point for the research to evaluate the effects of integrating AI tools into educational design processes on cognitive load, creativity, and metacognitive awareness. In particular, the ethical issues emphasized by the Beijing Consensus and the potential benefits of AI in education shaped the methodological approach of the chapter. Ethical considerations were carefully observed, and approval was obtained before conducting the research. Informed consent was also obtained from all participants. The privacy of the participants was protected, and the data were anonymized. In addition, the purpose and process of the research were communicated to the participants [32]. Confidentiality of the data and the voluntary participation of the participants were prioritized. The participants were fully informed throughout the research process.

3.1 Participants

A total of 37 (n = 37) participants from various academic levels, including undergraduate and graduate students in architecture, participated in the research. Participants were selected by registering within a certain registration period through platforms where the workshop program was announced. All individuals who applied during the registration period were included in the workshop; no pre-selection criteria were applied during this process, and participation was open to everyone. This method ensured that participants had various demographic and academic backgrounds and increased the generalizability of the research findings [33].

The participants' ages ranged from 20 to 35, with a balanced gender distribution (n male = 12, n female = 15). In addition, the participants' previous experiences with AI were diverse. This diversity ensured that different levels of knowledge and experience were reflected in the research findings. The participants' educational background and familiarity with AI tools contributed to the comprehensiveness and multidimensionality of the research's findings.

3.2 Procedure: Workshop Program

The increasing impact of technology on today's architectural design processes and the central role of AI in these processes reveal the necessity of new skills for designers. In this context, the workshop aimed to provide participants with the ability to create architectural concepts and scenarios using AI tools and to integrate these concepts into form-creation processes. The program was shaped by practical sessions that support creative thinking, critical evaluation, studies and technology-focused tools. Using AI tools such as AI Concept Generator and Adobe Firefly, participants experienced the processes of text-based idea development (text-to-text), creating visualized forms from abstract concepts (text-to-image) and improving the forms they produced (image-to-image). Thanks to these processes, creative thinking, visual communication, spatial awareness, technology integration and critical thinking skills were developed (Fig. 1).

The first day of the workshop focused on supporting the creative thought processes of the participants. For this purpose, interactive sessions and individual studies were organized, and the participants experienced the text-based idea development processes using the AI Concept Generator tool (Fig. 2). Participants generated a design concept using keywords they had decided on before. At this stage, it was emphasized that design is not only a visual action but also a process of constructing an intellectual structure. By working on certain conceptual themes, the participants learned to transform abstract ideas into a structured and expressible form.

The second day focused on the process of transforming scenarios into visual forms. Participants experienced text-to-image transformation by visualizing abstract concepts with the Adobe Firefly tool (Fig. 3). Participants used their design concept, which they created with AI Concept Generator. This process was supported by individual practical sessions. Participants transformed their scenarios into spatial forms using the visual design language and discovered the potential of using technology in design processes.

The third day was dedicated to participants critically evaluating and developing the forms they produced. Supported by Adobe Firefly's image-to-image features, this stage provided participants with the opportunity to reconsider and improve their designs (Fig. 3). Participants used a reference image, which they produced by taking inspiration from text-to-image models to create new images. The presentation and discussion sessions held at the end of the day encouraged not only individual development but also collective learning. Participants had the opportunity to learn from each other and add new dimensions to their designs with different perspectives.

At the end of the workshop, participants completed an online survey consisting of eight open-ended questions to share their experiences and feedback on the process. This survey evaluated the effectiveness of the workshop and analysed in detail the awareness, skills and ways of thinking that participants gained throughout the process. The workshop highlighted the opportunities that technology-supported creative design processes offer designers and offered participants the opportunity to be a part of this transformation.

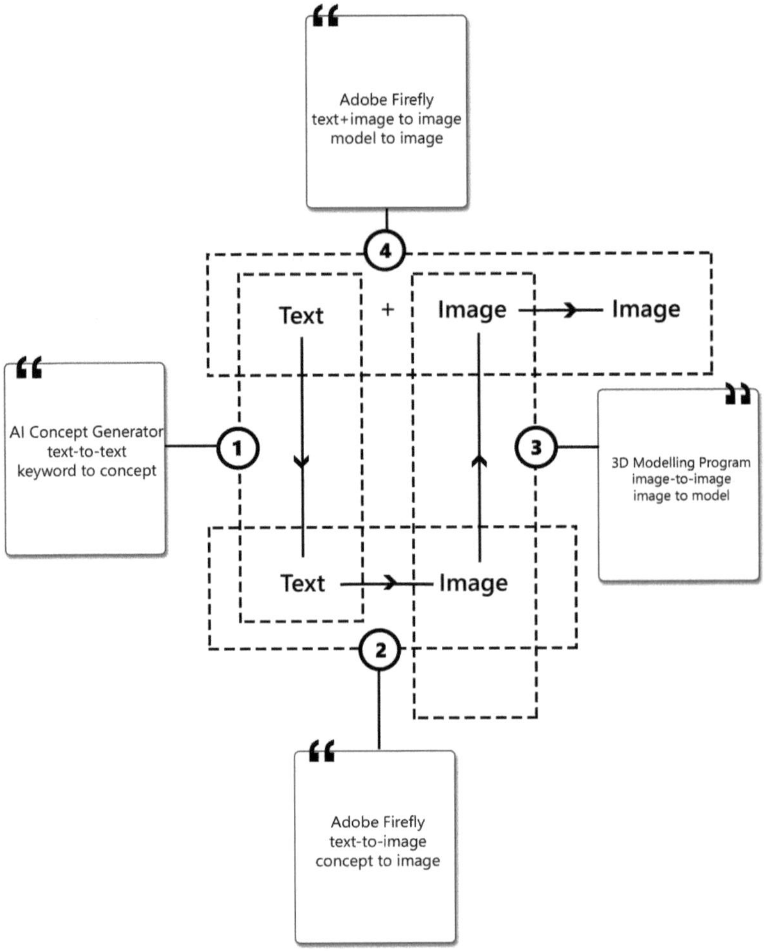

Fig. 1 Creating concepts and forms through AI tools in the design process

3.3 Data Collection Tool

In this research, an online form consisting of eight open-ended questions, which was sent to the participants at the end of the workshop, was used as a data collection tool. The form was designed to allow participants to evaluate the effects of AIAD processes on cognitive load, creativity, and metacognitive awareness. The questions were grouped under four main themes, with two questions in each theme. The themes were determined as follows:

- Cognitive Load Reduction: This theme aimed to understand the potential of AI tools to reduce the cognitive load of participants in design processes.

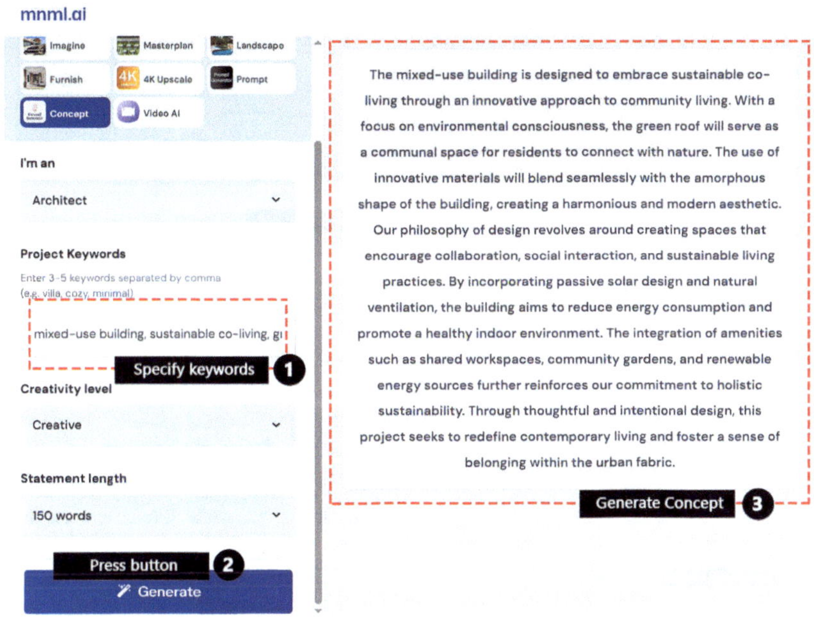

Fig. 2 Concept generation with keywords through by AI Concept Generator

- Creativity: The questions in this group were prepared to evaluate how AI tools contribute to the creative processes and innovative solution-generation capacity of the participants.
- Rapid Prototyping and Feedback: This theme examined the effects of AI-supported rapid prototyping and feedback processes on design results.
- Development of Metacognitive Processes: The last group aimed to investigate the effects of AI tools on metacognitive skills such as planning, monitoring, evaluating and controlling the design processes of the participants.

Table 1 lists the open-ended questions for each theme. These open-ended questions aimed to reveal the participants' experiences in AIAD processes through qualitative data. The answers obtained were examined using the thematic analysis method and formed the basis of the findings of the research.

3.4 *Data Analysis*

The data obtained within the scope of the research was analysed by the sentiment analysis method using MaxQDA software. Sentiment analysis is considered an effective method for understanding the emotional reactions of participants towards AIAD processes. This method provides a suitable analytical framework for making sense of

Fig. 3 Text-to-image and image-to-image generation process by Adobe Firefly

the participants' responses and examining the data set more systematically. During the data analysis process, the participants' responses were classified according to five sentiment tone categories: positive, somewhat positive, neutral, somewhat negative, and negative. The analysis process included the following steps:

1. Question-Based Analysis: All participants' responses for each question were categorized one by one according to their emotional tones. This step allowed for the examination of individual responses and detailed analysis of participant experiences at the emotional level.
2. Thematic Grouping: Responses were also grouped under the basic thematic framework of the research. These themes were determined as: Cognitive Load Reduction, Creativity, Rapid Prototyping and Feedback, and Development of Metacognitive Processes.

The findings were interpreted in detail to gain a more comprehensive understanding of the participants' experiences and to demonstrate the originality of the research. Quotations from participant responses were presented with the analysis results to increase the reliability of the research and support the findings. This process

Table 1 Open-ended questions and themes

No	Question	Theme
1	While using AI tools, can you briefly explain which tasks in your design process became easier and how these tools helped reduce your mental workload?	Cognitive load reduction
2	Do you think AI tools have alleviated repetitive or complex tasks in your design process, enabling you to save time and energy?	
3	In what ways do you think AI tools have contributed to your creative process? Have they provided new sources of inspiration or introduced different perspectives?	Creativity
4	How have the design suggestions generated by AI influenced your creative thinking? Can you briefly explain how these tools have supported you in developing innovative solutions?	
5	How do you think using AI tools for rapid prototyping and instant feedback has impacted your design process and outcomes?	Rapid prototyping and feedback
6	How would you describe the influence of rapid prototyping and feedback processes on your designs compared to traditional methods?	
7	Have you experienced instances where AI tools enhanced your ability to plan, monitor, evaluate, and control your design process?	Development of metacognitive processes
8	Can you explain how AI tools have improved your ability to plan, monitor, evaluate, and control your thought processes? Have these tools helped you make more strategic decisions?	

was carried out under the validation strategies recommended in qualitative analysis [34].

As a result, this analysis method provided an in-depth understanding of the participants' experiences and emotional reactions to the AIAD processes. Thematic grouping allowed us to compare and interpret the participants' different experiences at the emotional level, in line with the main objectives of the chapter. Thus, the innovative contribution of the research and the originality of its findings were guaranteed by the meticulous examination and thematic classification of the obtained data.

4 Results

The data obtained in this chapter was examined by sentiment analysis method using MaxQDA software to analyse the emotional reactions of the participants towards the AIAD processes. The responses given by the participants were classified according to positive, somewhat positive, neutral, somewhat negative and negative sentiment tones. This analysis aimed to understand the effects of AI tools on the design processes in detail.

Considering all questions and themes, it was seen that 88.18% of the responses were positive, 6.08% were somewhat positive, 1.35% were neutral, 4.05% were somewhat negative, and 0.34% were negative (Table 2). These findings show that the participants generally had a positive approach to AIAD processes and that these processes made significant contributions in cognitive, creative and metacognitive dimensions.

When the results are reviewed by question, only positive or somewhat positive responses were given to the first question. For the second, third, fourth, and sixth questions, no neutral or negative responses were observed. The fifth and seventh questions included not only negative but also other emotional tones, while responses to the eighth question reflected all emotional tones. Sample responses are shown in Table 3.

The analysis results were also evaluated thematically (Table 4). The findings reveal the effects of the AIAD process on the participants under certain themes. Accordingly, the most significant contribution of AI stands out as rapid prototyping and providing feedback (93.2%). This is followed by the development of metacognitive processes (89.2%), support for creativity (87.8%) and reduction of cognitive load (82.4%). These findings reveal the role of AI-based tools in the design process and how they contribute to the learning and thinking processes of the participants.

Table 2 Analysis results

	Positive		Somewhat positive		Neutral		Somewhat negative		Negative	
	Response	%	Response	%	Response	%	Response	%	Response	%
Q1	36	97.30	1	2.70	0	0.00	0	0.00	0	0.00
Q2	25	67.57	8	21.62	0	0.00	4	10.81	0	0.00
Q3	34	91.89	1	2.70	0	0.00	2	5.41	0	0.00
Q4	31	83.78	4	10.81	0	0.00	2	5.41	0	0.00
Q5	34	91.89	1	2.70	1	2.70	1	2.70	0	0.00
Q6	35	94.59	1	2.70	0	0.00	1	2.70	0	0.00
Q7	34	91.89	0	0.00	2	5.41	1	2.70	0	0.00
Q8	32	86.49	2	5.41	1	2.70	1	2.70	1	2.70
Total	261		18		4		12		1	
%	88.18		6.08		1.35		4.05		0.34	

Table 3 Sample responses from participants

Tone	Sample Response	Q
Positive	It allowed me to more easily visualize the finished version of the initial draft idea in my mind, along with similar concepts. In this stage, I was able to access multiple options more comfortably without spending a long time deliberating	Q1
Somewhat positive	With my current level of knowledge and limited understanding of AI programs, I believe they can be highly beneficial in design research and sketching phases. By creating various prototypes and collecting user feedback throughout this process, I can achieve more effective results. Using tools and methods offered by evolving technology, I aim to make my projects more innovative and user-friendly. The dynamic nature of the design world requires continuous learning and adaptation; therefore, I consistently challenge myself to develop further in these areas and acquire new knowledge	
Positive	Yes, we can access a broad dataset shaped by our own ideas. This way, we don't have to search for examples of what we envision or model and try everything during the design process	Q2
Somewhat positive	It is not particularly creative when it comes to generating concept texts from keywords, but it can be useful and time-saving in organizing the complex sentences we have in mind	
Somewhat negative	Sometimes, AI suggestions can be too general, which occasionally requires additional adjustments	
Positive	It provides several different solutions within a scenario and allows us to compare their visuals or the proposed solutions. Based on this, we can generate similar solutions in our design	Q3
Somewhat positive	I usually stick to the ideas I already have in mind and try to improve them with minimal changes. However, these tools allowed me to clearly see different perspectives as well	
Somewhat negative	Sometimes, AI suggestions deviate from what I expect, which can make my creative process more challenging	
Positive	It presents me with mass ideas that wouldn't have occurred to me. In this way, it shortens my design process and compensates for my shortcomings in creativity	Q4
Somewhat positive	Although it often provides suggestions unrelated to the concept, there is beauty in this irrelevance. It can be seen as a mind-opening exercise during moments when our idea generation becomes stagnant	
Somewhat negative	In a sense, it was constructive; however, on the other hand, following the pre-generated visuals instead of thinking could be considered limiting for creativity	
Positive	Receiving instant feedback allows me to identify design flaws early and simplifies the correction process	Q5
Somewhat positive	It significantly saves time and reduces mental clutter, though it still requires further development	
Neutral	The effort spent during the conceptual design phase can be utilized in other stages of the design process	

(continued)

Table 3 (continued)

Tone	Sample Response	Q
Somewhat negative	Sometimes, we need to experiment with multiple ideas through trial and error. At this point, obtaining quick, finished results increases the time we can dedicate to the idea we want to focus on	
Positive	A prototype that previously took hours to create can now be completed in minutes, resulting in significant time savings	Q6
Somewhat positive	Its speed has a positive impact, but the lack of complete accuracy is a drawback	
Somewhat negative	It accelerated the process but limited my role as a designer	
Positive	Yes, timing is a crucial factor. Since it generates the design idea immediately, the work shifts more intensely toward development	Q7
Neutral	Not yet	
Somewhat negative	I couldn't manage my data very well, but from what I observed with my friends, it's possible to progress incrementally. However, offering something new each time can be exhausting when trying to build upon the project	
Positive	Yes, because we saved a significant amount of time, allowing us to dedicate more time to other areas	Q8
Somewhat positive	In the initial stage, it helped me make more strategic decisions (such as the positioning of the mass). However, in the experiments I conducted by uploading mass images, I did not achieve the same level of efficiency. Abstract and general imagery was more helpful to me. In the final stage, I received very limited and disconnected design suggestions, and I realized that what I had in mind did not translate into the visuals	
Neutral	I believe that writing a concept text is more important than creating visuals when it comes to making strategic decisions. The concept text plays a crucial role in defining the overall framework of the project, and I see the process of refining visuals as an improvement phase that revolves around it	
Somewhat negative	I have concerns that it might limit the creative process. Therefore, I believe it would be more logical and beneficial to use it after creating some initial designs, as it would be much more helpful at that stage	
Negative	I have not yet gained enough experience or adaptation to answer this question	

The findings regarding the reduction of cognitive load show that AI tools significantly reduce the cognitive load in the design process. The results obtained by analysing the responses to question 1 and question 2 were categorized as 82.4% positive, 12.1% somewhat positive, and 5.4% somewhat negative. Participants emphasized that AI tools provide mental relaxation by facilitating repetitive tasks. These findings reveal that AI tools play a supporting role in cognitive processes and enable users to focus on more complex and creative tasks. While participants stated that AI tools are effective in certain tasks, they stated that the functionality of the tools is limited in some cases. However, in general, AI tools seem to have a positive effect on cognitive load.

Questions 3 and 4 under the theme of creativity examine how AI tools contribute to creative processes. The responses under this theme were measured as 87.8% positive, 6.7% somewhat positive and 5.4% somewhat negative. These rates show that most participants think that AI adds innovative ideas and new perspectives to their designs. In addition, it reveals the potential of AI tools to support creative thinking and develop innovative solutions. The somewhat negative response rate indicates that some participants find these innovative contributions limited. This situation reveals that AI is perceived as an effective tool in creative processes, but there are some differences depending on individual experiences.

Questions on rapid prototyping and feedback processes (Question 5 and Question 6) assess the capacity of AI to accelerate and improve these processes. According to the results, 93.2% were grouped as positive, 2.7% somewhat positive, 1.3% neutral and 2.7% somewhat negative. These rates show that most participants think that AI optimizes rapid prototyping processes. Providing instant feedback in feedback processes supported the design process in terms of both time and quality.

The findings regarding the development of metacognitive awareness and processes reveal the positive impact of AI tools in this area. The findings included 89.2% positive, 2.7% somewhat positive, 4% neutral, 2.7% somewhat negative and 1.3% negative responses. Positive and somewhat positive responses emphasized that AI tools improved the skills of planning, monitoring, evaluating and controlling design processes. Somewhat negative and negative responses indicate that some participants found the impact of these tools on metacognitive processes limited. However, in general, these findings emphasize that AI tools play an important role in the development of metacognitive skills (Table 4).

5 Discussion

The findings of this chapter show that AIAD processes offer significant contributions in terms of reducing cognitive load, encouraging creativity, optimizing rapid prototyping and feedback processes, and increasing metacognitive awareness. The chapter is largely consistent with existing studies in the literature but offers new perspectives compared to studies conducted in different geographical and cultural contexts. The results of this research provide theoretical and practical insights into how AI tools can be used in design processes and provide an important basis for future research.

The findings reveal that AI tools significantly reduce the cognitive load in the design process (82.4% positive response). This situation is directly related to the CLT proposed by Chandler and Sweller [3] and Paas and van Merriënboer [4]. According to CLT, effective management of cognitive resources improves the learning process and problem-solving skills. This chapter shows that AI tools, especially by automating repetitive and procedural tasks, enable designers to focus on more creative and analytical processes. This result is also consistent with the study by Zhang et al. [10] that AI-based tools reduce the extraneous load and allow designers to spend more time on creative processes. However, the fact that some participants

Table 4 Analysis results according to the themes

	Positive		Somewhat positive		Neutral		Somewhat negative		Negative	
	Response	%	Response	%	Response	%	Response	%	Response	%
Cognitive load reduction	61	82,4	9	12,1	0	0	4	5,4	0	0
Creativity	65	87,8	5	6,7	0	0	4	5,4	0	0
Prototyping feedback	69	93,2	2	2,7	1	1,3	2	2,7	0	0
Metacognitive awareness	66	89,2	2	2,7	3	4,0	2	2,7	1	1,3

(5.4%) stated that the functionality of AI is limited indicates that AI may not be equally effective in every scenario. This situation can be explained by Sweller and Chandler's [3] distinction between intrinsic load (task complexity) and extraneous load (unnecessary cognitive load); the possibility that AI tools may create additional cognitive load in some tasks should be considered.

The chapter provides strong evidence that AI tools support creativity (87.8% positive response). This finding is consistent with the theories of Amabile [2] and Csikszentmihalyi [5] explaining the creativity processes. When considered within the scope of the Creative Problem Solving (CPS) Model of Finke, Ward, and Smith [15], it is seen that AI tools support creative problem-solving by optimizing divergent (different thinking) and convergent (focused thinking) processes. However, some participants (5.4%) found the creative contributions offered by AI to be limited. This situation shows that the contribution of AI to creative processes may be related to the level of perceived control, as emphasized by Boden [9] in his studies on creative AI. In other words, the extent to which users adopt and integrate the ideas generated by AI may be a critical factor in determining the effectiveness of the process.

One of the most striking findings of the chapter is that AI significantly improves rapid prototyping and feedback processes (93.2% positive response). This result is consistent with the findings of McCormack et al. [7] that AI accelerates design processes and strengthens feedback mechanisms. AI tools offer great advantages in design education and professional practice as they enable designers to quickly materialize their ideas and strengthen iterative design cycles by accelerating feedback processes. However, some participants stated that they experienced limited difficulties in feedback processes (2.7% somewhat negative, 1.3% neutral response). When this situation is evaluated within the framework of Schön's [17] concept of reflective practice, it reveals that the feedback provided by AI may differ in terms of its capacity to deepen designers' insights. Therefore, it is recommended that feedback processes be optimized in a user-centred manner and that designers be encouraged to use these tools consciously and critically.

The chapter reveals that AI helps designers improve their metacognitive awareness (89.2% positive response). This finding supports the studies conducted by Yazici and Dogan [19, 35, 36] and Ball & Christensen [18] on the critical role of metacognitive awareness in design processes. Especially when evaluated in the context of Schön's [17] concept of reflective practice, it can be said that AI improves designers' ability to monitor, control and organize their own thought processes. However, some participants (5.4%) stated that the effect of AI on metacognitive processes is limited. This situation shows that Cross's [37] approach of designers discovering and developing their own ways of thinking with the "designerly ways of knowing" approach may not fully coincide with the opportunities offered by AI.

6 Conclusion

This chapter has brought a new perspective to the existing literature by examining the effects of AIAD processes on cognitive load management, creativity, and metacognitive awareness. The findings reveal that the integration of AI into design processes improves the cognitive and creative capacities of designers and has the potential to create significant transformations in design practices.

The data obtained show that AI tools reduce cognitive load by automating repetitive tasks, support designers' creative problem-solving processes, and increase their metacognitive awareness. However, it was also determined that the effects of AI tools may vary depending on the participants' experiences. This situation reveals that pedagogical and methodological approaches need to be developed for more effective integration of AI into design processes.

The chapter has some limitations. Firstly, the limited number of participants and the fact that the research was conducted only in the field of architecture limit the generalizability of the findings. In future studies, including participants from different design disciplines and with various professional backgrounds, will allow for the evaluation of the effects of AI tools from a broader perspective. Secondly, the characteristics of the AI tools used and the participants' familiarity with these tools have caused the results obtained to differ. User-friendly interfaces, the quality of the suggestions provided by the algorithms, and the way AI is integrated into the design processes directly affect how designers evaluate these tools. Therefore, in future studies, a comparative analysis of different AI systems should be conducted to examine which tools support designers' cognitive and creative processes more effectively. Thirdly, this chapter was conducted online with volunteers and within the scope of a three-day workshop. This requires evaluating the effect of participants' motivation levels and the duration of interaction with AI tools on the findings. Studies spanning longer periods should be conducted to understand the long-term effects of AI-assisted design processes.

Despite all its limitations, this chapter makes significant contributions to the field by systematically addressing the effects of AIAD processes on cognitive load management, creative problem-solving, and metacognitive awareness. In future studies, it is recommended that comprehensive analyses be conducted on how AI tools are used in different design disciplines, new models be developed on how these technologies can be integrated more effectively in education, and more in-depth studies be conducted on how AI transforms not only production processes but also the way designers think.

In conclusion, this chapter shows that the integration of AI into design processes offers a new and original perspective on how it can transform designers' cognitive and creative processes. In this context, it is of great importance to increase interdisciplinary research to make AIAD systems more inclusive, user-centred and pedagogically integrated.

References

1. Sweller J (1988) Cognitive load during problem solving: effects on learning. Cogn Sci 12:257–285. https://doi.org/10.1207/s15516709cog1202_4
2. Amabile TM (1983) The social psychology of creativity: a componential conceptualization. J Pers Soc Psychol 45:357. https://doi.org/10.1037/0022-3514.45.2.357
3. Chandler P, Sweller J (1991) Cognitive load theory and the format of instruction. Cogn Instr 8:293–332. https://doi.org/10.1207/s1532690xci0804_2
4. Paas FGWC, Van Merriënboer JJG (1994) Instructional control of cognitive load in the training of complex cognitive tasks. Educ Psychol Rev 6:351–371. https://doi.org/10.1007/BF02213420
5. Csikszentmihalyi M (1997) Flow and creativity. NAMTA J 22(2):60–97. Retrieved 10 April 2024, from https://archives.montessori-ami.org/do/8021477b-18d4-41de-949a-346a773fe798
6. Flavell JH (1979) Metacognition and cognitive monitoring: a new area of cognitive–developmental inquiry. Am Psychol 34:906–911. https://doi.org/10.1037/0003-066X.34.10.906
7. McCormack J, Hutchings P, Gifford T et al (2020) Design considerations for real-time collaboration with creative artificial intelligence. Organised Sound 25:41–52. https://doi.org/10.1017/S1355771819000451
8. Gero JS (2002) Computational models of creative designing based on situated cognition. In: Proceedings of the 4th conference on Creativity & cognition, pp 3–10. https://doi.org/10.1145/581710.581712
9. Boden MA (1998) Creativity and artificial intelligence. Artif Intell 103:347–356. https://doi.org/10.1016/S0004-3702(98)00055-1
10. Zhang H, Chen P, Xie X et al (2024) A hybrid prototype method combining physical models and generative Artificial Intelligence to support creativity in conceptual design. ACM Trans Comput-Hum Interact 31:1–34. https://doi.org/10.1145/3689433
11. UNESCO (2019) Beijing consensus on artificial intelligence and education. In: International conference on Artificial intelligence and education, Planning education in the AI Era: Lead the Leap. Retrieved 12 Jun 2024, from https://unesdoc.unesco.org/ark:/48223/pf0000368303
12. Paas FGWC, Van Merriënboer JJG, Adam JJ (1994) Measurement of cognitive load in instructional research. Percept Mot Ski 79:419–430. https://doi.org/10.2466/pms.1994.79.1.419
13. Cross N (2007) From a design science to a design discipline: understanding designerly ways of knowing and thinking. In: Michel R (ed) Design research now: essays and selected projects. Birkhäuser Basel, Basel, pp 41–54. https://doi.org/10.1007/978-3-7643-8472-2_3
14. Lawson B (2006) How designers think: the design process demystified, 4th edn. Routledge, London. https://doi.org/10.4324/9780080454979
15. Finke RA, Ward TB, Smith SM (1992) Creative cognition: theory, research, and applications. MIT Press, Cambridge. https://doi.org/10.7551/mitpress/7722.001.0001
16. Guilford JP (1967) Creativity: yesterday, today and tomorrow. J Creat Behav 1:3–14. https://doi.org/10.1002/j.2162-6057.1967.tb00002.x
17. Schön DA (1983) The reflective practitioner: how professionals think in action. Temple Smith, London. https://doi.org/10.4324/9781315237473
18. Ball LJ, Christensen BT (2019) Advancing an understanding of design cognition and design metacognition: progress and prospects. Des Stud 65:35–59. https://doi.org/10.1016/j.destud.2019.10.003
19. Yazici G, Dogan F (2024) Fostering metacognition in the design studio: the effect of minimal interventions on architectural students' metacognitive awareness. Think Ski Creat:101651. https://doi.org/10.1016/j.tsc.2024.101651
20. Clemente V, Tschimmel K, Vieira R (2017) Why a logbook? a backpack journey as a metaphor for product design education. Des J 20(Suppl 1):S1530–S1542. https://doi.org/10.1080/14606925.2017.1352677
21. Hargrove RA (2013) Assessing the long-term impact of a metacognitive approach to creative skill development. Int J Technol Des Educ 23(3):489–517. https://doi.org/10.1007/s10798-011-9200-6

22. Chandrasekera T, Hosseini Z, Perera U (2024) Can artificial intelligence support creativity in early design processes? Int J Architect Comput. https://doi.org/10.1177/14780771241254637
23. Adeleye IO (2024) The impact of artificial intelligence on design: Enhancing creativity and efficiency. J Eng Appl Sci 3(1):1–13. https://doi.org/10.70560/vvsfej12
24. Pei Y, Wang L, Xue C (2024) Human–AI co-drawing: studying creative efficacy and eye tracking in observation and cooperation. Appl Sci 14(18):8203. https://doi.org/10.3390/app14188203
25. Wadinambiarachchi S, Kelly RM, Pareek S, et al (2024) The effects of generative AI on design fixation and divergent thinking. In: CHI 2024—Proceedings of the 2024 CHI conference on human factors in computing systems. Association for Computing Machinery, Inc., pp 1–18. https://doi.org/10.1145/3613904.3642919.
26. Chandrasekera T, Hosseini Z, Perera U, Bazhaw Hyscher A (2024) Generative artificial intelligence tools for diverse learning styles in design education. Int J Architect Comput. https://doi.org/10.1177/14780771241287345
27. Kim J, Lou MM (2023) The effect of AI-based inspiration on human design ideation. Int J Des Creat Innov 11:81–98. https://doi.org/10.1080/21650349.2023.2167124
28. Marocco S, Talamo A, Quintiliani F (2024) Applying design thinking to develop AI-based multi-actor decision-support systems: a case study on human capital investments. Appl Sci 14:5613. https://doi.org/10.3390/app14135613
29. Zimmerman BJ, Schunk DH (2011) Self-regulated learning and performance: an introduction and an overview. In: Schunk DH, Zimmerman BJ (eds) Handbook of self-regulation of learning and performance. Routledge, New York, pp 1–12. https://doi.org/10.4324/9780203839010
30. Khotimah K, Rusijono AM (2024) Enhancing metacognitive and creativity skills through AI-driven meta-learning strategies. Int J Interact Mob Technol 18(5):18–31. https://doi.org/10.3991/ijim.v18i05.47705
31. Steyvers M, Kumar A (2024) Three challenges for AI-assisted decision-making. Perspect Psychol Sci 19(3):722–734. https://doi.org/10.1177/17456916231181102
32. Orb A, Eisenhauer L, Wynaden D (2001) Ethics in qualitative research. J Nurs Scholarsh 33:93–96. https://doi.org/10.1111/j.1547-5069.2001.00093.x
33. Patton C, Sawicki D, Clark J (2015) Basic methods of policy analysis and planning, 3rd edn. Routledge, New York. https://doi.org/10.4324/9781315664736.
34. Creswell JW (2013) Steps in conducting a scholarly mixed methods study. DBER Speaker Series 48. https://doi.org/10.32873/unl.dc.dberspeakers.48.
35. Yazici G, Dogan F (2023) Shortcomings in design education: Exploring metacognitive activities. In: Jones D, Borekci NA, Clemente V, Corazzo J, Lotz N, Nielsen LM, Lesley-Ann N (eds.), The 7th international conference for design education researchers, 29 November–1 December 2023, London. https://doi.org/10.21606/drslxd.2024.060
36. Yazici G, Dogan F (2024) An inquiry into the relationships among design learning, metacognitive awareness, and academic goal orientation. In: Gray C, Ciliotta Chehade E, Hekkert P, Forlano L, Ciuccarelli P, Lloyd P (eds.) DRS2024: Boston, 23–28 June, Boston. https://doi.org/10.21606/drs.2024.319.
37. Cross N (2001) Designerly ways of knowing: design discipline versus design science. Des Issues 17(3):49–55. https://doi.org/10.1162/074793601750357196

Comparing Conventional and AI-Aided Approaches to Architectural Visuals: A Pedagogical Evaluation

Faruk Can Ünal and Derya Karadağ

Abstract The rise of AI in architectural design is changing traditional workflows, especially in tasks related to visualization and the refinement of incomplete architectural visuals. This chapter presents a comparative study between conventional digital editing methods and AI-aided completion tools within the context of architectural education. The aim is to understand how the integration of AI tools influences the process and its outcomes, specifically in terms of efficiency, accuracy, and usability for the production of architectural visuals. The research involves two groups of students completing the same tasks under identical constraints and objectives. The first group manually edits and completes architectural visuals using conventional Photoshop editing tools. The second group utilizes Photoshop Generative Fill tool that automates parts of the image completion process using AI-driven algorithms that generate detailed and contextually refined outputs. A reflexive thematic analysis-driven evaluation framework is employed to assess each group's results, emphasizing qualitative insights while incorporating quantitative user ratings for complementary analysis. This comparative study explores how these tools impact students' learning, creative decision-making, and ability to tackle complex real-world design challenges. By examining the outcomes of this pedagogical experiment, the study contributes to ongoing discussions about the role of AI in architectural education and practice.

Keywords Architectural visualization · Architectural education · Digital visualization · Manual editing · Generative AI

F. C. Ünal (✉)
Yeditepe University, Istanbul, Turkey
e-mail: farukcan.unal@yeditepe.edu.tr

D. Karadağ
Işık University, Istanbul, Turkey
e-mail: derya.karadag@isikun.edu.tr

M. Sedrez and G. Pitts (eds.), *Artificial Intelligence-Aided Design for Sustainability*, Urban Sustainability, https://doi.org/10.1007/978-981-95-1349-9_4

1 Introduction

In recent years, the rapid advancement and widespread availability of artificial intelligence technologies have begun to transform established practices across design disciplines, prompting renewed attention to their pedagogical implications. The integration of AI in architectural visualization is reshaping design workflows, offering automation-driven efficiencies while redefining traditional editing methods. Conventional digital visualization techniques, such as manual 3D modelling and post-processing, require extensive time and expertise, whereas AI-aided tools simplify these processes by generating and refining architectural visuals with minimal user involvement. While these advancements enhance efficiency, they also raise questions regarding their impact on learning, skill acquisition, and engagement in architectural education.

The emergence of artificial intelligence in architectural design represents not merely a technological shift, but a paradigmatic transformation in how design is conceived, represented, and critiqued. As Leach [1] argues, the adoption of AI in architecture signals a departure from human-centric authorship toward computational authorship, raising profound questions about the nature of creativity and aesthetic judgment. Similarly, Chaillou [2] underscores how diffusion models and GAN-based systems are recalibrating the architectural canon, proposing speculative yet operational alternatives to traditional visual styles. These developments transcend mere software advancement, revealing new epistemologies of design rooted in algorithmic reasoning. This study builds on these perspectives by empirically examining how such paradigmatic shifts manifest in educational contexts.

From an educational perspective, the integration of AI into visualization workflows also reframes how students engage with core design tasks. Rather than modelling creativity per se, AI tools like Generative Fill function as creative support systems, mediating decision-making and reducing technical barriers. Recent empirical studies have also explored how generative AI tools support conceptual thinking and creative experimentation in design studios [3]. As students interact with these tools, questions arise not about the generation of novel design knowledge, but about how such tools influence engagement, learning trajectories, and the development of representational competency in early design education. Del Campo and Leach [4] extend this argument by suggesting that AI acts as a co-designer an entity capable of proposing novel spatial configurations that provoke human reflection rather than simply executing predetermined outcomes. Meanwhile, Bernstein [5] cautions against the wholesale substitution of human skill with automation, advocating for a hybrid model that preserves human agency while leveraging AI's computational power.

Despite this growing body of theoretical literature, the pedagogical implications of AI tools in architectural education remain underexplored. While researchers emphasize the creative potential of generative tools, empirical investigations of their impact on learning -particularly in the context of foundational visualization skills- are limited. This study addresses this gap by comparing conventional Manual Editing

(ME) with Generative Fill (GF) workflows in architectural visualization, examining how these approaches influence students' engagement, skill development, and perception of creative autonomy.

To investigate these dynamics, this research is guided by the following questions:

- How do conventional and AI-aided workflows impact students' engagement with architectural visualization?
- What are the key advantages and limitations of AI-aided visualization tools in comparison to conventional digital editing techniques?

By examining these aspects, the study contributes to discussions on the role of AI in architectural pedagogy, offering insights into balancing automation with fundamental design competencies. As AI continues to integrate into educational frameworks, the need for structured pedagogical strategies becomes essential to ensure that future architects develop both technical proficiency and critical design awareness.

2 Background

Architectural visualization plays a crucial role in conveying architectural design. The focus on the visual in architecture is deliberate, as human interaction with both natural and built environments is primarily visual. A broad range of human activities, from aesthetic appreciation to the planning of actions, depends on visual information to assess situations and make decisions. Visualization has been a significant aid to the understanding of environment [6]. The designed space should be represented in a way that meets users' needs and desires by enriching with natural and cultural data [7]. Before the advent of digital technologies, architectural visualization relied heavily on hand-drawn sketches and perspective drawings to communicate design ideas. These traditional methods, while highly artistic and expressive, were time-consuming, labour-intensive, and limited in terms of flexibility. Any design revisions required redrawing or rebuilding, making the process less iterative compared to digital workflows. Architectural visualization underwent a significant transformation, shifting from manual techniques to computer-generated images with the digital technologies [8]. In architectural visual production, modelling and rendering are fundamental steps in creating realistic and detailed representations of architectural designs. In fast-paced architectural workflows, taking a quick model render and completing it with digital editing tools has become a common practice to save time while creating visuals. This conventional approach allows for greater flexibility, as adjustments to lighting, textures, entourage (people, furniture, landscape elements, etc.), and atmospheric effects can be efficiently made in post-production.

The integration of AI-aided tools in architectural visualization has fundamentally transformed how architects and designers create, refine, and enhance visual outputs [9, 10]. AI-driven methodologies have introduced new approaches for processing raw architectural visuals, enriching them with natural and cultural elements while streamlining complex workflows. As AI continues to reshape architectural design and education, it enables more efficient generative design processes and automation strategies that optimize visualization techniques [11, 12]. In practice, AI tools enhance architectural workflows by automating rendering, intelligent image processing, and material adaptation, significantly improving efficiency and realism. AI-powered features such as automated image completion, sky replacement, and texture enhancements not only accelerate production but also provide greater creative control, allowing designers to experiment more freely with composition, atmosphere, and narrative elements. By integrating these advancements into architectural education and practice, AI fosters new possibilities for exploration and innovation while redefining conventional design methodologies.

As technological advancements continue to reshape various fields, emerging technologies must be harnessed to strengthen education systems and enhance learning outcomes [13–15]. The integration of artificial intelligence in education has been recognized as a key driver for sustainable development, as highlighted in the Beijing Consensus on Artificial Intelligence and Education [16]. This global policy framework emphasizes the importance of AI-enhanced education in promoting accessibility, innovation, and interdisciplinary learning while ensuring ethical and sustainable implementation. In alignment with Sustainable Development Goal 4 (SDG 4)—which advocates for inclusive and equitable quality education and lifelong learning opportunities—AI tools in architectural education have the potential to streamline visualization processes, optimize resource efficiency, and foster new pedagogical approaches that align with sustainability goals.

To effectively integrate AI into architectural education, educators must develop the necessary competencies to navigate, teach, and evaluate AI-assisted design tools [17]. The UNESCO AI Competency Framework for Teachers [18] provides a structured guideline that supports teachers in adopting human-centred, ethical, and pedagogically sound AI applications. In the context of architectural visualization, these competencies empower educators to critically assess AI-aided tools, integrate them into design curricula, and guide students in understanding their impact. By incorporating AI literacy into pedagogical strategies, educators can ensure that AI is used not merely as a productivity tool but as a medium to enhance creativity, sustainability, and critical engagement with technology. For students, the transition from conventional architectural visualization to AI-aided methods requires a redefinition of core competencies. The UNESCO AI Competency Framework for Students [19] provides a structured approach to AI literacy across four dimensions: human-centred thinking, ethical awareness, AI techniques, and AI system design. As generative AI tools become increasingly prevalent, students must not only learn their technical applications but also develop a critical perspective on their implications for authorship, design originality, and sustainability. This framework supports the development of AI fluency in architectural education, ensuring that students acquire both the technical skills to leverage AI effectively and the ethical awareness to address its challenges. By embedding these competencies into the curriculum, educators can equip students with the necessary tools.

The integration of artificial intelligence in education has the potential to enhance accessibility, efficiency, and engagement, particularly in design-based fields such as architecture [20, 21]. AI-driven tools can facilitate personalized learning experiences, provide real-time feedback, and streamline complex workflows, allowing students to focus more on creative and analytical thinking rather than repetitive technical tasks. Beyond improving design education, AI has broader implications for the future of work and skills development, reshaping the core foundations of teaching and learning. As automation and intelligent systems redefine professional practices, architectural education must evolve to equip students with critical digital competencies, adaptability, and problem-solving skills. By incorporating AI into curricula, institutions can bridge the gap between emerging technologies and industry expectations, fostering a generation of architects who are proficient in both traditional design principles and AI-aided methodologies. In this way, AI not only transforms the learning process but also prepares students for the dynamic and technology-driven nature of architectural practice in the future.

3 Methodology

This study investigates the impact of AI-aided design tools on architectural workflows by comparing conventional Manual Editing (ME) with Generative Fill (GF) methods. The research employs a mixed-methods approach, integrating quantitative analysis of user ratings with qualitative insights from thematic analysis. The primary aim is to assess efficiency, user experience, and creative engagement in AI-assisted versus conventional digital workflows. Data collection involves structured task completion followed by surveys with open-ended reflections to capture both measurable performance indicators and subjective experiences. To analyze the qualitative data, Braun and Clarke's [22] reflexive thematic analysis was employed, providing a flexible and interpretive approach to identifying patterns in students' interactions with these methods. This approach facilitated the extraction of key themes reflecting participants' engagement, including their perception of how each approach impacted their architectural visualization process. Themes were categorized by frequency: high frequency (appearing in over 50% of responses), moderate (25–50%), and low (less than 25%). The analysis was conducted based on the following headings: 'Experience of Completing the Task,' 'Thematic Analysis of Students' Experiences,' 'Challenges Encountered While Completing Tasks,' and 'Conveniences of the Tools While Completing Tasks.'

The experimental setup utilized Photoshop as the primary digital environment to ensure consistency across both conventional and AI-aided methods. The conventional approach was conducted using Photoshop's standard editing tools, while the AI-aided approach was implemented through its Generative Fill feature. By using the same software environment, external variables influencing the results were minimized. Participants were tasked with completing two unfinished architectural visualization tasks -one exterior and one interior scene- each containing architectural

Fig. 1 Selected architectural visuals of the exterior and interior spaces for the experiment [23, 24]

elements but lacking specific visual details (Fig. 1). The inclusion of both scene types was intended to explore potential differences in perceived difficulty and workflow efficiency across varying spatial contexts. The design of the experiment aimed to replicate real-world architectural visualization scenarios by requiring students to refine and complete architectural renderings using either ME or GF. The experiment was integrated into regular course hours to align with students' learning experiences while ensuring methodological rigor. Workflow efficiency and user perceptions were assessed through structured documentation of participants' experiences, providing valuable comparative data on the impact of both approaches in architectural visualization.

The participants in this study were 55 first and second year architecture students enrolled in the 'Computer-Aided Presentation Techniques' course at institutions in Turkey that use English as the language of instruction. Although the participants' native language was different, the study was conducted in English so no language-related barriers were encountered. This introductory-level course was deliberately chosen to ensure a participant group with minimal prior bias toward either method, thereby enhancing the relevance of the findings. The students had no prior experience with artificial intelligence tools and had only been trained in conventional digital presentation techniques, primarily using Photoshop. To ensure comparable skill levels, participants were selected based on their shared stage in architectural visualization training. They were divided into two groups according to their assigned method: one using ME and the other using GF. Both groups worked on identical tasks, completing unfinished architectural visuals. To capture students' experiences, a survey was conducted following task completion. Ethical considerations were carefully observed: participants' informed consent was obtained and documented, their anonymity was maintained, and the study was conducted under the approval of the institutional ethics committee. This methodological structure allowed for an unbiased evaluation of both approaches, highlighting their respective strengths and challenges in architectural visualization.

4 Findings and Discussion

In this study, we explore how participants interact with two distinct workflows in Photoshop: the conventional manual editing techniques and AI-supported Generative Fill (Fig. 2). By featuring a thematic analysis, we aim to uncover differences in user responses, focusing on their experiences, satisfaction, and the challenges they faced. The data comes from two groups of participants: 26 who relied on manual editing techniques and 29 users who utilized the Generative Fill feature. This comparison seeks to shed light on how AI tools are reshaping creative workflows and influencing user perceptions in digital visualization.

The findings, summarized in Fig. 3, compare user experiences with the Manual Editing (ME) and Generative Fill (GF) for completing the given tasks. User perceptions are categorized into five levels of difficulty: Very Difficult, Difficult, Neutral, Easy, and Very Easy. The data reveals distinct patterns: ME received mixed responses, with 11 users rating it as Neutral and 11 as Easy, followed by 2 users selecting Very Easy. Difficult and Very Difficult ratings were minimal, with 1 user for each, suggesting that while the method is generally manageable, it may lack intuitiveness for some. Conversely, GF received primarily positive ratings, with 16 users selecting Easy and 9 selecting Very Easy, while only 2 users rated it as Neutral and 2 as Difficult. Notably, no respondents found GF Very Difficult, emphasizing its

Fig. 2 Student visualizations using manual editing (left) and generative fill (right) in Photoshop

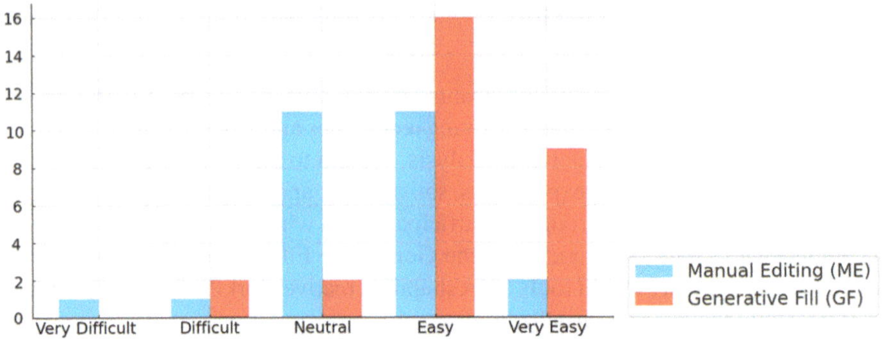

Fig. 3 Comparison of responses from ME and GF users on the overall experience of completing the task

user-friendly nature. Overall, GF outperforms ME in ease of use, as reflected by the higher concentration of favourable responses.

The thematic analysis of ME for students who rated their experience as Easy or Very Easy highlights meaningful connections between engagement, learning, and design (Table 1). Process Simplicity stood out as the most common theme, with many students emphasizing how practice transformed the manual approach into a smooth and even enjoyable process. This demonstrates how hands-on repetition builds both skill and confidence, revealing the ME's potential to foster growth through effort. Moderate-frequency themes like Exceeded Expectations, and Exploration and Curiosity show how students found unexpected value in mastering manual methods. Many were surprised by their ability to achieve results they initially thought were beyond their capabilities. These moments of discovery underline how manual techniques engage students in deeper problem-solving and exploration, encouraging creativity and adaptability. Unchanged Expectations appeared less frequently, reflecting a small group of students who found the manual approach aligned closely with their initial expectations. Ultimately, these findings reveal the value of manual techniques in cultivating practical skills and creativity. ME uniquely offer students the chance to connect more deeply with their work through practice and iteration, making them a vital component of the design learning process.

The thematic analysis of students' experiences with GF, as reported by those who rated the tool as Easy or Very Easy, identified five key themes: Exceeded Expectations, Process Simplicity, Exploration and Curiosity, Creative Empowerment, and Minor Process Challenges (Table 2). High-frequency themes, such as Exceeded Expectations and Process Simplicity, appeared in over 50% of responses, with students admiring the tool's ability to deliver high-quality, realistic results and its intuitive ease of use. Moderate-frequency themes, including Exploration and Curiosity and Creative Empowerment, emerged in 25–50% of responses. Students described discovering unexpected capabilities of the tool and experiencing increased engagement and creativity. GF also facilitated design refinement, allowing students to visualize their creations more effectively. Low-frequency challenges, noted by

Table 1 The thematic analysis of students' experiences with ME, specifically from those who rated the tool as easy or very easy

Theme	Description	Frequency	Examples of student expressions
Process simplicity	The process was easier or less complex than anticipated, often becoming smoother with practice	High	*–It seemed complicated at first, but as I practiced, it became enjoyable and straightforward* *–I thought it was harder at the start, but once I understood, it was easy*
Exceeded expectations	Students were surprised by their ability to achieve results or learn more than they initially expected	Moderate	*–I was expecting to be less good than I did, but with practice, it gets easier.* *–I didn't expect to gain this much; I completed the task quickly and effectively*
Exploration and curiosity	Students engaged in learning new methods, discovering unexpected possibilities, or adapting their techniques	Moderate	*–I didn't think I could change the angles of objects, but once I learned, I could* *–I explored filling spaces in a way I liked and found it easier than expected*
Unchanged expectations	Students noted that their expectations remained unchanged, indicating alignment between their initial perceptions and the experience	Low	*–I didn't think it would be difficult, and it wasn't.* *–My expectations didn't change because we were prepared for it.*

fewer of respondents, involved minor technical difficulties, such as needing multiple attempts to achieve desired outcomes. However, these did not weaken significantly from the overall positive experience. In summary, GF demonstrates strong potential for enhancing students' design processes by fostering creativity, simplifying workflows, and exceeding expectations, with minor improvements needed to address occasional challenges.

The thematic analysis reveals key differences between students' experiences with ME and GF, particularly among those who rated the tasks as Easy or Very Easy. ME initially appeared more challenging, requiring repeated practice for students to achieve fluency and confidence. It encouraged deeper exploration and curiosity, as students discovered new methods and refined their designs through iterative problem-solving. Additionally, ME presented a gradual learning curve, emphasizing the development of technical mastery over time. While both methods were generally well-received, minor challenges were more noticeable in ME, reflecting the additional time and effort required for mastery. In contrast, GF stood out for its ability to simplify complex tasks through automation, offering immediate ease of use. It

Table 2 The thematic analysis of students' experiences with GF, specifically from those who rated the tool as easy or very easy

Theme	Description	Frequency	Examples of student expressions
Exceeded expectations	Students were surprised by the tool's ability to surpass their expectations, providing high-quality or efficient results	High	*–I didn't expect it to look this realistic, but the result was beautiful*
Process simplicity	The process was easier than anticipated, requiring minimal effort to achieve results	High	*–I thought it would be harder, but it was easy and very realistic* *–I got the hang of it quickly*
Exploration and Curiosity	Users were curious about the tool and discovered unexpected capabilities or enjoyed the process	Moderate	*–I was curious about how AI works and how it would look* *–It was fun and educational*
Creative empowerment	The tool enabled users to imagine or transform designs, enhancing their creative capabilities	Moderate	*–I could imagine myself in the space I created, which helped me see its flaws* *–Transforming an image completely with AI was inspiring*
Minor process challenges	Users faced small difficulties but still perceived the experience as smooth and manageable	Low	*–It took a few attempts, but the process was smooth overall*

fostered creative empowerment by enabling students to produce realistic and high-quality results with minimal effort, often exceeding expectations. GF also provided an intuitive and accessible experience, reducing the cognitive load associated with design tasks. Overall, the findings highlight how these two approaches offer different ways of working, each contributing to the design process in its own way. While GF excels in delivering rapid results and streamlining workflows, ME enhances practical skills and deeper engagement with the design process.

Table 3 highlights the challenges encountered by the ME users while completing tasks. The most frequent difficulties were related to perspective and positioning, as well as finding and formatting images. Students often struggled to adjust angles, align perspectives, and position visuals realistically within their compositions. Similarly, sourcing suitable visuals in appropriate formats, like PNG or JPG, proved challenging, with issues such as website restrictions and format incompatibility frequently mentioned. Moderately frequent challenges included adding realistic shadows and removing image backgrounds. Many students found it difficult to create cohesive shading or adjust lighting to match the composition, while others struggled with cleaning up image edges during background removal. These issues reflect the detailed and precise nature of manual editing tasks, which require significant technical effort. Time and workflow management emerged as a low-frequency challenge, with some students expressing difficulty in managing the task within the given timeframe. However, a small number of students reported no significant difficulties, indicating that prior experience or familiarity with manual techniques made the process smoother for them. The findings emphasize the demanding nature of ME, which requires users to master technical precision and resourcefulness to achieve desired results. Challenges related to perspective, realism, and formatting point to areas where additional training or streamlined tools could support students.

Table 4 illustrates the challenges encountered by the students while using GF during the tasks. The most reported difficulty was articulating ideas to the AI, particularly when students needed to describe complex or specific visuals. This often resulted in frustration when prompts failed to generate the desired outcomes. Similarly, issues with selection and adjustment, such as aligning objects, resizing layers, and maintaining consistent perspectives, were also frequent. These technical challenges highlight the need for improved tools to support precision in user workflows. Moderately frequent challenges included difficulties in finding or generating suitable visuals that matched the task's requirements, with mismatched scale or perspective often being cited as barriers. Additionally, some students faced an initial learning curve, particularly those unfamiliar with Photoshop or AI tools. However, as they practiced, many overcame these obstacles and adapted to the workflow. A smaller group of students reported no significant difficulties, finding the tool intuitive and effective for their needs. These findings underscore the dual nature of GF as a highly capable tool with some technical aspects to navigate. While it excels in offering creative opportunities, its effectiveness joints the user's familiarity with the platform and ability to navigate its technical aspects.

The comparative analysis of Tables 3 and 4 highlights distinct patterns in the challenges faced by students using ME and GF, reflecting differences in technical demands, workflow complexity, and user experience. ME challenges were more manual and resource-intensive, with frequent struggles in aligning perspectives, sourcing compatible visuals, and achieving realistic shading. These tasks required significant technical effort, underscoring the labour-intensive nature of manual workflows. In contrast, GF challenges largely centred on the interaction between users and AI, with difficulties in articulating precise prompts and handling selection and adjustment tasks. These issues stemmed from limitations in AI's interpretation of user input and its capacity to integrate changes, highlighting the need for improved algorithmic structure. Findings suggest that while ME offers unparalleled opportunities for customization and skill development in more advanced contexts, GF excels in efficiency and accessibility for basic to intermediate tasks.

Table 5 outlines the conveniences reported by the ME users when using manual editing techniques, highlighting key aspects of their experience with familiar tools and workflows. The most frequently mentioned convenience was Background Removal, with students emphasizing the ease of removing image backgrounds using tools. Many appreciated how this functionality allowed them to quickly clean up visuals and eliminate unnecessary elements with minimal effort. Another major

Table 3 Responses from ME users on the challenges encountered while completing tasks

Theme	Description	Frequency	Examples of student expressions
Challenges in perspective and positioning	Students struggled with adjusting angles, perspectives, and positioning images to fit the composition realistically	High	*–I struggled with creating perspectives and positioning images* *–Aligning perspectives was challenging*
Difficulties in finding and formatting images	Locating suitable visuals in appropriate formats (e.g., PNG, JPG) and ensuring compatibility with the task posed significant challenges	High	*–Finding a PNG image was difficult* *–It was hard to find JPG files without signing up on websites* *–Finding photos with the right format was tricky*
Shadows and realism	Students reported challenges in adding realistic shadows or adjusting lighting to make the final composition cohesive and natural	Moderate	*–Making shadows was difficult because they didn't fit naturally* *–Creating realistic shadows was tiring* *–Shading was hard to match the composition*
Background removal	Removing backgrounds from images cleanly and effectively was a recurring challenge for many students	Moderate	*–Removing the background from images was tricky* *–Cleaning the edges were a challenge*
Time and workflow management	Some students found managing time and completing all steps within the given timeframe to be a significant difficulty	Low	*–Time management was tough* *–I struggled to manage my time efficiently while completing the task*
No significant challenges	A subset of students reported no significant difficulties, indicating familiarity or comfort with manual techniques	Low	*–No challenges at all* *–None, the process was smooth*

Table 4 Responses from GF users on the challenges encountered while completing tasks

Theme	Description	Frequency	Examples of student expressions
Difficulty communicating ideas to AI	Students struggled with accurately describing their desired outcomes in prompts, especially for specific or complex visuals	High	*–Writing prompts for specific detail changes in the image* *–I struggled to describe what I wanted with words* *–It's hard to convey ideas for exact objects*
Challenges with selection and adjustment	Issues with selecting areas, adjusting layers, and aligning objects within the composition were frequently mentioned	High	*–I had trouble aligning objects to the same perspective.* *–When layering, absurd visuals occurred, and resizing was difficul*
Difficulty finding desired visuals	Students encountered problems locating suitable visuals or generating realistic images to match their project needs	Moderate	*–The AI generated unrealistic results, and I struggled to find suitable visuals.* *–The scale and perspective of objects didn't match*
Adaptation and familiarity	Initial struggles with adapting to new tools and understanding the workflow, particularly for those unfamiliar with Photoshop or AI tools	Moderate	*–At first, understanding the steps was hard, but with practice, I overcame it.* *–I'm new to Photoshop, so using selection tools was challenging*
No significant challenges	A subset of students reported no significant difficulties in completing the task, suggesting that the tool was intuitive for them	Low	*–I didn't face any challenges.* *–None, it was smooth and enjoyable*

benefit was Efficiency and Speed, where students noted that manual tools enabled relatively fast execution of common tasks, such as duplicating, scaling, and adjusting elements. Familiarity with these tools allowed for a practical and streamlined workflow once users understood the details. Moderate themes included Precision and Control, and Accessibility and Familiarity. Students valued the detailed control manual techniques offered, particularly for perspective adjustments, scaling, and achieving design cohesion. Additionally, the familiarity of tools like Photoshop shortcuts and common functionalities provided convenience for those already acquainted with the software. A small subset of students reported No Significant Convenience, indicating uncertainty or a lack of specific benefits when using manual tools. Overall, Table 6 highlights ME's strengths in providing precision, control, and familiarity, which many students found beneficial for achieving cohesive designs. However, the responses also reflect a reliance on prior experience, suggesting that less experienced users may not derive the same level of convenience from manual tools. Enhancing accessibility and providing additional guidance could improve the user experience for a broader audience.

Table 6 summarizes the perceived conveniences provided by GF, as reported by the users, emphasizing its impact on workflow efficiency, usability, and creativity. The most frequently mentioned benefit was Time Efficiency, with students highlighting how GF reorganised rendering, visual generation, and spatial organization, significantly reducing the time required to complete tasks. Respondents noted that it shortened the process and avoided time-consuming activities like manual perspective adjustments, marking a clear advantage over traditional methods. Another prominent theme was Ease of Use, as students found the tool's intuitive interface and ability to generate visuals based on simple prompts highly accessible. Many appreciated how GF simplified the visual creation process, making it easier and faster to generate images that aligned with their project needs. Moderately frequent themes included

Table 5 Responses from ME users on the conveniences of the tools while completing tasks

Theme	Description	Frequency	Examples of student expressions
Background removal	Students frequently mentioned the ease of removing backgrounds from images	High	*–Removing the background with one click was very helpful* *–Background removal was very useful*
Efficiency and speed	The manual tools allowed for relatively fast execution of tasks like copying, scaling, and adjusting elements	High	*–Quickly duplicating and scaling images made the process faster* *–It was practical; Photoshop is easy to use once you know the details*
Precision and control	Manual techniques provided more control over details, such as perspective adjustments, scaling, and achieving design cohesion	Moderate	*–Perspective-making tools and shadowing helped make the design cohesive* *–The precision allowed for proper adjustments in angles and shadows*
Accessibility and familiarity	Common Photoshop features were convenient for students who were already acquainted with them	Moderate	*–Photoshop tools made the process easier* *–Shortcuts were very handy* *–The tools were familiar and simple to use*
No significant convenience	A few respondents indicated that they were unsure or did not find any specific convenience in the manual tools	Low	*–I'm not sure* *–I don't know*

Table 6 Responses from GF users on the conveniences of the tools while completing tasks

Theme	Description	Frequency	Examples of student expressions
Time efficiency	GF reduced the time required for rendering, finding visuals, and organizing spaces, allowing students to complete tasks faster	High	*–It saved time and prevented wasting time on perspectives* *–It shortened the process and made it much simpler than manual work*
Ease of use	The tool's simplicity and intuitive interface allowed students to easily generate and select visuals that fit their project needs	High	*–AI creating visuals based on my words made it very easy* *–It's easier and faster to provide some source of visual rendering*
Creative freedom and flexibility	Students appreciated the ability to translate their ideas into visuals, explore multiple alternatives, and make quick modifications	Moderate	*–I could reflect my imagination onto my project easily* *–It created several alternatives and helped me choose*
Realism and Quality	The tool provided realistic images with high-quality rendering that aligned with students' expectations	Moderate	*–I created realistic images in a short time* *–It was great to have visuals created fast and in the right dimensions*
Convenience in searching and organizing	GF eliminated the need to search for resources manually or adjust perspectives, saving effort and improving the workflow	Moderate	*–It saved me from searching for the right photo or image for my project* *–Instead of searching objects one by one on Google, it saved time*
No Convenience	A few respondents indicated that GF did not provide significant benefits for their tasks	Low	*–It didn't provide any convenience because I couldn't use the tools I wanted*

Creative Freedom and Flexibility, and Realism and Quality. Students valued the tool's ability to translate ideas into visuals, explore multiple alternatives, and make quick modifications, enabling them to reflect imagination and choose from various options. Additionally, the quality of outputs, such as realistic images with accurate dimensions, met students' expectations and contributed to the overall convenience. Convenience in Searching and Organizing was another moderate theme, with GF reducing the manual effort required to search for and format resources, thus enhancing workflow efficiency. Respondents specifically noted how the tool eliminated the need to search for the right photo or image or individually adjust perspectives, saving both time and effort. A small subset of students, however, indicated No Convenience, citing difficulties in using specific tools or aligning GF's functionalities with their needs. In summary, Table 6 highlights GF's strengths in saving time, simplifying tasks, and enabling creativity, while also identifying isolated instances of unmet expectations. These insights suggest that while GF excels in efficiency and usability, addressing specific user needs and limitations could further enhance its effectiveness in design workflows.

The comparison between ME and GF highlights distinct strengths that cater to different aspects of the design process. ME's precision and control offer unmatched flexibility for detailed adjustments, making it well-suited for tasks that demand refined customization. However, this level of control comes with a trade-off, as ME requires significant technical knowledge and manual effort. Users must carefully manage elements like perspective, shadows, and alignment, often resulting in longer workflows but allowing for deeper engagement with the design process. In contrast, GF's automation streamlines workflows, enabling students to save time on repetitive tasks like rendering and visual generation. While GF reduces cognitive and technical load, making it accessible even to those with limited experience, its reliance on accurate prompt articulation can sometimes lead to outputs misaligned with expectations. Another key distinction lies in how each method handles workflow challenges. ME places the burden of efficiency on the user, relying on their familiarity with tools to navigate tasks effectively, which can lead to tiredness in complex projects. Meanwhile, GF eliminates many time-consuming manual steps, offering an efficient solution for rapid ideation and iterative exploration. While both methods have their advantages, integrating GF's automation with ME's precision could provide a balanced approach, allowing designers to optimize both efficiency and customization in their workflows.

While the thematic findings offer structured insights into students' reported experiences, several unstructured observations from the process provide additional illumination of the pedagogical dynamics. Within both groups, students exhibited moments of self-correction, experimentation, and informal reflection, particularly when they encountered limitations or discovered shortcuts. ME participants, for instance, often verbalized their problem-solving strategies during the session, demonstrating how challenges catalyzed active learning. Meanwhile, those using GF displayed surprise and enthusiasm at the immediacy of the results, with some expressing curiosity about the AI's decision-making process despite the absence of direct control. These interactions revealed distinct learning dispositions shaped by the tools themselves: ME encouraged meticulous iteration and strategic adjustment, while GF fostered a more intuitive, exploratory approach.

5 Conclusion and Future Directions

This study investigated the comparative impact of conventional ME and AI-Aided GF workflows in architectural visualization, focusing on efficiency, user experience, and creative engagement. The findings reveal distinct advantages and limitations associated with each method. GF demonstrated clear strengths in enhancing workflow efficiency, reducing cognitive load, and generating high-quality visual results with minimal effort. However, challenges related to prompt articulation and occasional misalignment between generated visuals and user expectations indicate areas for improvement. Conversely, ME, while more labour-intensive, fostered deeper engagement with the design process, supporting iterative refinement and skill development.

The results reflect the experiences of learners with relatively minimal prior exposure to both methods. Future research should expand the participant group to include a more diverse range of users, including professionals with advanced expertise in digital visualization. Testing these workflows in professional settings could provide further insights into how these approaches function in real-world architectural practice. Further research should explore the refinement of AI-aided tools to enhance accuracy, contextual awareness, and realism in visual generation. Additionally, examining how professionals engage with these methods could help understand their applicability beyond educational settings. Investigating the impact of repeated manual editing processes on skill retention and expertise development would also be valuable, particularly in contrast to the convenience-driven automation of AI tools.

Another essential area for future exploration is the ethical dimension of AI integration in design education. The integration of AI raises critical questions regarding authorship, creative autonomy, and the potential over-reliance on automated tools. Understanding its implications in pedagogical contexts, particularly in terms of fostering independent design thinking, is essential. While GF streamlines workflows, it is crucial to ensure that students still develop foundational design skills that are traditionally honed through manual processes.

A balanced integration of both methods -leveraging GF's automation with ME's precision–could enhance the design process by addressing their respective limitations while maximizing their distinct strengths. While AI facilitates efficiency and rapid iteration, the iterative nature of manual workflows plays a crucial role in skill acquisition and expertise development. Equipping students with proficiency in both methodologies may better prepare them for professional practice, fostering both technological adaptability and hands-on design competency. By strategically integrating AI and manual editing into design education, educators can cultivate a learning environment that promotes efficiency while safeguarding the integrity of creative skill.

Reflecting on the broader educational context, this study offers partial alignment with the competencies outlined in the UNESCO AI Competency Framework for Teachers and Students, particularly in fostering awareness of AI tools, promoting human-centred thinking, and initiating critical reflection on technological integration. However, several challenges emerged that underscore the complexity of actualizing these competencies in architectural education. From the educators' perspective, the rapid evolution of AI tools outpaces formal pedagogical training, raising concerns about their ability to guide students through both technical use and critical discourse. For students, while tools like Generative Fill facilitate creative exploration, the lack of transparency in AI decision-making and the risk of over-reliance on automated processes can hinder the development of foundational visualization skills. These tensions suggest that while UNESCO's guidelines offer a valuable normative framework, their full realization requires institutional commitment, iterative curriculum development, and support for educators navigating the intersection of technology and design pedagogy.

Acknowledgements We would like to express our gratitude to Önder Aras for his support during the research process. We also sincerely thank the students who participated in the study for their engagement and contributions.

References

1. Leach N (2021) Architecture in the age of artificial intelligence. Bloomsbury Publishing. http://digital.casalini.it/9781350165533
2. Chaillou S (2022) Artificial intelligence and architecture: from research to practice. Birkhäuser.
3. Karadağ D, Ozar B (2025) A new frontier in design studio: AI and human collaboration in conceptual design. Front Arch Res. https://doi.org/10.1016/j.foar.2025.01.010
4. del Campo M, Leach N (2022) Machine Hallucinations: artificial imagination and the future of design. John Wiley & Sons, In Designing with Artificial Intelligence
5. Bernstein P (2018) Architecture | design | data: practice competency in the era of computation. Birkhäuser.
6. Koutamanis A (2000) Digital architectural visualization. Autom Constr 9(4):347–360
7. Ünal FC, Demir Y (2020) Local data: a new approach to data in architecture. A|Z ITU J Fac Arch 17(1):13–23 https://doi.org/10.5505/itujfa.2019.45220
8. Tong H, Türel A, Şenkal H, Ergun S, Güzelci OZ, Alaçam S (2023) Can AI function as a new mode of sketching: a teaching experiment with freshman. Int J Emerg Technol Learn 18(18):234–248
9. Muslu O, Koçyigit RG (2025) Artificial intelligence in the context of photorealism in architectural visualization. CONTEXTO. Rev Fac Arquit Univ Autónoma Nuevo León 19(29):15–32
10. Güzelci OZ, Karadag I (2024) Revisiting the key components of creativity through generative AI. In: Making art with generative AI tools. IGI Global, pp 1–16
11. Swift IP, Chattopadhyay D (2024) A value-oriented investigation of photoshop's generative fill. arXiv preprint arXiv:2404.17781
12. Montenegro N (2024) Integrative analysis of Text-to-Image AI systems in architectural design education: pedagogical innovations and creative design implications. J Archit Urban 48(2):109–124
13. Ceylanlı Z, Aktaş E (2021) A critical assessment of an extended learning environment in interior design studio. J Des Stud 3(1):71–81
14. Narin ZS, Ünal FC (2023) An evaluation of sketching processes in physical and immersive environments. In: Digital design reconsidered-proceedings of the 41st conference on education

and research in computer aided architectural design in Europe, vol 1, Graz, pp 20–22. September 2023, 201–210
15. Fonseca D, Redondo E, Valls F, Villagrasa S (2017) Technological adaptation of the student to the educational density of the course. A case study: 3D architectural visualization. Comput Hum Behav 72:599–611
16. Unesco (2019) Beijing consensus on artificial intelligence and education. In: International conference on artificial intelligence and education, planning education in the AI era: lead the leap. Beijing, pp 16–18. May 2019
17. Liu X (2024) Integrating generative AI into landscape architecture education: methodologies, applications, and ethical considerations. J Digit Landsc Arch:937–945
18. Unesco, Miao F, Cukurova M (2024) AI competency framework for teachers. https://doi.org/10.54675/ZJTE2084
19. Unesco, Miao F, Shiohira K (2024) AI competency framework for students. https://doi.org/10.54675/JKJB9835
20. Copper C, Harrison PH, Yang Z (2024) Artificial intelligence literacy: collaborating to support image research in architecture education. Paper presented at the ACSA 112th Annual meeting: disrupters on the edge, Vancouver, pp 14–16. March, 2024
21. Gallega RW, Sumi Y (2024) Exploring the use of generative AI for material texturing in 3D interior design spaces. Front Comput Sci 6:1–21. https://doi.org/10.3389/fcomp.2024.1493937
22. Braun V, Clarke V (2006) Using thematic analysis in psychology. Qual Res Psychol 3(2):77–101
23. Evermotion (2025) Suburbs 3d models. Archexteriors, vol 13 https://evermotion.org/shop/show_product/archexteriors-vol-13/7374
24. Evermotion (2025) Modern office spaces 3d models. Archinteriors, vol 33 https://evermotion.org/shop/show_product/archinteriors-vol-33/11224

Design Practice with Generative Artificial Intelligence for Sustainability: An Experimental Study with Undergraduate Designers

Eduardo Napoleão and Alice Demaria Silva

Abstract The use of Generative Artificial Intelligence (GenAI) has implications across creative industries, including fashion design, architecture, gaming, and music. Considering literature research and a practical workshop, this study analyzes the design process based on the relationship between designers and GenAI tools, examining the creative process. The educational and data collection methods used included slide presentations, group discussions, and a guided project. The practical component of the methodology focused on a two-day workshop. The study concludes that there are three main challenges to the design process: originality of creations, technological dependence, and transformations in the professional domain. These changes highlight the human centrality of the designer as a creative and critical agent, who interprets and uses technology for creation and is aware of the creative process. The future of design must balance human and technological capabilities and ensure that the effects of this interaction produce results based on human, social and sustainable values. In conclusion, we don't see GenAI tools as replacements for designers, but as tools that require human guidance to generate design solutions. They need to go beyond aesthetics and reach functional and symbolic values.

Keywords Creativity · Design practice · Design project · Generative tools · Sustainability · Artificial Intelligence

1 Introduction

Creativity can be viewed and defined in many ways, as there is no consensus among authors. One way of defining it is based on the ability to produce new, useful and surprising ideas, which is considered an important aspect of human cognition. If

E. Napoleão (✉) · A. D. Silva
University of Vale Do Itajaí, Itajaí, Brazil
e-mail: eduardonapoleao@univali.br

A. D. Silva
e-mail: alice.silva@univali.br

M. Sedrez and G. Pitts (eds.), *Artificial Intelligence-Aided Design for Sustainability*, Urban Sustainability, https://doi.org/10.1007/978-981-95-1349-9_5

we analyze creative studies from a human perspective in relation to the evolution of technology and work, we can affirm that not only can human beings be observed from a creative perspective, but there is also the possibility of generating creativity through machines. With the advancement of Generative Artificial Intelligence (GenAI) models, there is a growing interest in exploring the creative capacities of both, especially in linguistic and artistic contexts. These areas of study seek to understand the extent to which Artificial Intelligence (AI) can contribute to creative solutions, given its limitations in abstract problem solving, critical thinking, and originality [1].

The creation of images through text represents a paradigm shift in the design process. These systems, which provide the ability to transform natural language prompts into high-quality digital art, raise questions about the concept of creativity, changes in the design process, and ethics and intellectual property. The product-centric paradigm is no longer sufficient to evaluate the creativity involved in generating images from text, and the analysis must be extended to the creative process, considering the human-machine interaction through GenAI. This perspective is supported by an online survey of AI-generated art communities, such as those associated with the Midjourney tool, which highlighted the role of prompt engineering as a creative practice [2]. The use of GenAI tools significantly improves the efficiency and quality of processes, enables early error detection, optimizes production processes, and facilitates resource management [3].

There are studies that relate the characteristics of the use of generative AI to creativity, design, and the workplace. Some suggest that AI can facilitate the creative process, fostering creative solutions and critical reflection, especially when the results are analyzed deliberately and with appropriate mediation [4]. However, these technologies also face criticism and ethical challenges, such as the use of personal data to train models and concerns about replacing human functions in the marketplace. Collaboration between humans and AI can mitigate these issues, promote better creative outcomes, and explore new possibilities [5].

On the other hand, the use of AI tools in the professional education of future designers is controversial. Researchers defend the inclusion of generative AI technologies in design curricula, while others point to limitations related to human intuition, contextual understanding, and true creativity. However, it is suggested that collaborative practice between humans and AI can overcome these limitations [5, 6].

This study seeks to explore the intersection between the creative process applied to design practices and the use of generative AI, with a focus on sustainable contexts applied to the education of designers at the undergraduate level. This research is based on an experimental and exploratory study that investigates how the interaction between design students and generative AI tools can foster more creative and sustainable design practices. Based on the reflections presented in the literature and a practical workshop, the study proposes to analyze the design process based on the relationship between designers and the use of generative AI tools, considering not only the results, but examining the creative, interactive and collaborative processes.

The central question guiding this research is: What are the contributions and implications of integrating generative AI tools in promoting creative and sustainable design practices in the education of designers at the higher education level? To answer this question, the study adopts an experimental and qualitative approach, focusing on the interaction between design students and generative AI tools. To this end, the study analyzes the ethical, methodological and practical implications of this integration, seeking to contribute to the field of design mediated by new technological and creative possibilities, organizing their interactions and practices through a two-day workshop.

The originality of this research lies in the implementation of an experimental workshop with design students, allowing for an analysis of their perceptions regarding the integration of generative artificial intelligence tools in the field of design, with a focus on the creative process. With the recent popularization of this technology, which has shifted from a restricted innovation to an—relative—accessible resource, this study contributes by detailing students' practical experiences, reflections, and the challenges they face in adopting these tools. Furthermore, this research provides a significant contribution by presenting a structured workshop model, highlighting methodologies and discussions aimed at preparing future designers for an evolving job market. Thus, this study not only expands the understanding of the impact of generative AI in design education but also suggests pathways for critically integrating this technology into professional practice.

2 Creativity, Generative AI and the Designer's Formative Process

Understanding the role of design in today's world is necessary to promote innovation. Designers and creative professionals must be aware of today's conditions and challenges to create meaningful solutions, and they must seek to understand how creativity and the design process can collaborate with innovation processes to foster change. This intersection is necessary to address existing problems and foster new realities in society. When people are involved in the design process, the results are generally more satisfying and relevant to their specific needs [7].

Artificial intelligence can outperform humans in some tasks, such as image classification and text comprehension, but not in complex problems, such as those involving mathematics and common sense. Looking at the quantitative characteristics of AI, we can say that industry produces more machine learning models than academia, with high training costs, such as the estimated $78 million for GPT-4 and $191 million for Gemini Ultra between 2023 and 2024. In addition, the number of base models is growing, with a tendency for them to be open source. At the same time, GenAI tools such as ChatGPT are becoming more common, with 63% of respondents to a survey aware of their existence and half using them regularly. However, the global perception of AI is still marked by doubt and uncertainty, with 52% of people expressing concerns about its impact [8].

In terms of creativity, co-creation, and iteration within the design process, we understand that generative AI tools can enhance designers' creativity by allowing them to explore a wider range of ideas and generate multiple design variations. This collaborative approach allows designers to experiment with different concepts more quickly [9]. In this case, the concept of creativity is related to the number of ideas generated and considers that generative AI tools can accelerate the development of new combinations and results. The author suggests that AI should be seen as a collaborative partner, not a replacement for human designers. This perspective is important for the future vision of the sustainable design process, as generative AI tools can potentially enhance rather than reduce the creative process. The balance between human intuition and AI capabilities can be the necessary foundation for achieving appropriate design outcomes.

However, it is important to note the distinction between generative AI and human reasoning and language use [10]. It is argued that generative AI lacks the capacity for genuine critical analysis, being unable to distinguish between what is possible and what is not, and operating without any moral or ethical principles to guide its outputs.

Regarding the design process and workflow, AI tools automate repetitive tasks such as image resizing and color matching, saving time and allowing designers to focus on more complex and strategic aspects of their work [9]. In terms of ethical considerations, there are issues related to authorship, intellectual property rights, and the potential for bias in AI-generated designs.

The use of AI tools is also related to the potential increase in the imaginative capacity of creators. When we created a textual prompt in the generative image tool Leonardo.ai in December 2024, we received four different images related to the requested result in just a few seconds, something that would have taken a long time to generate using other methods (Fig. 1).

Fig. 1 Image created by Generative Artificial Intelligence tool Leonardo.ai showing horses playing cards on Mars

The following text prompt was used in the Leonardo.ai AI generative imaging tool: "A surreal and imaginative scene of a group of horses sitting around a circular table on the Martian surface, playing cards. The red, rocky terrain of Mars surrounds them, with dramatic orange and pink hues in the sky. The table is set with colorful cards and poker chips, and the horses each have distinct personalities, wearing fun accessories like cowboy hats or sunglasses. In the background, distant Martian mountains and a glimpse of Earth in the sky add to the otherworldly atmosphere. The mood is playful and lighthearted, blending the unique Martian setting with the whimsical concept of horses playing a card game". The text prompt used was created with the help of the ChatGPT tool, based on another prompt: "I need to create an image of horses playing cards on Mars. This image will be generated by Leonardo.ai. Create a suitable prompt for the tool".

Therefore, we can describe the characteristics of text-driven image generation systems—such as Midjourney, ChatGPT and Leonardo.ai—and establish some aspects of their relationship between creativity, humans and digital tools in a design context. Questions can be asked, such as whether the product generated by generative AI can be considered creative. Given the contexts presented, initiatives that consider the creative process rather than focusing only on the outcome can be beneficial, as there are difficulties in assessing creativity in text-generated art due to the opacity of the systems, the confidential nature of the prompts, and the lack of knowledge about the creative processes used [2].

Thus, discussions arise on how to integrate AI tools into projects that involve human creativity, such as design or architecture—particularly in urban planning projects, where there is a need to operate at scale while seeking more effective and efficient solutions for clients. The central issue lies in the effectiveness of creativity, that is, the ability of AI to generate innovative and functional solutions. Moreover, it is also essential to assess aspects related to creativity awareness and the appreciation of design outcomes, highlighting the critical role of human interpretation and validation in the solutions created with AI assistance [11].

The education of future design professionals at the undergraduate level requires knowledge and awareness regarding the United Nations Sustainable Development Goals (SDGs). Beyond Goal 4—Quality Education, which is the most evident—the SDGs 8, 9, and 12 stand out as key pillars for exploring the use of generative AI in design education, fostering decent work, innovation, and sustainability. SDG 8 aims for inclusive economic growth; SDG 9 promotes infrastructure development and sustainable industrialization; and SDG 12 emphasizes the importance of sustainable consumption and production. The integration of these goals into the training process of designers, who will eventually use generative AI as a professional tool, must consider aspects such as productivity and the search for innovative solutions, optimization and adaptability, as well as waste reduction and the minimization of environmental impacts [12].

3 Methods

From an educational standpoint, the practical component of the methodology was centered on a nonconsecutive two-day workshop held on a university campus. Conducted over two afternoons, the workshop totaled eight hours in duration. It was facilitated by two professors and involved ten participants, including current undergraduate students from both graphic design and game design programs. The main objective was to explore the intersection between the creative process applied to design practices and the use of generative AI, with a focus on sustainable contexts applied to undergraduate design education. The workshop was organized into four distinct phases: (1) An introduction to generative AI for designers; (2) a hands-on exploration of generative design tools; (3) practical applications in design; and (4) creative experimentation and a project showcase (Fig. 2). These phases were executed through practical application, integrated with periods of group discussion. This structure enabled participants to articulate their understanding of the topic, investigate the functionalities of the tools, produce tangible outcomes, and subsequently engage in a comprehensive reflection on the entire process. Group discussions were initiated by the facilitating instructors, based on questions presented in the presentation slides: "How can the application of generative AI foster creativity, optimize workflows, and extend the boundaries of graphic design?" and "What are the potential positive and negative impacts on the design field?". These discussions were conducted informally, encouraging active participation and attentive listening among attendees. The practical sessions incorporated a theme based on the United Nations Sustainable Development Goals, serving as a framework for both creative and critical project development. The educational and data collection methods employed during the workshops included slide presentations, group discussions, and guided practical projects. The discussions were audio-recorded, uploaded to YouTube and transcribed using ChatGPT's Video Summarizer app. The data gathered throughout the workshops was analyzed and is presented in the next section of this document. Also, the ethical considerations necessary for the development of the book chapter were carefully taken into account, as all participants had to sign a contract on the rules of conduct and participation with the university where the workspace was used to participate in the event.

Fig. 2 The four stages of the workshop

4 Data Collection and Analysis

The first day of the workshop on the use of GenAI was organized with a focus on initial presentations on the main topic, its history and general applications in products and services. The general characteristics of the design process were also presented. The aim was to provide interactions between the group of participants, as well as presentations and practical experimentation with generative AI tools. The importance of clarity and objectivity in writing prompts was discussed, both in professional and academic contexts. During the activities, strategies and tools were presented to help produce effective prompts. To do this, they practiced writing formal electronic messages, such as emails in a work or academic organization, considering aspects such as voice tone, personas and background.

In a structured way, Part I—Introduction to Generative AI for Graphic Design, involved an open conversation with the participants about their perceptions of the relationship between design and the use of generative AI tools in the design process. This discussion addressed initial perceptions of the potential positive and negative impacts of using generative AI tools in students' daily lives and later in their professional careers. The data considered this first approach, including the initial presentation by the facilitators, before experimenting the tools.

At the same time, an informative presentation was given on topics such as the applications of AI throughout the ages, its logic of use and training, current studies and applications, such as the case of Madonna's 2024 tour [13]. A more prominent approach took place during the topic "Generative AI for designers", where insights were discussed such as the existence of a new era in human creative evolution, similar to the transition from analog to digital era in cinema and phonography, where co-creation between humans and machines is increasing, although it is already present in tools such as automatic correctors and software such as Photoshop and Canva. In this context, it was discussed that the popularization of generative AI tools could be compared to the invention of photography, which eventually transformed other mediums, such as painting, by redefining creative processes. In the design industry, its use can enable automation, personalization and customization, allowing for exploration and access to new aesthetics and styles. It also promotes collaboration, inclusion and accessibility, while raising fundamental ethical questions. Despite the positive tone, some participants were concerned about how the tools might change their workflows and even eliminate their roles in teams.

Also, on the first day of activities, Part II—Hands-on Exploration of Generative Design Tools explored the first practical approach with the students, through the exploration of generative AI tools focused on text. The initial activity consisted of properly structuring a response to a inquiry by preparing text to be used as a response to a request received by email. The exploration of generative AI began with the creation of texts in tools designed for this purpose, such as ChatGPT, Gemini and Copilot. The main task was to generate different versions of the text to be used in emails that were supposed to respond to an everyday professional question. The workshop facilitators emphasized that the message must be tailored to the target audience, which requires a prior understanding of who will receive it, since in different situations, the ability to communicate clearly is essential. For example, when addressing a hierarchical superior, a formal and respectful tone may be used, while messages to colleagues may allow for informality. The media through which messages are transmitted also change their format [14]. Thus, this was the basic context where generative AI tools were tested and discussed among the participants.

Thus, the main question of the activity was: "Tell your superior, via text message, that you won't be able to attend an event because of a holiday". Participants were asked to create prompts with requests for results in the GenAI tools mentioned above. In addition, Google's Prompting Guide 101 was used as a guide for the creations. The document indicates the use of four blocks to create effective prompts: Persona, Task, Context and Format, and indicates that they do not necessarily have to be used at the same time [15]. Different prompts were created using the proposed tools, and the textual results were discussed and analyzed in groups. The comparative analysis showed that the answers varied according to the training base of each tool, which highlighted the need for adjustments and customization on the user's part. In addition, the group discussions indicated that constant refinement of the message throughout the process would be appropriate to achieve an adequate voice tone, including elements such as acknowledgements, justifications and alternative contacts.

The discussions highlighted the importance of iteration when using these tools. Throughout the workshop, the facilitators emphasized that continuous refinement of the resulting texts is key to keeping them in line with the purpose of the output. This refinement process involves experimenting with different formats, such as limiting the number of words or turning a text idea into a bulleted list. In this way, it was possible to link the fundamentals of creativity and design, such as iteration, prototyping, and organizing around appropriate goals, with GenAI tools in a loop of continuous learning, where results improve gradually as prompts are refined, tested, and discussed again.

In addition, the comparative analysis explored in the workshop highlighted the benefits of using different tools at the same time. By using them in combination, a variety of outcomes can be explored and the range of choices available to participants can be broadened. It was recommended that students try more than one generative AI tool, both in the context of text and image generation. It was also suggested that they experiment with different approaches and formatting styles. In terms of developing prompts, it is expected that this approach will allow more precise texts to be produced that are tailored to the specific needs of each design project, thus producing more appropriate results. In Fig. 3, we can see the participants exploring different prompts and tools while discussing the results of the activity.

At the end of the second stage and the first day of activities, it was possible to see that the strategic use of generative AI tools in the design process, combined with constant iteration, can contribute to the development of better prompts, increasing the possibility of more appropriate and effective results. In addition, the facilitators recommended that participants use generative AI tools between the two days of the

Fig. 3 Exploring different prompts and Generative AI tools

workshop. This was intended to encourage the integration of these tools into participants' daily routines, encouraging experimentation, exploration of their potential, and the development of greater familiarity with the technology.

The second day of the workshop took place after a two-week break and focused on hands-on experimentation with GenAI tools applied to design processes, with the aim of generating small design projects and then discussing the creation process and the results generated. Here, the research and creation of projects related to the United Nations Sustainable Development Goals (SDG) was added as a general theme as part of the brief for generating projects [16]. This was followed by a critical discussion of the impacts observed after the practical application of the tools in the design process. The main objective of this phase of the workshop was to encourage reflection on the negative and positive impacts of integrating GenAI in the design process, highlighting the creative, ethical, professional, operational, educational and sustainable implications during the development of a design project.

During Part III—Practical Applications in Graphic Design, it was possible to see how generative AI tools, whether used in the context of generating images or text, can enhance designers' creative output by supporting the creative process. However, this also raised some significant concerns in different dimensions among the workshop participants (Fig. 4). Among the challenges discussed were issues related to (1) originality of productions, (2) technological dependence, and (3) changes in the job market. Students reported feelings of ambiguity regarding the authorship of creations, as the role of GenAI in developing specific solutions often blurred the distinction between human and automated contributions.

One of the points of criticism discussed during the practical development of the project was the superficiality of some of the proposed solutions generated by the GenAI tools. Although the model can produce results that appear adequate at first glance, participants highlighted the lack of conceptual depth in many of the results obtained. This reinforced the importance of understanding GenAI not as a substitute for the designer, but as a tool that requires human guidance to generate solutions that transcend aesthetic aspects and achieve functional and symbolic values.

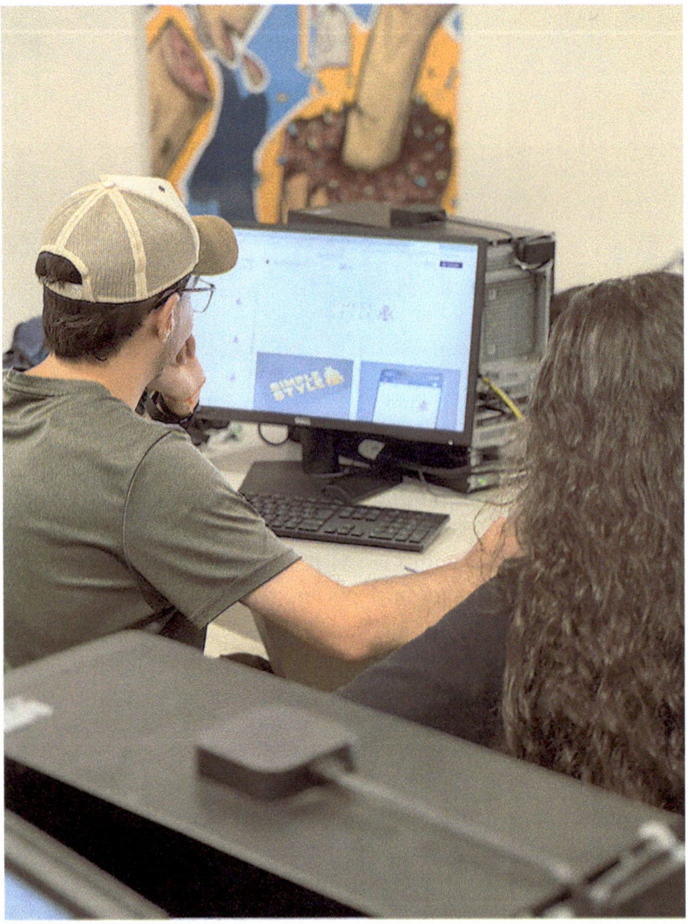

Fig. 4 Design project development using generative AI tools

Participants also discussed the impact of the use of these tools on the professional market. The automation of processes and the speed with which GenAI can generate alternatives raises concerns about reducing opportunities for early-career designers, especially in repetitive or operational roles. However, workshop participants also recognized that the need to adapt is an ongoing issue. For them, therefore, the ability to integrate the use of these tools and manage their use in the process is as important as the critical interpretation of their results. This balance between the operational and the strategic in design will be essential both during their education and in their professional careers.

The ethical dimension of the implications of using GenAI in the design process was another issue raised by participants. Questions about the use of databases to train AI models led to discussions about intellectual property and consent to use data to train models. The perception that AI-generated creations are based on pre-existing work developed by third parties raised doubts about the intersection of technological innovation and copyright, and the legal implications of the results suggested by tools based on text prompts.

One of the benefits reported by participants was the ability of GenAI tools to automate repetitive tasks, allowing designers to spend more time on strategic and conceptual activities. In the workshop activities, generative tools were used to create sketches of illustrations, drawings, logos, and text, exploring different compositional variations and developing visual solutions in a short period of time. This flexibility not only optimized the workflow, but also broadened the scope for experimentation, encouraging the exploration of ideas that might not have been considered in a design process that didn't consider the use of GenAI tools, given the time required to generate new ideas.

The use of GenAI tools in the process was also highlighted as an educational strategy to help learn and organize complex content. Participants reported using AI platforms to learn new skills or concepts related to design, such as graphics software or theories related to the field. The ability of these tools to provide quick answers tailored to the user's individual needs was mentioned as an advantage, especially for students at the beginning of their careers. The problem of the reliability of the information generated by these tools was presented as a counterpoint to the positive educational perception, since errors, inaccuracies, or biases can affect learning and lead to misinterpretations. This aspect reinforces the importance of critical analysis on the part of users when using AI in educational contexts.

Students also raised questions about the impact of using GenAI tools on the creation of contemporary visual culture. They recognized that although these tools depend on the existence of pre-existing data to perform, they have the potential to introduce new graphic and textual perspectives based on the different possibilities of combination. This reinforces the role of the designer as a curator and mediator of technology, capable of directing significant results, be they aesthetic, functional or symbolic. In addition, the tools were seen as a means of democratizing access to creative tools, allowing professionals with different levels of experience to achieve high quality results.

Another topic discussed was the balance between using generative AI and preserving the creative identity of the designer. While the technology offers practical and rapid solutions, participants emphasized the importance of maintaining control over the creative process and ensuring that the decisions made reflect a clear human intent at the end of the process. In this sense, GenAI was described as a "creative assistant" that complements, but does not replace, the designer's work. In this context, we can consider that a creative assistant would be a facilitating tool for the continuity of the design process, which can, for example, propose different perspectives for solving the same problem. Thus, it can quickly expand the capacity to generate solutions and propose results in a project, which is a desirable characteristic in different stages of the design process, such as ideation and prototyping.

In the end, GenAI was recognized as having significant benefits for the field of design, from simplifying operational processes to broadening the creative portfolio. However, participants emphasized the need to integrate this technology critically and ethically, considering both its potential and its limitations. In addition to technical and process adaptations, the sustainable integration of GenAI into design requires a review of the ethical and creative paradigms that guide the profession, with the formulation of clear guidelines that protect the human values inherent in the practice of design and the rights of the people involved in the creative process. Finally, participants discussed the relevance of the designer as a central actor in a context where technology is an available resource, but creativity and originality remain essential elements of the design process.

The discussions in the third phase of the workshop took place continuously in conjunction with the activities in Part IV—Creative Experimentation and Project Showcase, which presents the technical and practical results of the event. Although not the focus of the research, it was necessary for the discussions among the participants to take place. As a result, one of the outcomes of the creative phase of the workshop was chosen as a way of presenting the visual results developed by the participants, with the aim of demonstrating the steps they had worked out and supporting the reflections discussed with the group on the results achieved at the end of the workshop.

The Brief[1] for the "Veggies" project was to create a variety of unspecified graphic communication elements for a beauty product related to the following characteristics: Tech Aesthetics, Cyberpunk, Dystopian, Futuristic, Organic, Vegan, DIY Enthusiast and located in Dublin, Ireland. The design process involved iterating prompts in ChatGPT to expand the briefing and generate the first set of ideas, in addition to the discussions reported earlier in this chapter. The process was documented using the Miro tool, which was also useful for organizing communication between the group participants, as they had only 60 min to complete the project. From the first results selected from the textual iteration of prompts in ChatGPT, results related to the brief were created for use in the "Ideogram.ai" tool to generate visual results. It is possible to see the continuous result of the flow, different GenAI tools and a linear record of how the iterations took place—both textual and graphical, in Fig. 5. Much of the text

[1] The briefings were randomly created using the tool https://www.brandbrief.io/.

cannot be read in the image, as the authors' intention is for it to be observed as a creative flow. Most of the text is in Portuguese, except for the prompts used. These are: "Design a brand identity for 'Veggies', an organic beauty products company targeting DIY enthusiasts. The concept should feature items that customers can combine, inspired by the interactive style of 'Little Alchemy' games. The aesthetic should blend organic elements with a casual, clean, and futuristic cyberpunk vibe. Use natural tones like greens and browns, accented with white and light blue, and incorporate simple packaging. Highlight the DIY aspect with modular components that users can mix and match to create their own beauty products"; and "Design a brand identity for 'Veggies', an organic beauty products company targeting DIY enthusiasts. The concept should feature items that customers can combine, inspired by the interactive style of 'Little Alchemy' games, with a futurist aesthetic, cyberpunk, focus on white and denim. Highlight the DIY aspect with modular components that users can mix and match to create their own beauty products."

The same process described in the previous paragraph was repeated after the briefing to refine the prompt and generate visual results that were more applicable to the group's interests. As the results were generated, the participants searched for alternatives in other tools, trying to compare the results and select the most appropriate ones, as shown in Figs. 6, and 7. New visual results were generated in the Leonardo.ai tool.

Textual options were also created for the " voice tone" of the product. Finally, the ability to iterate the design process and experiment with the results was extended to other tools. The Veggies project participants decided to use ChatGPT to create tone-of-voice possibilities for the visual communication of the product and the Suno.com tool to generate musical soundtracks. The iterative process, including the prompt used, is shown in Fig. 8. It was used the following prompt: "a song that's feminine and fancy, but also cyberpunk and jeans, no lyrics, vegan also".

5 Discussion

Regarding the central question of this study, it was observed that the implications of integrating GenAI tools into the design process lie in the combination of technical skills, professional ethics, and a broad understanding of contemporary challenges. Among the contributions, the study emphasizes that iteration and prototyping allow a continuous learning cycle between the designer and the technological tool, leading to progressively more refined results. It was observed that continuous iteration between tools and designers, within an organized process, is more effective when there is a constant dialogue between the participants, both human and non-human. To improve the process, the use of mind maps and semantic boards can be indicated to find more appropriate keywords to optimize the generation of prompts and improve the results. Other strategies observed were the use of generative text AI tools to refine prompts for image creation, as well as exploring results in different tools using the same prompts.

Fig. 5 Linear representation of visual and textual iterations of the Veggies project

Fig. 6 Visual iterations using the "Leonardo.ai", "Ideogram.ai", "Flux-ai.io" and "ChatGPT" tools

Fig. 7 Other Visual iterations using the "Leonardo.ai" and "Ideogram.ai" tools

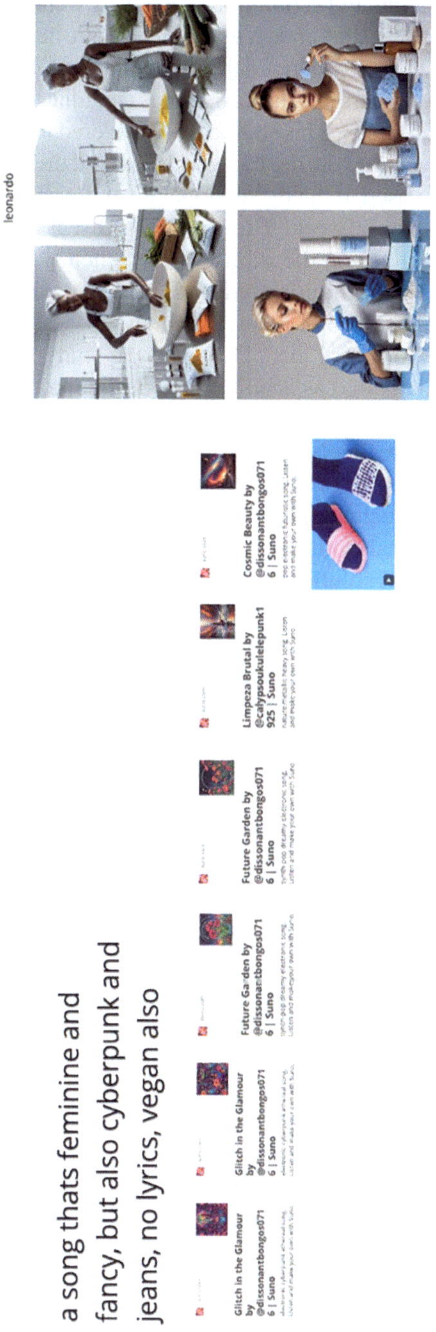

Fig. 8 Visual and sound iterations during the design process of the Veggies project

The process of prompt design highlights that the human factor remains essential in the design process, curating and refining the results as well as providing creative direction, supporting the view that creativity in GenAI art lies more in the collaborative process than in the final product [2]. This collaboration not only enhances creative capabilities but also positions the designer as a mediator between technological possibilities and human needs. This should be considered in educational projects, such as the one tested and applied during the workshop, and there are indications that it could also be a guideline used at the professional level. For this, it should be tested in other experiments and future work. The workshop discussion facilitated a shift in the participants' perspectives on the subject. Consequently, they adopted a more active and confident approach, exhibiting reduced fear and anxiety, and a diminished concern about the future relevance of their roles in the face of GenAI's potential to automate design tasks and displace jobs. However, the study shows that the use of GenAI tools should be understood as an ally of the design process and the designer, not a substitute, and highlights the need to adapt design practices to incorporate new technologies ethically and creatively.

The study concludes that there are three main challenges to the design process: the originality of creations, technological dependence in the digital field, and transformations in the professional domain. Originality requires the critical mediation of the designer to ensure that solutions are not only aesthetically pleasing but also functionally and symbolically relevant since the final quality of a design project is related to the balance between these characteristics. Dependence on and changes in technology are constant, as seen, for example, with the popularization of the personal computer and the Internet. However, their introduction has always generated doubts and anxieties related to changes in working paradigms, as discussed by the event participants. The importance of studies on creativity, culture, prototyping, history, and design management in the education of future designers was also noted, as a form of discernment and design guidance during the creation process.

Future studies must balance human and technological capabilities and ensure that the effects of this interaction produce results based on human, social and sustainable values and needs. In conclusion, we don't see GenAI tools as substitutes for designers, but as tools that require human guidance to generate design solutions. According to their specificity, these solutions must go beyond aesthetic aspects and reach functional and symbolic values. The future of AI-mediated design also requires analyses of the social and sustainable impacts of this technology. The study indicated that aligning the creative characteristics outlined in the project briefing with the SDGs enabled discussions among participants that included ethical applications to both the technical project and the design process. This also fostered a more appropriate reflection on inclusion and accessibility in the context of GenAI tools. In design education environments, the critical integration of these themes with the use of GenAI in the creative process can create spaces for discussion on the sustainable integrity of the profession and provide mechanisms for training designers who are aware of issues related to global urgencies.

References

1. Ismayilzada M, Paul D, Bosselut A, Lonneke, Van der Plas L (2024) Creativity in AI: progresses and challenges. ArXiv preprint arXiv:2410.17218. https://doi.org/10.48550/arXiv.2410.17218
2. Oppenlaender J. (2022) The creativity of text-to-image generation. In: Paper presented at the 25th international academic Mindtrek conference, Tampere University, Tampere, 16–18 November 2022. https://doi.org/10.1145/3569219.3569352
3. Kayıran HF (2024) The applicability of artificial intelligence in the fields of engineering and design. Advances in human and social aspects of technology book series, pp 41–64. https://doi.org/10.4018/979-8-3693-4318-0.ch003
4. Vartiainen H, Tedre M, Jormanainen I (2023) Co-creating digital art with generative AI in K-9 education: socio-material insights. Int J Educ Art 19(3):405–423. https://doi.org/10.1386/eta_00143_1
5. Alvarado KMV (2023) La Perspectiva del Diseño Gráfico ante la Inteligencia Artificial. Rev Investig Form: Innovación Y Apl Técnico-Tecnológicas 5(2)
6. Osadcha K, Osadcha M (2023) Generative artificial intelligence vs humans in the process of creating corporate identity elements. Inf Technol Learn Tools 98(6):212–230. https://doi.org/10.33407/itlt.v98i6.5494
7. Cabral SCR, Obregon RFA, Zandomeneghi ALO (2023) ERGODESIGN & USIHC 2023 & JOP'Design 2023, Blucher Design Proceedings, 11, 1475-1483, ISSN 2318-6968. https://doi.org/10.1016/ergodesign2023-100
8. HAI Stanford University Human Centered Artificial Intelligence (2024) AI index annual report. https://aiindex.stanford.edu/report/. Accessed 13 Dec 2024
9. Adeleye IO (2024) The impact of artificial intelligence on design: enhancing creativity and efficiency. J Eng Appl Sci 3(1):1–13. https://doi.org/10.70560/vvsfej12
10. Chomsky N, Roberts I, Watumull J (2023) Noam Chomsky: The false promise of ChatGPT. https://www.nytimes.com/2023/03/08/opinion/noam-chomsky-chatgpt-ai.html. Accessed 18 Apr 2025
11. Leach N (2022) AI and the limits of human creativity in urban planning and design. Artificial intelligence in urban planning and design: technologies, implementation, and impacts book, pp 21–37
12. Mossin N, Stilling S, Bøjstrup TC, Larsen VG, Lotz M, Blegvad A (2018) An architecture guide to the UN 17 sustainable development goals. https://www.uia-architectes.org/en/resource/architecture-guide-to-the-un-17-sustainable-development-goals-english/. Accessed 15 Jan 2025
13. Runway (2024) Generating new worlds for Madonna's Celebration Tour with Runway. https://runwayml.com/customers/generating-new-worlds-for-madonnas-celebration-tour. Accessed 17 Dec 2024
14. McLuhan M. (2013) Understanding media: the extensions of man. Electronic ed. Berkley
15. Google (2024) Prompting guide 101. https://services.google.com/fh/files/misc/gemini-for-google-workspace-prompting-guide-101.pdf. Accessed 15 Dec 2024
16. United Nations Development Programme (2024) Sustainable development goals. https://www.undp.org/sustainable-development-goals. Accessed 17 Nov 2024

Sustainability of Architectural Design Processes in the Metamodern Age: The Value of AI-Human Collaboration

Belinda Humphris

Abstract We are living in an era, oscillating between the rigid objectivity of modernism and the cynicism of postmodernism, while simultaneously embracing meaning and value. This is the age of metamodernism, driven by rapid technological advancements, among which, artificial intelligence (AI) presents a critical challenge to the sustainability of architectural practice: adapt or risk obsolescence. Exploration of the rise of metamodernism and the application of AI elicits key questions: How is architecture expressed in this metamodern age? How is AI used in architectural practice, and what role can it play in metamodern design? Will AI replace architects in the design process, or is there a sustainable way for designers to retain control? In response, integration of Large Language Models (LLMs) within a metamodern process is explored through a case study. Conducting a Research By Design methodology, the goal of the study is to engage with LLMs and physical design through a metamodern process of iterative experimentation, oscillating between deconstructions and reconstructions to refine production outputs. Resulting in an example of what metamodern architecture and architectural design processes could be, this chapter aims to illustrate how AI was able to be integrated sustainably, retaining meaning and creative control over the design.

Keywords Artificial Intelligence (AI) · Large Language Model (LLM) · Architectural design · Metamodernism

1 Introduction

An emerging cultural shift has seen the rise of metamodernism, characterised by oscillation between modernist ideals and postmodern scepticism, creating new opportunities for architectural expression [1]. Through developments in the technological revolution artificial intelligence (AI) is increasingly being incorporated into the design process. This chapter explores a hybrid approach to sustainably utilise large

B. Humphris (✉)
Deakin University, Geelong, VIC, Australia
e-mail: Belinda.Humphris@gmail.com

M. Sedrez and G. Pitts (eds.), *Artificial Intelligence-Aided Design for Sustainability*, Urban Sustainability, https://doi.org/10.1007/978-981-95-1349-9_6

language models (LLMS), a form of AI, within a metamodern design process. The goal, to ensure that creative engagement and control resides with the designer and not the AI.

Metamodernism: The rise of a new age, considers how cultural theory informs design, providing a foundation for understanding architecture in a metamodern era. Characteristics of metamodernism are based on its oscillating nature, described by cultural theorists Timotheus Vermeulen and Robbin van den Akker in 2010, as well as Brendan Graham Dempsey's attempt to combine several theories into one overarching definition [1, 2]. Dempsey describes metamodernism as iterations of deconstruction and reconstruction [2]. Vermeulen and van den Akker point to Herzog & de Meuron's Elbphilharmonie in Hamburg as an example of metamodern architecture and how it is expressed in this metamodern age [1].

Focusing on LLMs, *The role of artificial intelligence in architectural design,* investigates how AI is used in architectural practice. Their adoption within the industry and potential as a design tool are discussed. Questions arise over the use and associated risks and benefits of LLMs when incorporated into architectural design processes.

Building on this, *Comparison of typical design process using LLMs with process for metamodern case study,* presents a case study extract from a Master Class Thesis at Deakin University's School of Architecture and Built Environment [3]. This case study is presented as a sustainable approach to integrating AI within a metamoden design process through the adaptive reuse of an industrial shed to become the Maloppio Metamodern Museum [3]. The case study considers LLMs to be a metamodern tool due to significant development and uptake in the metamodern age and their iterative nature, which aligns with metamodern characteristics [3].

Exploration into the use of LLMs within a metamodern design process is undertaken, applying a Research By Design methodology [4], to a hybrid design approach in which both LLMs and human design processes are utilised.[1] A (human-created) model is used as the basis for LLM iterations and is inputted into Midjourney multiple times throughout iterative cycles. This ensures outputs retain key conceptual elements of an arch over a shed form. In contrast to common AI workflows, LLM outputs are not considered as an endpoint of the process, undergoing further steps of deconstruction and reconstruction through human-led design to produce unique, human-created/AI-assisted, rather than solely AI generated, outcomes. Crucially, this allows creative control to be retained by the human designer throughout the process.

Concluding with advocacy for integration of LLMs within a human-centered design approach, this chapter argues that by maintaining authorship at every stage, architects can take advantage of AI's capabilities for rapidly generating multiple options without surrendering control to the AI. In doing so, an iterative metamodern design process may become more sustainable, with AI remaining as a design tool rather than a substitute for human intelligence and creativity.

---

[1] A loose summary of Hauberg's description of Research by Design, developed by the Dutch practice at The Faculty of Architecture in Delft, is a process of inquiry, whereby research is undertaken by means of design activity. In other words, research is primarily conducted through design rather than other forms of investigation.

2 Metamodernism: The Rise of a New Cultural Age

How do we define an era where grand narratives are reemerging and existential issues like war, economic instability, environmental crises and the rapid change of the technological revolution will impact our jobs and our daily lives? In 2010, cultural theorists Timotheus Vermeulen and Robbin van den Akker noted a cultural shift beyond postmodernism, calling it metamodernism [1]. They asserted that metamodernism was a "structure of feeling," describing it as a new cultural sensibility that values meaning and sincerity [1]. Stoev linked structures of feeling both emotional effect and affect [5].

In Notes on Metamodernism, Vermeulen and van den Akker defined the term 'meta' from the original Greek term, 'metaxis', as a dynamic state of oscillation with metamodernism existing between many opposing poles, holding multiple truths at once, for example, oscillation between modernism and postmodernism [1]. Researchers Dina Stoev and Jimenez Lei, further discuss this oscillating state as a duality, and simultaneously existing between extremes [5, 6]. Such dualities could be, for example, the metamodern concept of meta-irony, which simultaneously holds irony in a dual state of existence with sincerity [5, 6].

Building on this foundation, author Brendan Graham Dempsey put forward an overarching framework in an attempt to unify theories from many of the key researchers in this field, including the work of Vermeulen and van den Akker [2]. Dempsey refuted Stoev and Lei's idea of dipolar oscillation, expanding the meaning of the Greek word 'metaxis' as a state of being simultaneously beyond and between many poles [2]. Highlighting the iterative nature of metamodernism, he defines a dynamic state of oscillation between conditions of deconstruction and reconstruction [2, 7].

Why is metamodernism relevant to design? Architecture is shaped by the cultural ethos of its time. Metamodernism is a cultural theory describing an all-encompassing worldview affecting everything we do, including architectural design [2]. The need to understand the ethos underpinning the thinking of today, how this influences the architecture that we build, how we deal with the tools that we build it with, and the way we view its history and its meaning, is foundational to the practise of creating metamodern architecture and examples of this are provided by Kadagishvili [8].

Vermeulen and van den Akker noted that an emergent metamodern style, separate to postmodernism, was yet to be defined [1]. However, many aspects of metamodernsim seem to align with architectural design theory, such as the concept of 'structure of feeling', and consideration for the importance of history, meaning, and concern for environment [1, 9].

So how is metamodernism expressed in the age of metamodernism? Vermeulen and van den Akker cite Herzog & de Meuron's Elbphilharmonie in Hamburg as an example of metamodern architecture, identifying its oscillation between nature and culture, as well as past and present [1]. The design incorporates adaptation of an historic industrial structure into a concert hall while adding a striking glazed iceberg form above, which reflects the surrounding context [1].

The case presented in this chapter attempts to apply a metamodern process that oscillates between deconstruction and reconstruction, while utilising the iterative nature of LLM technology. This process is employed to create a solution for the adaptive reuse of an historical industrial shed, resulting in an oscillating metamodern form that considers contextual elements of the site, such as its industrial history and representation of impending submersion from rising sea levels through form.

3 The Role of Artificial Intelligence in Architectural Design

LLMs, a type of generative AI, are already being adopted by the architectural profession and their use in design will most certainly become ubiquitous [9]. Examples of LLMs include Midjourney, Dall E and ChatGPT [10, 11]. Early adopters of these technologies are experimenting with their use in generating design ideas [12, 13]. Recognisable names such as Design Specialist, Olly Thomas (BIG), and Tim Fu (Studio Tim Fu, previously at ZHA) have indicated that they use LLMs to generate ideas as a basis for design development [12, 13]. In an interview with Rishabh Wadhwa, Fu explained this approach as human inputs processed through an LLM to generate outputs over a number of iterative cycles until a human-specified criteria is met [13]. The selected output is developed into a refined design solution.

The potential benefit of integrating LLMs into architectural design is their ability to efficiently produce infinite design options at speeds unattainable by humans [10, 14, 15]. However, their true power lies in their iterative nature [15]. With each evolution, the user has control to decide which outputs to select as inputs for the next cycle (Fig. 1). Results can be reprocessed with additional parameters or other inputs, such as imagery, in an exponential and unlimited process. However, user influence is limited to the start and end of each iteration to add inputs, select outputs and decide whether to continue or discontinue the process. Figure 1 illustrates the typical process for the generation and selection of LLM outputs.

It is important not only to consider the benefits of using LLMs in design, but also the pitfalls. Of concern is over-reliance on technologies that produce instant results, bypassing hundreds of hours of human design effort. LLMs are just another design tool, however results can be generated faster and in greater quantity than through traditional design methods [9, 11, 16, 17]. Rather than accepting machine-generated outputs as a final design solution, designers can rework or further modify outputs after completing the LLM process where they can create unexpected and unconventional outcomes as an extension of their creative process. The case study shows how maintaining control over this process (rather than handing the reins to AI) allows architects to utilise LLMs as a powerful asset to enhance rather than replace human creativity [3].

One design application of LLMs is to create outputs for a specific style. This can be problematic. Stylistic definitions are based on datasets with content limitations, and the pool of data available to LLMs impacts the types of outputs that are generated [10,

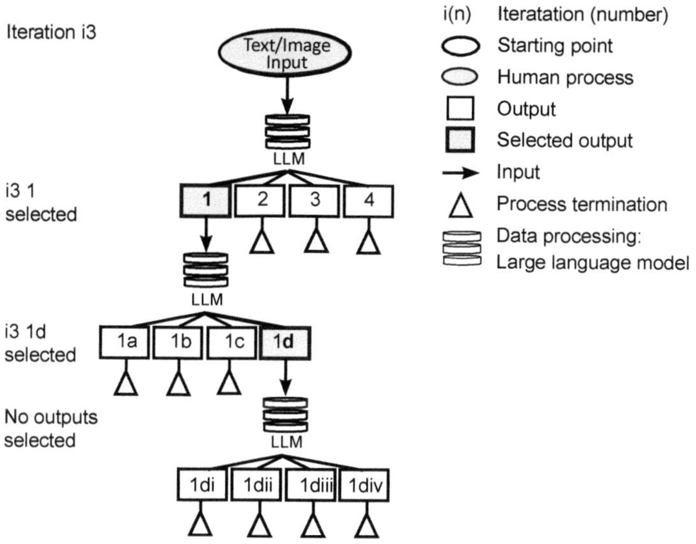

Fig. 1 Output generation and selection process for output 1

11, 18]. Due to content limitations, LLMs may also preference certain architectural styles over others, reducing diversity in output content [10, 17].

Unless explicitly specified LLMs cannot discern which preferred theoretical model/s to use when requesting outputs that express a specific style. For example, ChatGPT-4 generated the following description of metamodernism:

> Metamodernism is a cultural and philosophical movement that oscillates between the skepticism of postmodernism and the sincerity of modernism. It embraces contradiction, blending irony with genuine emotional engagement, and often navigates between optimism and doubt. Unlike postmodernism, which deconstructs grand narratives and revels in cynicism, metamodernism seeks to reconstruct meaning while acknowledging complexity and uncertainty. This results in a cultural ethos that is both self-aware and earnest, mixing nostalgia with innovation, sincerity with playfulness, and depth with ambiguity. Metamodernism is evident in contemporary art, literature, architecture, and media, where creators balance critical distance with heartfelt engagement, crafting works that are both reflective and hopeful.

In this example, ChatGPT was asked to define 'metamodernism'. The output was drawn from Vermeulen and van den Akker's definition and did not include other frameworks such as Dempsey's [1, 2]. This demonstrated that over-reliance on information generated by LLMs is ill-advised [10, 17, 18].

Generating images using Midjourney based on the style, 'metamodernism', produced images without context and it was difficult to determine how they relate to metamodernism (Fig. 2). Results for 'metamodernist building', prompt c (Fig. 2) produced a series of unusual (perhaps oscillating) forms that do not convey other key concepts such as, Structure of feeling [1]. This demonstrates the importance of human participation throughout the process rather than relying on the LLM to generate a style.

Fig. 2 Interpretations of metamodernism: Images generated using Midjourney (v.5)

Rather than replicating a style, Fu and Thomas describe a process in which an LLM, such as Midjourney, generates images as the basis for a design solution [12, 13]. This approach has several drawbacks. Replication of LLM outputs as a final solution risk minimising the role of human design thinking and potentially places AI in a lead role rather than the architect [15, 16]. LLM outputs are confined by the scope and content of user inputs, and since output generation relies on datasets, this affects how styles are interpreted, limits diversity of outputs and creates the possibility of producing outputs that closely resemble copyrighted work [17, 18].

So, what role can AI play in a sustainable metamodern design approach without compromising the role of the designer? This is addressed through exploration into the use of AI in a metamodern design process outlined in *4. Comparison of typical design process using LLMs with process for metamodern case study*. As an emerging design technology in the metamodern era and due to their iterative nature, which aligns with metamodern characteristics, an attempt to create a metamodern case study is provided, employing LLMs as a metamodern design tool [3]. A key aim is to ensure the designer retains control whilst utilising the many benefits provided by incorporating LLMs into a metamodern design process.

4 Comparison of Typical Design Process Using LLMs with Process for Metamodern Case Study

The following pages present a case study exploring the use of LLMs to undertake a project that would be considered as metamodern and utilises metamodern design processes [3]. This process applies Hauberg's Research by Design methodology [4] to investigate the use of LLMs to create an architectural design with theoretical underpinnings in metamodernism. Specifically an LLM (Midjourney) is incorporated into a metamodern design process to explore: The application of a metamodern approach to design process; The use of LLMs as a metamodern tool for design; and the sustainable use of LLMs in the design process. This process leads to the design of an elevation which has been provided as an attempt at what metamodern architecture, and design process, could be. It also aims to demonstrate sustainable utilisation of LLMs to allow full creative control over the entire process for the development of an elevation [3].

The case study presents the development of the Maloppio Metamodern Museum at Point Henry, Geelong, a final thesis project for the Master of Architecture at Deakin University, Australia [3]. The project aimed to re-establish Point Henry, a former industrial complex, as a tourist destination through adaptive reuse of a 260m long steel portal framed shed, formerly housing a series of aluminium smelters. Incorporating a significant iterative process using LLMs, metamodern theory, and traditional design practices, this section explores the development of the north-east façade of the museum through a hybrid design approach.

A metamodern process of deconstructions and reconstructions involved development of a concept model followed by the use of LLMs to generate content options, further human-led exploration using traditional design processes, and finally human refinement. The objective was to create a metamodern design, employing a metamodern process that oscillated between human design activities and Midjourney iterative outputs. The intention was to offer an example of a metamodern design process which utilised LLM technology to create form and program (not covered in this chapter).

During early project development the elevation concept was identified as a single axially related arch over the portal-frame museum (formerly the shed) (Fig. 3). This concept carried through to the end design (Fig. 5). LLM outputs are probabilities based on initial inputs, meaning that it is difficult to prevent outcomes from deviating from the initial input [19]. The main issue in the use of LLMs for the development of the case study was how to ensure the concept was retained through LLM processing [3].

In the case study the design process itself was used to research how a metamodern design and process could be achieved [3]. To produce an outcome that might be considered as a metamodern design, metamodern concepts such as oscillation, achieving a state of between, and introduction of meaning and emotion were explored through form and composition. Conversely, this research attempted to develop a metamodern process that oscillated between method and process to produce the design outcome.

Design exploration involved a series of six iterations (i1-i6) oscillating between processes of deconstruction, reconstruction and reconstructions of reconstructions to produce a final design (Fig. 4). Iteration series 2–3 were based on an initial i1 model developed in iteration series 1, derived from the basic concept of the arch over the

Fig. 3 Basic concept for North-east elevation: Arch over former industrial shed (adapted to become the museum)

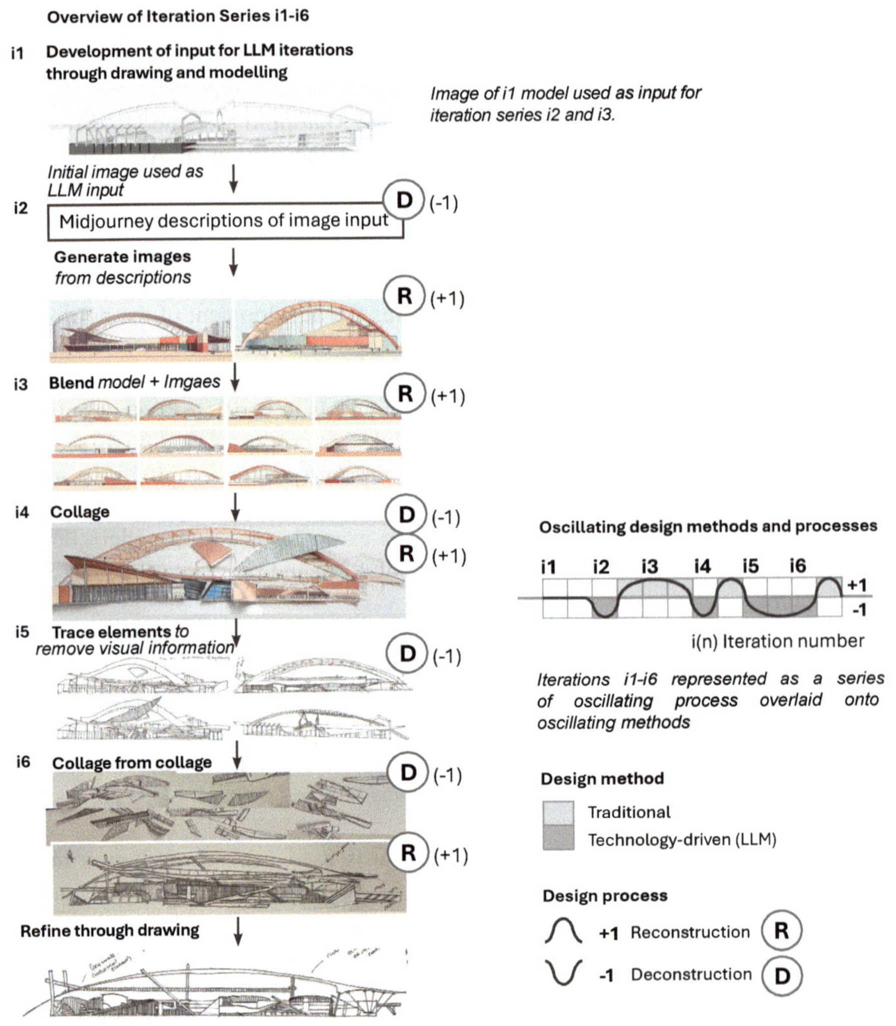

Fig. 4 Iterative design process: Selected outputs from iterations i1-i6, derived from a series of oscillating processes and methods (images i2-3 from Midjourney 6; diagrams and other images [3])

museum, prior to any use of AI. This model was used as input to ensure iteration outputs retained these key conceptual elements.

Figure 4 demonstrates how iterations i1-i6 became part of an oscillating metamodern design method and process (Fig. 4). Beginning with a traditional design process, iteration 1 (i1) involved development of a 3D model from drawings and modelling. This model became input for iterations i2 and i3 in Midjourney. Iterations i4-i6 involved human design processes, oscillating between deconstructions and reconstructions of i2 and i3 outputs through collaging and drawing. Overlaying deconstructions and reconstructions of the design processes onto traditional or LLM design methods revealed the oscillating nature of this approach.

A typical use of an LLM such as Midjourney is for idea generation [12, 13]. With each iteration more outputs are generated in an expansive process that diverges further from the initial input as more generative cycles are undertaken. Aside from the selection of outputs and slight adjustments of input instructions users have little control over the process or the end result [15, 20].

A different approach was undertaken in the case study with the initial (human-designed) input fed back into the generative cycle during multiple iterative sequences, to become a self-referential, rather than expansive, design approach (Fig. 5). Generation of LLM outputs became part of a metamodern process of iterations of deconstruction and reconstruction combined with traditional design methods (Fig. 4).

Figure 5 demonstrates the integration of the initial input across multiple iterative cycles rather than only at the start of the process. This approach, applied to iteration series 2 and 3, generated outputs that remained aligned with the original elevation concept rather than diverging from it, enabling key design elements to be retained throughout successive iterations.

After development of the model in Iteration series 1 through traditional processes, Iteration series 2 and 3 involved the generation of LLM content using Midjourney. The objective was to generate variations on design elements to be used for collaging. The use of LLMs was discontinued once enough variations were generated to create a large amount of content for collaging in the next series. Taking the process from

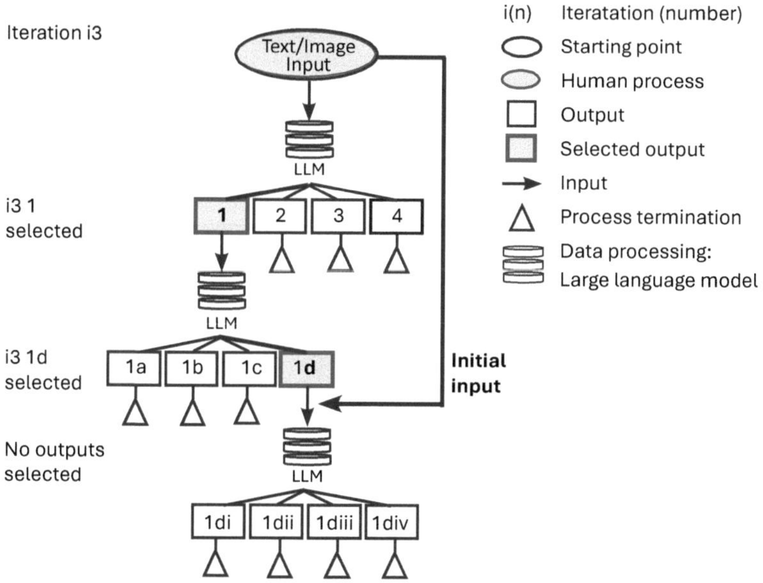

Fig. 5 Alternative approach for an LLM generative cycle in the case study: Re-entering initial input at iteration i3 1d

the AI back into human hands meant there was no question of (human) authorship or ownership.

Iteration series 2 consisted of two parts. The first generation of iterations, i2 1 to i2 5 involved generating text descriptions of the i1 model, with i2 5 selected as input to generate images i2 5a-d (Diagram 1). In the second generation of iterations, i2 5e-h (Diagram 2), images were generated from the i2 5 description and the i1 model as input.

In Iteration Series 2, images 5a-5d were generated from the description for output i2 5 (Diagram 1). At i2 5bi-iv the iterative process was discontinued as outputs began to deviate significantly from the elevation concept of the initial i1 input (an arch over a shed form), indicating a change to the initial input was required. Therefore, a second round of iterations were performed. Rather than only starting with the i1 image as the first input, both the i2 5 text and the i1 image were used as the first input to generate iteration sequence i2 5e-h images (Diagram 2). The aim of reinputting the

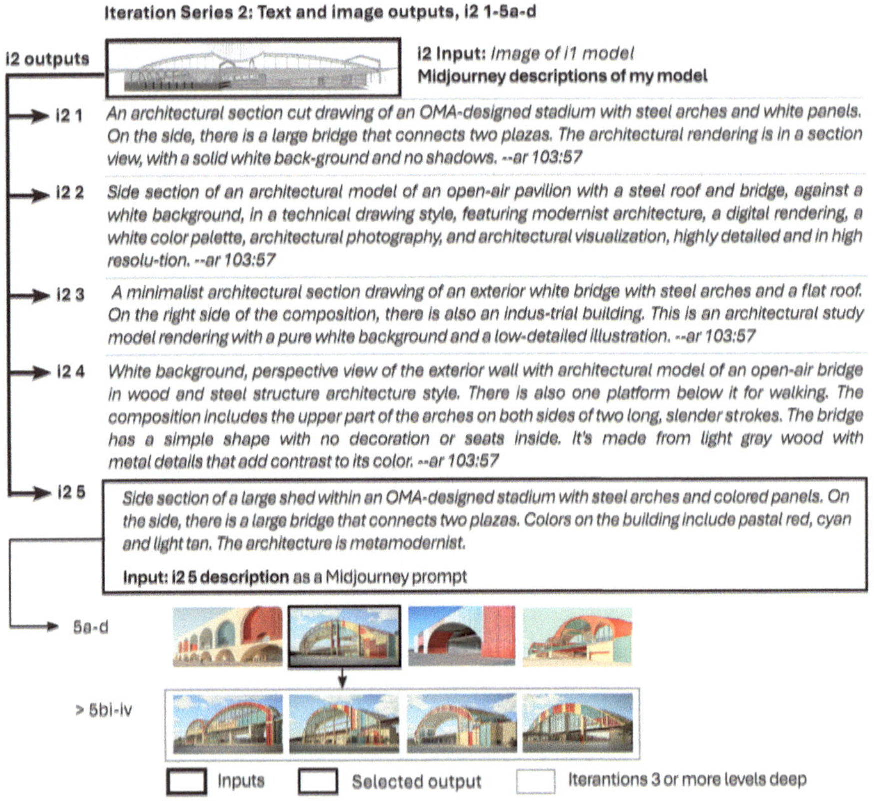

Diagram 1 Iteration Series 2: Outputs from LLM iterations 1–5 (text and images, Midjourney version 6)

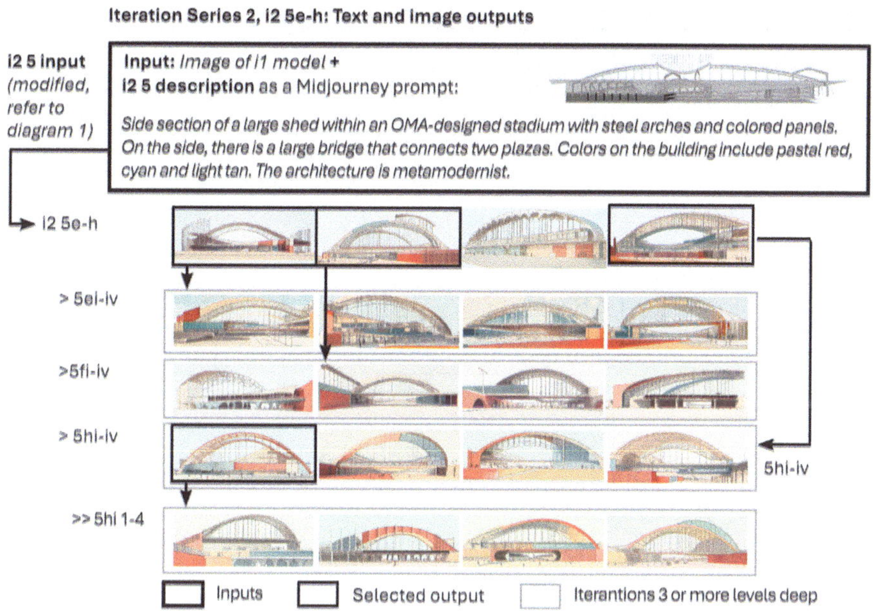

Diagram 2 Iteration Series 2: Outputs from LLM iteration 5e-h (text and images, Midjourney version 6)

model was to better align outputs with the original elevation concept by reinstating the conceptual elements of an arch over a shed.

Further iterations were then generated for i2 5e, 5f and 5h, with one final round of iterations generated for 5hi. Unlike previous iterations (i2 5a-d), i2 5e-h produced images that for the most part depicted an arch over a shed, demonstrating the impact of inputting the i1 model back into the iteration sequence to reinstate conceptual elements. In this series output selection was based on personal preference, focusing on variety in how elements were expressed.

Midjourney allows the use of multiple images as inputs, so iterations 5e and 5hi were selected as inputs for Iteration Series 3 (Diagram 3), in conjunction with the i1 model. i1 was used again as an initial input at the start of i2 5e-h to increase the likelihood of the outcomes continuing to have the required elements of an arch and a shed.

Outputs from series 3 iterations (Diagram 3) were generated from the i1 image combined with two outputs from iterations series 2. This resulted in images closely resembling a shed-form beneath an arch. This again helped to ensure outputs did not have major variations, such as the exclusion of the shed form or the arch, or inclusion of new elements that did not align with the elevation concept. As with series 2 sequence i2 5 (Diagram 2) selection was based on personal preference, focusing on variety in how elements were expressed.

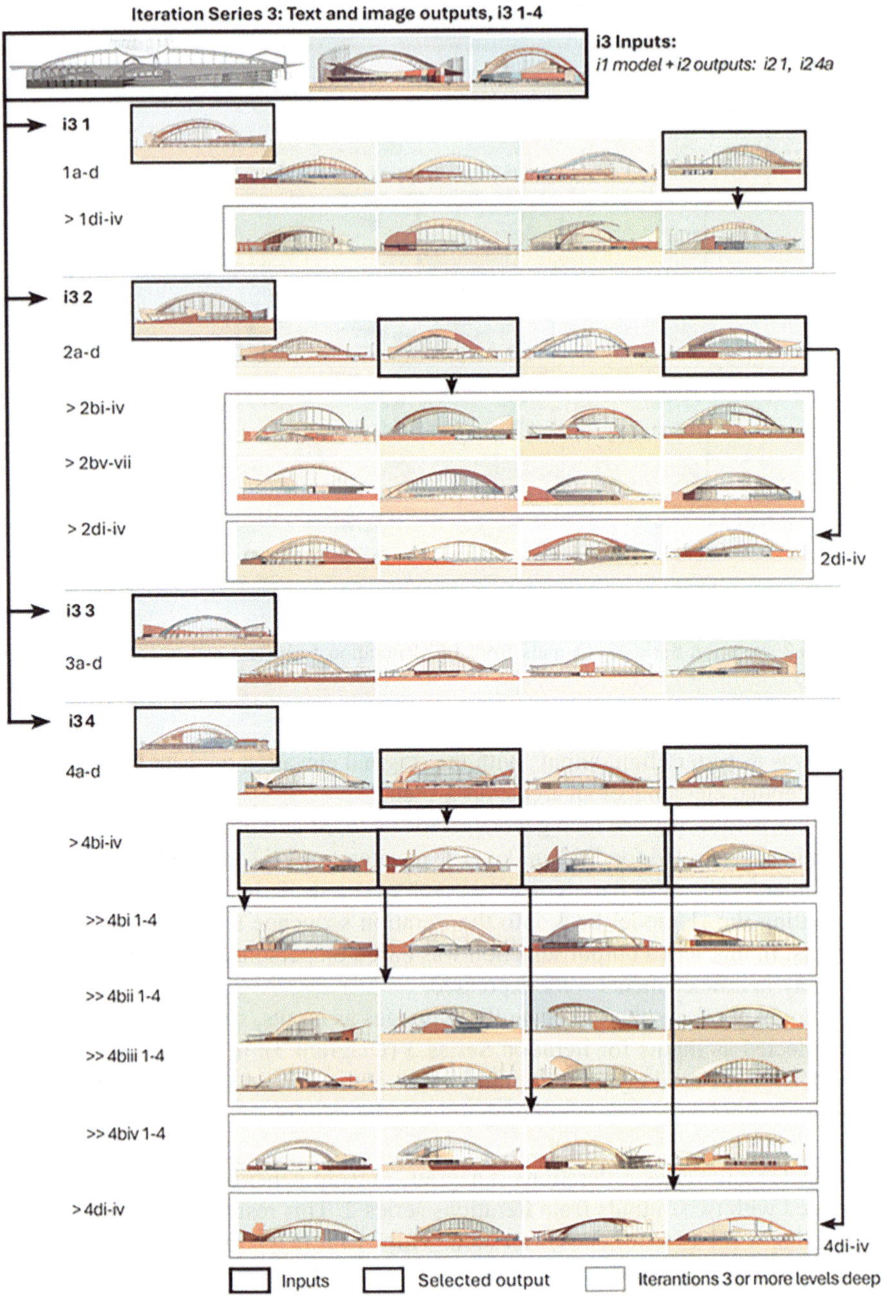

Diagram 3 Iteration Series 3: Outputs from LLM iterations 1–4 (text and images, Midjourney version 6)

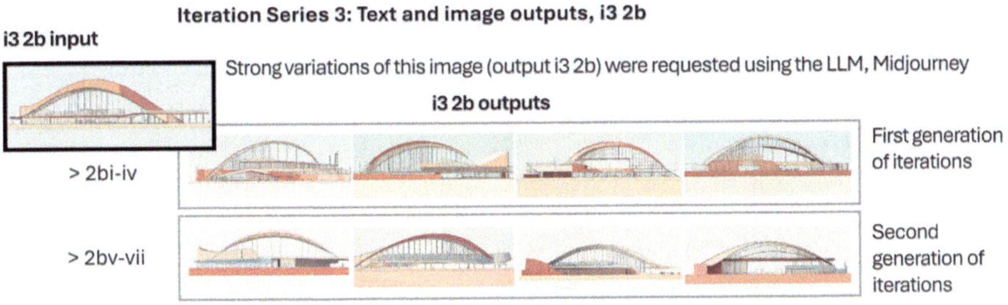

Diagram 4 Repetition of Iteration Series 3, 2b outputs (images, Midjourney version 6)

Under the same conditions, with the same inputs, outcomes in Diagram 3 differed and were not repeatable, demonstrating the probabilistic nature of LLM iterations. The same issue occurred with other LLMs, such as ChatGPT. In cases where repeatable and identical results are required this may be problematic.

To further demonstrate this, iterations for sequence i3 2b were generated twice from the same input, producing iteration 2bi-iv, and iteration 2bv-vii (Diagram 4). Again, outputs had similarities, however they were all unique.

The design objective of iterations series 4–6 (Diagram 5) was to create a considered solution applying a creativity that extended beyond the recreation of an LLM-generated image. Maintaining physical engagement throughout the design process was considered important to avoid relinquishing full control to the LLM.

Iteration Series 4 to 6 involved physical collaging methods involving processes of deconstruction and reconstruction of LLM outputs generated in Iteration Series 2 and 3. So why use collage to intercept the iterative process of generative LLM outputs? The rapid pace of generative design using LLMs resulted in a high volume of outputs. As a designer, it became increasingly difficult to keep track of such rapid generation. At this stage LLM outputs had become too similar with little variation between elements, whereas the collaging process required variety in the expression of elements to be successful.

In Iteration Series 3 (Diagram 3) this was addressed by identifying a stopping point. Outputs were generated until there was enough variety in the material to produce a number of different collages for Iteration Series 4–6 (Diagram 5). Collaging enabled creation of new design options through deconstruction and reconstruction of LLM outputs, resulting in a hybrid design approach. This allowed human control to be regained over the quality and quantity of outcomes, and to physically and creatively reconnect with the design.

The design exploration process ended after iteration series i6 (Diagram 5). The last drawing in Diagram 5 is a composite drawing derived from iteration 6 elements, programming requirements, site context and the original elevation concept, and this became the basis for the final elevation solution.

Figure 6 shows a render of the north-east elevation for the final design. Neither the original human-created model (i1) or the AI generated content encompassed the richness of the final design. The metamodern process, incorporating LLM technology,

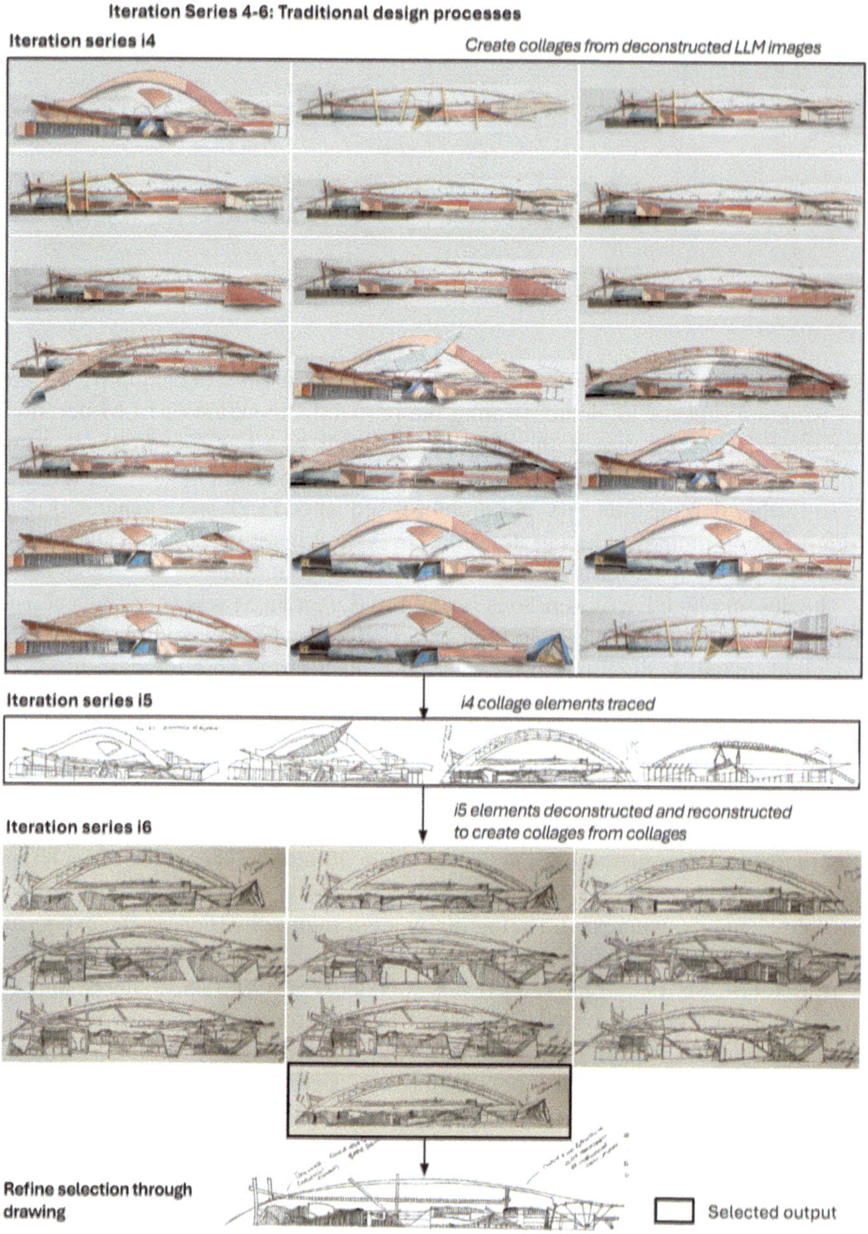

Diagram 5 Iteration Series 4-6: Outcomes from traditional design processes [3]

Fig. 6 Final design resolution: Render of north-east elevation [3]

produced a spectacular and unique form, demonstrating how integrating AI with traditional design methods can create results that surpass those achieved by either approach alone. The original elevation concept was still evident and key conceptual elements, incorporating a single axially related steel arch positioned over the length of the shed, were retained (Fig. 6). It could be argued that the final elevation sits between postmodern deconstruction, and modern rationalisation of structure. This was overlaid with a metamodern consideration for context and meaning, with the arch referencing the waves of the surrounding ocean and its eventual inundation of the land due to climate change.

5 Findings: Metamodern Design Process, AI and (Human) Design Sustainability

The case study demonstrated a mixed design approach to create a design solution derived from LLM generated content and traditional processes [3]. Applying a design by research methodology to develop a metamodern process, six iteration series oscillating between human-led and AI-led processes were undertaken [4]. Series 1 entailed the (human) development of a model used as input for series 2 and 3. Series 2 and 3 used LLMs to generate content, while series 4 to 6 involved human-led processes of collaging, modelling and drawing (Fig. 4).

Iterative series 2 and 3 demonstrated that LLM output generation is unlimited with the LLM generating many more variations on the initial i1 input then through traditional methods alone. New material was generated by the LLM to be used in further design activities, generate new and unexpected outputs, and combine ideas in novel ways (Diagrams 1, 2, and 3). This content was based on single or blended inputs and generated without the creative and practical constraints or considerations that may limit human design thinking (Diagrams 1, 2, and 3).

At iteration series 3, the issue of when to stop generating iterations became apparent. With the ability to rapidly generate hundreds of outputs this process had the potential to quickly get out of hand. The risk of becoming overwhelmed by large quantities of outputs was mitigated by using LLMs to generate material, rather than a solution, for iterations 4 to 6 which involved collaging and modelling (Diagram 5).

This also allowed human control over design outcomes to be retained and meant that authorship of the design was not in question.

Due to the probabilistic nature of LLM's, iteration series 2 revealed that over time outputs started to diverge from the initial input (Diagrams 1 and 2). At iteration 5 (series 2) it became apparent that the only way to retain conceptual elements in outputs (the arch over the museum) was to reintroduce the i1 model as input both at this point and throughout subsequent iterative cycles (Diagram 2). A common approach to the use of LLMs is for the generation of a design solution, however, several issues were observed during this process that would make the use of LLM outputs untenable as final solutions. This included a lack of detail, a lack of control over the detail, and a lack of consideration for important contextual factors such as structural integrity and materiality, culture, environment, longevity, feasibility, siting, and history.

A key question was, how much of the project would be developed using AI? This impacts authorship and human control. Discontinuing the design process after iterative series 3 (Diagram 3) would have meant selecting an LLM output as a final design solution and without the benefit of further development using other, human-based processes, this would have resulted in poorer design outcomes. Although outputs included key conceptual elements, they lacked the complexity and the richness of the final design solution and were devoid of metamodern characteristics such as meaning imbued through human creativity, critical thinking, and contextual understanding. Reliance on LLMs to drive design outcomes without human oversight and accountability was found to be problematic, underutilising the value of human input and reducing their role to a passive act.

Continuing the design process after discontinuing the use of LLMs at the conclusion of iteration series 3 (Diagram 3) enabled re-engagement with traditional processes in iteration series 4 to 6 (Diagram 5) and greater human control over the development and refinement of the design. This engagement included modelling, collaging and drawing based on LLM content developed in the previous iterative series. At this stage, the benefits of human-led design were utilised including, consideration of context, human creativity and critical thinking, human insight and accountability, knowledge and experience.

Comparison of the initial human-developed model (i1) with the final elevation developed from AI and human processes (Fig. 6) demonstrates how this mixed approach, oscillating between design methods and processes, increased the complexity and richness of the design outcome (Fig. 7). Further exploration could also have occurred by using two LLMs concurrently (e.g. Midjourney for images, and ChatGPT for text), as was the case for the development of program and spatial outcomes in other stages of the project.

Fig. 7 Initial i1 model (left) and final elevation developed through a series of oscillating processes and methods, utilising both AI-led and human-led design processes (right) [3]

6 Conclusion

A hybrid design approach to create a metamodern architectural response using Large Language Models (LLMs) was presented through a case study [3]. This chapter focussed on one of the design processes used in the development of an elevation for a hypothetical final Master Class project with the School of Architecture and Built Environment at Deakin University [3]. The design proposition was the adaptive reuse of a single 260m long steel portal framed industrial shed to accommodate the 'Maloppio Metamodern Museum' at Point Henry, Geelong. Key design elements were a single axially related steel arch, positioned over the length of the shed, explored through a significant iterative design process using LLMs. Key issues explored through engagement with a design by research methodology were: when and how to use AI in the project; the use of AI as a tool or to create a design solution; switching between AI-led versus human-led design processes; significance of AI in the project and how much of the project is developed from AI; authorship and who is in control of the design and its processes; and finally, can AI be used in a metamodern process to create a metamodern design?

Applying a design by research methodology enabled experimentation to create a metamodern design process that oscillated between LLM generative cycles and manual design activities, and between traditional and contemporary approaches. This oscillating approach allowed the designer to switch between AI and human-led processes to generate new design options. Greater human engagement was achieved with continuous human influence over the entire process, producing outcomes that were strongly linked to human creativity and the original design intent. Defining when and how LLMs were used in the project created clear boundaries around the role of AI versus the role of a human designer and the extent to which LLMs influenced the final design. Although LLMs can be used as a tool to directly derive and extrapolate a final design solution, the case study revealed that this was not a sustainable option to address the many complexities of design, and resulted in poorer, less considered results.

The use of LLMs ended once a large enough pool of diverse elements had been generated in iteration series 3. However, creative exploration continued through several iterations of manual design activities involving drawing and collaging. Although the application of AI in the project was significant, it was only utilised in a portion of the design process. Out of six iterative series, LLMs were used in two (series 2 and 3), with the larger portion of the design work undertaken through human-led processes. Restrictions around the role and significance of LLM outputs over the final design outcome enabled a tenable means to sustain human influence and control over the design process. This chapter presented an alternative design approach within a metamodern context, where LLM outputs were not treated as an endpoint to the creative process, but rather a technology-driven generator of materials within a meaningful, human-centred process. This approach could provide benefits to the design process, including aiding creativity and enriching the design process by producing more detailed outcomes with consideration of theoretical, contextual and other requirements, demonstrating how AI can assist, rather than replace, the role of architects.

References

1. Vermeulen T, van den Akker R (2010) Notes on metamodernism. J Aesth Cult 2(1):1–14. https://doi.org/10.3402/jac.v2i0.5677
2. Dempsey BG (2023) Metamodernism or the cultural logic of logics. ARC Press, Minesota USA
3. Humphris B (2024) Maloppio metamodern museum. [Master's thesis, Deakin University]. School of Architecture and Built Environment
4. Hauberg J (2011) Research by design: a research strategy. Archit Educ J Rev Lusófona de Architectura e Educacao 5:46–56. https://www.researchgate.net/publication/279466514_Research_by_design_a_research_strategy
5. Stoev D (2022) Metamodernism or metamodernity. Arts 11:91:1–26. https://doi.org/10.3390/arts11050091
6. Lai J (2019) Between irony and sincerity. Log Summer 46:23–32. https://www.jstor.org/stable/26759025. Accessed 23 July 2023
7. Dempsey BG (2023a) After postmodernism | 8, Metamodern Synthesis [online interview]. Brendan G Graham Dempsey. YouTube, 13 July 2023. https://www.youtube.com/watch?v=jKwUDh35q-8. Accessed 22 July 2023.
8. Kadagishvili D (2013) Metamodernism as we perceive it (Quick Review). Eur Sci J Dec 2:559–565
9. Zhang Z, Fort JM, Mateu LG (2023) Exploring the potential of artificial intelligence as a tool for architectural design: a perception study using Gaudí's works. Buildings (2075–5309) 13:7:1863–1882. https://doi.org/10.3390/buildings13071863
10. Russell P, Hall I (2024) Generative artificial intelligence, large language models, and JID innovations [editorial]. JID Innov 4(2):1–4. https://doi.org/10.1016/j.xjidi.2024.100256
11. Van Noorden R, Perkel JM (2023) AI and science: what 1,600 researchers think. Nature 621(7980):672–675. https://doi.org/10.1038/d41586-023-02980-0
12. Uszkurat O (2023) The future of architecture: a look beyond the AI hype, Olly Thomas, the Design Specialist Manager at BIG London [online interview]. Upstairs. YouTube, 25 Oct 2023. https://www.youtube.com/watch?v=HQrnROCUjBQ. Accessed 22 July 2023
13. Wadhwa R (2022) Using AI to design: Tim Fu from Zaha Hadid Architects [online interview]. The Blessedarch Podcast. YouTube, 8 December 2022. https://www.youtube.com/watch?v=Fm1ZEETe7wc&t=1408s. Accessed 22 July 2023.
14. Fan L, Li L, Ma Z, Lee S, Yu H, Hemphill L (2024) A bibliometric review of large language models research from 2017 to 2023. ACM Trans Intell Syst Technol 15(5):1–25. https://doi.org/10.1145/3664930
15. Leach N (2022) Architecture in the age of artificial intelligence: an introduction to AI for architects. Bloomsbury Publishing USA, 2021. ProQuest Ebook Central. https://ebookcentral.proquest.com/lib/deakin/detail.action?docID=6729618. Accessed 28 July 2023

16. Çeken B, Aköz B (2024) Yapay Zekanin Tasarima Etkisi ve Dall-E Örneği (The Impact of Artificial Intelligence on Design: The Example of Dall-E). Anadolu University Journal of Art & Design / Sanat & Tasarım 14:1:374–397. https://doi.org/10.20488/sanattasarim.1506116
17. Stokel-Walker C, Van Noorden R (2023) What ChatGPT and generative AI mean for science. Nature 614(7947):214–216. https://doi.org/10.1038/d41586-023-00340-6
18. Chaillou S (2022) Artificial intelligence and architecture: from research to practice. Berlin, Boston: Birkhäuser
19. Radford A, Wu J, Child R, Luan D, Amodei D, and Sutskever I (2019) Language models are unsupervised multitask learners. OpenAI. https://d4mucfpksywv.cloudfront.net/better-language-models/language_models_are_unsupervised_multitask_learners.pdf
20. Bordas A, Le Masson P, Thomas M, Weil B (2024) What is generative in generative artificial intelligence? A design-based perspective. Res Eng Design 35:427–443. https://doi.org/10.1007/s00163-024-00441-x

Hybrid Design Process: A Sustainable Approach Based on Analog Tool and Generative Artificial Intelligence

Sara Regiane Chornobai, Angelica Paiva Ponzio, and Fabiano de Vargas Scherer

Abstract Generative Artificial Intelligence (GenAI) models—more specifically text-to-image generation models, can serve as creative tools in the conceptual phase of the architectural design process, assisting in the generation of visual references and the communication of ideas. This support can be beneficial in academic contexts, where architecture students are in a phase of conceptual learning and the formation of their foundational design references. The conceptual phase is also responsible for key decision-making that influences the creation of the architectural design, including aspects related to sustainability. In this context, this study presents a theoretical proposal for a hybrid design process that combines the use of a GenAI text-to-image model—currently integrated within ChatGPT 4o—with an analog tool—Analog Tool for Bioclimatic Architecture (ATBA)—developed to guide the analysis, synthesis (generation), and evaluation of images in a way oriented toward sustainable architecture (bioclimatic characteristics and strategies for residential dwellings). Preliminary research findings indicate that the hybrid design process can stimulate creativity and innovation in academic contexts by articulating fundamental bioclimatic references with generative conceptual explorations aimed at the design of architectural projects aligned with the principles of environmental sustainability.

Keywords Design · Process · Creativity · Bioclimatic architecture · Artificial intelligence · Sustainability

S. R. Chornobai (✉) · A. P. Ponzio · F. de Vargas Scherer
Program of Research and Postgraduate Studies in Architecture (PROPAR), Postgraduate Program in Design (PGDesign), School of Architecture, Federal University of Rio Grande Do Sul, Porto Alegre, Brazil
e-mail: sara.chornobai@ufrgs.br

A. P. Ponzio
e-mail: 00008997@ufrgs.br

F. de Vargas Scherer
e-mail: fabiano.scherer@ufrgs.br

M. Sedrez and G. Pitts (eds.), *Artificial Intelligence-Aided Design for Sustainability*, Urban Sustainability, https://doi.org/10.1007/978-981-95-1349-9_7

1 Introduction

Recently, Generative Artificial Intelligence (GenAI) models have become available to the general public through accessible platforms with intuitive interfaces. Among them, text-to-image generation models have attracted the attention of artists, architects, and designers. This is because such models can generate original images in a simplified manner (based on natural language descriptions), quickly (within seconds), and with high graphic quality (comparable to photographs, artistic illustrations, and rendered images). This scenario has sparked various discussions regarding the responsible use of these technologies. While these concerns remain important, it is still necessary to understand how these models can serve the architectural discipline.

Although architecture is not limited to images or even three-dimensional models, images are part of the architect's toolkit [1]. Sketches, diagrams, drawings, figures, and photographs serve to communicate ideas, represent designs, illustrate theories, record spatial concepts, and reference built works. Within the design process, images are essential from the early stages, during conceptual generation. In these phases, images assist both in the exploration of reference and related works and in the graphic expression and communication of ideas, which are linked to the reflective activity present in creative processes [1]. Images, therefore, can be understood as one of the means of supporting the creative process in architecture.

Thus, it is assumed that GenAI models that generate images from text can serve as creative tools in the conceptual phase of architectural design, assisting in the generation of original visual references and the communication of ideas. This support can be particularly beneficial in academic contexts, where architecture students are in a phase of conceptual learning and the formation of their foundational design references. This context is understood as a receptive space for process experimentation, allowing greater freedom for design explorations and enabling the adoption of new tools and investigations.

Considering the potential of text-to-image generative models to produce original images based on the description of characteristics and formally expressive concepts, it is understood that the use of these tools can extend beyond merely exploring existing references from reality, while still maintaining a commitment to it. In other words, this study argues that such models can be better utilized in the design process when employed with the purpose of assisting in imagining and creating possible and desirable futures.

Among these futures, sustainability emerges as a central concern. In this context, "Our Common Future" [2] lays the foundation for the concept of "Sustainable Development" which advocates for a more responsible approach by humanity in response to the climate emergency and the worsening negative impacts on the environment and its inhabitants [3]. In the field of architecture, the alternatives that align with this notion of sustainability vary according to specific perspectives, such as: the ecological crisis; the depletion of non-renewable natural resources; waste production and pollutant emissions; quality of life and the humanization of socio-economic relations;

water and wastewater management; environmental comfort; and energy efficiency and performance in buildings [4, 5].

Despite acknowledging the existence of different perspectives, there is a consensus that the principle of sustainability should be integrated, especially at the beginning of the design process [4, 5]. This is because the conceptual phase of the project is responsible for key decision-making that influences the creation of the overall design solution [6, 7]. It is at this stage that conceptual and contextual relationships should be considered to understand not only the constraints affecting the building's design but also the opportunities to positively impact its surroundings [9].

Given this framework, the following question arises: How can architectural design education foster creative processes that integrate text-to-image GenAI models to explore design concepts oriented toward environmental sustainability?

Therefore, this study presents a proposal for a hybrid design process that combines the use of a text-to-image GenAI model—ChatGPT—with an analog tool—Analog Tool for Bioclimatic Architecture (ATBA)—developed to guide the analysis, synthesis (generation), and evaluation of images in a way oriented toward sustainable architecture. This research is characterized as an investigative applied study and addresses the problem using a qualitative approach [10]. Furthermore, this study is structured as follows: it begins with a theoretical contextualization on generative models and AI, creative activity in the design process, and environmental sustainability in architecture; then, it presents the development of the analog tool and its possible applications; and finally, the considerations are discussed.

2 The Evolution of Artificial Intelligence

Artificial Intelligence (AI) has evolved rapidly since its conceptual origins in 1950, when Alan Turing [11], in his seminal paper, posed the question "Can machines think?" and introduced what later became known as the Turing Test. A fundamental concept in AI, the Turing Test was conceived as a method to evaluate whether a machine's behavior could be indistinguishable from that of a human. But it was only in 1955 that the term "Artificial Intelligence" was introduced in John McCarthy's Dartmouth proposal, to be formalized shortly thereafter at the 1956 Dartmouth Summer Research Project [12], an event widely regarded as the field's founding moment [13].

Following these early milestones, AI research experienced periods of stagnation—often referred to as "AI winters"—due to inflated expectations, funding challenges, and unmet technological promises. A major resurgence began in the early 2010s with the revival of neural networks, marking the third boom of AI, catalyzed by the success of deep convolutional architectures such as AlexNet [14]. Significant improvements in computing power and the availability of large datasets from the internet have marked a major advances era since the 2020s—these developments paved the way for more powerful and accessible AI models, accelerating breakthroughs across various domains [15].

2.1 GenAI and Its Technological Foundations

According to Bommasani et al. [16], "a foundation model is any model that is trained on broad data (generally using self-supervision at scale) that can be adapted (e.g., fine-tuned) to a wide range of downstream tasks." This category encompasses large-scale generative models capable of multimodal outputs—such as DALL·E [17, 18], Stable Diffusion [19], Midjourney, and Imagen [20]—all designed to generate novel outputs informed by extensive training data.

GenAI has emerged as a transformative technology. Unlike earlier AI models, which often relied on rule-based systems or classical statistical approaches, GenAI—powered by deep neural networks—can produce new data samples by learning patterns from their extensive datasets. This capability is rooted in deep learning [15], a subset of machine learning, and is fundamental to fields such as Natural Language Processing (NLP) and Computer Vision (CV). These technologies form the foundation of modern generative models, including large language models such as GPT-3 [21]—the seminal precursor to ChatGPT—for text generation, and text-to-image systems such as DALL·E [17, 18] for image synthesis.

One of the first generative models applied to image synthesis was the Variational Autoencoder (VAE), introduced by Kingma and Welling [22]. VAEs compress image data by encoding it into a latent space and later decoding it, allowing for both reconstruction and variation of inputs. Additionally, VAEs serve as a foundational component in later generative models, as they enable efficient data compression and latent space representation.

Shortly after, Generative Adversarial Networks (GANs) were introduced by Goodfellow et al. [23], using a generator–discriminator framework where the generator creates synthetic images, and the discriminator evaluates their authenticity. These models have gained significant attention not only for their high-quality outputs but also for their ability to generate unexpected and novel outputs, often referred to as "machine hallucinations", inspiring creative applications in design disciplines. In architecture, Conditional Gans and specialized variants such as pix2pix and StyleGAN, have improved controllability, enabling experiments in plan generation, style transfer, and architectural form exploration [24]. In particular, StyleGAN has demonstrated strong performance in image learning and reconstruction simulation, as explored by Eroglu and Gül [24]. Despite their strengths, GANs face challenges such as mode collapse and training instability, affecting the consistency of the generated outputs.

A major shift occurred in 2015 when Sohl-Dickstein et al. [25] introduced diffusion models, laying the foundation for GenAI based on thermodynamic principles. These models were later refined by Ho et al. [26], significantly improving their efficiency and effectiveness in image generation. Unlike GANs, diffusion models offer higher image quality, more stable training, and high-fidelity outputs. Inspired by physics, diffusion models gradually add noise to images, in a forward diffusion, to later remove it step by step by reverse diffusion, generating new images based on the training data.

A second—complementary—breakthrough was the introduction of Transformers by Vaswani et al. [27], whose self-attention mechanism enables models to dynamically weigh the importance of different input elements. This architecture has become a cornerstone in models like GPT for text generation and in text-to-image synthesis pipelines.

These innovations paved the way for Latent Diffusion Models (LDMs), which combine key features of VAEs, Convolutional Neural-Network-based U-Net architectures [28], Transformers and Diffusion processes. LDMs process images in a lower-dimensional space using VAEs, which helps reduce the amount of computing power needed while still maintaining high-quality outputs [19]. In LDMs, Transformers convert text prompts into embeddings—or numeric representations of the data—that guide the diffusion process through cross-attention layers, ensuring that the generated image aligns closely with the text. CNNs, originally developed for image recognition [29], improve image quality and structure. A U-Net architecture, based on CNNs, gradually removes noise from the image, while the VAE decoder reconstructs the high-quality final version.

2.2 AI in Creative and Architectural Fields

AI-powered generative models have attracted significant attention in creative disciplines such as graphic design, art, and architecture [30]. Although AI-generated 3D outputs—by the time this chapter is being written—are predominantly limited to representations such as meshes, point clouds, and Non-Uniform Rational B-Splines (NURBS) [31] —and not yet seamlessly integrated into Building Information Modeling (BIM) workflows [32, 33],—these outputs remain valuable for testing spatial configurations and conceptual explorations, including 3D printing and 2D image referencing [32]. Building on this premise, our study focuses on AI's contribution in the early design ideation phase by examining whether 2D text-to-image tools might enable designers to rapidly explore conceptual ideas, experiment with aesthetic directions, and generate contextual references before transitioning to detailed digital modeling and documentation [33].

2.3 ChatGPT

This research focuses on the integration of 2D artificial image making—initially via the stand-alone DALL·E models and subsequently within the ChatGPT large language mode (LLM). While DALL·E 1 relied solely on Transformer architectures, DALL·E 2 introduced CLIP guidance (Contrasting Language Image Pre-Training—responsible for associating image to text) alongside a diffusion decoder. DALL·E 3 has, instead, removed the separate CLIP component, employing an enhanced Transformer-based text-encoder coupled with diffusion for image generation [17, 18].

Uniquely, DALL·E 3 operated as a built-in image generation module within ChatGPT until it was discontinued in early 2025. Users engaged in a natural-language dialogue, and ChatGPT refined those inputs into detailed prompts for DALL·E 3. The resulting image was then displayed within the conversation. According to OpenAI [17, 18], this integration not only enhanced the quality of the generated images but also had improved bias mitigation and the prevention of disallowed content. Moreover, the interactive process allowed ChatGPT to retain context, enabling iterative refinement that aligns with the user's research objectives.

Following DALL·E 3, OpenAI introduced ChatGPT 4o, a fully multimodal LLM with native image generation within its autoregressive architecture [34]. Its enhanced LLM capabilities allow it to craft and refine detailed visual prompts directly from user dialogue. According to Chen et al. [35] GPT-4o outperforms prior diffusion-based systems in text rendering, composition, spatial reasoning, and image-to-image transformations.

3 Creative Activity

According to Kowaltowski, Bianchi and Petreche [36], creative activity has been acknowledged since antiquity—initially understood by philosophy as divine inspiration—evolving into more contemporary theories of cognitive processes that emerged prominently during the nineteenth century. Currently, creativity is approached from a systemic perspective, which considers the relationship from the brain level to interactions with the social and cultural environment. Therefore, the concept of creativity is influenced by the historical evolution of theories, reflecting the sociocultural and technical factors of their time [36].

Boden [37] argues that, for humans, creativity is based on cognition, association of ideas, perception, analogical and reflective thinking, motivation, memory, emotion, culture, and personality. In this sense, she understands that creativity can be classified into three types: combinatorial, when common ideas are combined in unexpected ways, such as analogies; exploratory, when new ideas are generated through an exploration within their conceptual space, creating unexpected ideas; and transformational, when they are generated through the transformation (expansion or removal) of the conceptual space structure, creating truly surprising ideas that were previously considered impossible [37].

In this research, it is considered that the creation of something new does not necessarily have to result in a product. According to Tschimmel [38], creativity is not only expressed when a new product or form emerges but also when new questions are raised, new problems are identified and previously unknown ideas, answers, and solutions are generated. Thus, this study acknowledges that creativity is understood to occur when originality is accompanied by purpose and contributes to a solution [36]; when, in addition to being new, the idea is surprising and holds value [37].

Although unique to each designer, Lawson [1] explains that the design process follows a structure that develops through cumulative phases, each consisting of different stages in which consecutive choices are made by the designer. These phases are characterized by sequences of analysis (understanding the problem), synthesis (generation of solutions), and evaluation (critical assessment of solutions) for each decision-making step [1]. In reality, however, these stages do not occur in such a logical and unidirectional manner; rather, the process resembles an iterative cycle between analysis, synthesis, and evaluation [1].

In the design activity, the creative process begins with a stimulus or a problem and, "through intricate networks of associations and mental interactions", is able to generate ideas that will be developed and validated through procedural methods [39]. According to Lawson [1], one of the key questions regarding the creative process is understanding how references are used by the designer, as some may be employed "in highly innovative ways that not only surprise us but also seem entirely relevant".

Therefore, to develop creative processes, the abilities to abstract and imagine must be combined with phases of convergent and divergent thinking, which are supported by the mastery and ability to manipulate knowledge within a given field [36]. By understanding these basic mechanisms, it becomes possible to train or cultivate creativity through creative methods and techniques [36].

In this context, various creativity methods and techniques have been developed primarily to stimulate fluency and flexibility in idea generation. Although often used as synonyms, methods are logical procedures that pre-organize the operations to be used in a creative process, whereas techniques are intermediate procedures that can be applied at specific moments and typically rely on tools (instruments such as tables and checklists). Among the creative methods that support the early stages of architectural design, idea generation methods serve as tools for understanding problems and expanding the possibilities for innovative solutions [36].

4 Environmental Sustainability in Architecture

In an attempt to align architecture with the concept of sustainability, Szokolay [8] argues that parameters related to environmental performance must regain a central role during the design phase. In this context, every decision has an impact, and therefore, it is increasingly necessary for decisions to be made in an integrated manner, with clear information and precise criteria [1]. According to Heywood [5], buildings account for half of the world's energy consumption, and therefore, reducing this consumption should be understood as one of the central issues in minimizing environmental impact. For Lamberts, Dutra, and Pereira [7], this is strongly related to how indoor comfort is provided, as the increasing reliance on active thermal conditioning and lighting systems continues to intensify, leading to a higher demand for operational energy.

Guiding the formal development of an architectural project, the architectural design is responsible for adopting general strategies that are directly related to environmental concerns, such as site placement, geometry, solar orientation, the positioning and sizing of enclosures and fenestrations, access points, and circulation flows, among others [6, 7]. Some of these decisions, although made in the early stages, are crucial for aligning with the concept of sustainability, as they establish long-term relationships with the climate and environmental context, critically affecting user comfort, and consequently, energy consumption and building durability [5–7].

Aiming for low-energy consumption architecture, Gurgel [6] argues that climate should define the initial parameters of the design, as projects adapted to local climatic conditions tend to be more environmentally successful. The climatic adaptation of architecture was also investigated by the Olgyay brothers in 1973, when they introduced the concept of bioclimatology—understood as the study of the relationship between climate and living beings—into architecture, coining the term "bioclimatic design". This concept refers to architectural projects that utilize natural strategies for building conditioning in order to promote thermal comfort for occupants [7].

To adopt these natural strategies, architects must understand the microclimatic scale, which refers to the immediate surroundings of the building. This is because local climate variables are influenced by other environmental factors, such as the presence of water, vegetation, topography, soil, and neighboring buildings, among others [6, 7]. Moreover, not only is the environmental comfort of a project affected by the microclimate, but the project itself also influences the environmental quality at this scale. Therefore, it is the designer's responsibility to develop architectural solutions that are integrated with the local context conditions [5].

A design that adapts to the local climate to promote environmental comfort is neither a novelty nor a contemporary approach, but rather a vernacular and disciplinary practice, as it has already been described in ancient architectural treatises [7, 9]. However, bioclimatic architecture is now being advocated as one of the foundational principles for aligning architecture with sustainability [5–8].

By understanding the influence of climate on architectural form and its constructive elements, Szokolay [8] identifies four climatic design archetypes: cold climates, temperate climates, hot and dry climates, and hot and humid climates. Their characteristics are formally expressive, and some of them can be identified in images. These archetypes are based on the nature of the human thermal challenge and are primarily observed in vernacular or indigenous architecture [8]. Heywood [5] argues that designing should be informed by the architecture produced by the region's indigenous peoples, as it is typically built in harmony with the local climatic conditions, making it well-suited to its environmental context.

5 Development of the Analog Tool

Following a thorough literature review, bioclimatic characteristics and strategies recommended by specialized literature were identified [5–8] to serve as guidelines and design parameters. The research focused on qualitative parameters applicable in the early stages of creative exploration and conceptual design. Residential buildings

Climate Archetypes		Cold Climate	Temperate Climate	Hot and dry Climate	Hot and humid Climate
Bioclimatic Strategies		Insulation (capacitive); Solar heating	Solar heating; Thermal inertia; Shading; Evaporative cooling; Cross ventilation	Thermal inertia; Shading; Evaporative cooling; Selective ventilation	Shading; Ventilation; Indirect evaporative cooling
Bioclimatic Elements		Chimney; Winter garden	Winter garden; Chimney; Eaves; Brises; Pergoled; Balcony; Light Shelves	Tank or water spout; Pergoled; Patio whit vegetation; Wind tower; Moucharaby; Water Mirror	Ventiled attic; Wind directors; Cobogo; Balcony; Roof
Materials		Wood; Stone; Brick	Wood; Concrete; Stone; Brick	Adobe; Earth; Stone; Textile	Wood; Bamboo; Fiber; Taipa
Form		Compact	Compact or fragmented	Compact	Fragmented
		Centralized	Linear or centralized; Solar house	Centralized; Patio house	Linear
Enclosures		Ground	Ground or elevated	Ground	Elevated
		Dark colors	Dark or light colors	Light colors	Light colors
		Heavy and insulated	Light and insulated	Heavy and insulated	Light and reflective
		Protected	Protected	Exposed	Procted
		Inclined coverage	Inclined coverage	Flat or slightly slopped coverage	Inclined coverage
Fenestrations	Solar Protection	Unshadowed	Shadowed in summer	Shadowed	Shadowed; Reflective and isolated windows
	Number	Few	Medium	Few	Many
	Ventilation	Small	Medium	Small	Large
	Orientation	Facing towards the Ecuador	Facing towards the Ecuador	Avoid faces facing direct solar radiation	Avoid east and west faces

Fig. 1 Analog Tool for Bioclimatic Architecture (ATBA)

were chosen as the object of study, given their suitability for the application of bioclimatic strategies and the classification of climatic archetypes.

After gathering and analyzing the data, it was concluded that there are general guidelines for specific climate types, as well as climatic parameters associated with these types. These can be easily applied in the early stages of the design process, guiding an architectural approach that is more responsive to the local climatic conditions. Despite the adoption of different climatic classifications among the analyzed references, it was observed that the archetypes can be summarized into four main climate types: cold, temperate, hot and dry, and hot and humid. As the final outcome of this research, a framework (Fig. 1) was developed, presenting the main bioclimatic guidelines and qualitative parameters for the four primary climatic archetypes identified.

Named the "Analog Tool for Bioclimatic Architecture" (ATBA), the framework has an informational character, and its visual is designed to facilitate the identification and application of guidelines and parameters at the academic level. The content is organized into the following categories: bioclimatic strategies; bioclimatic elements; materials; form; enclosures; and fenestrations—were further subdivided into solar protection, number of openings, ventilation, and orientation of openings.

5.1 *Application of the Analog Tool*

The development of the Analog Tool for Bioclimatic Architecture (ATBA) enabled the exploration of GenAI models in the design process in ways oriented toward environmental sustainability in architecture—particularly in terms of adaptation to local climatic contexts.

Before reaching the current stage of this research, the Analog Tool for Bioclimatic Architecture (ATBA) had already been tested in an earlier study by Ponzio et al. [40], which combined analog criteria from the Archetypal Climatic Board (ACB) with AI image generation and parametric modeling to optimize architectural forms. A reference image was created using DALL·E 2, selected for its alignment with ACB parameters (e.g., facade view, orientation, material clarity), and refined using Stable Diffusion inpainting. This image informed the creation of a parametric 3D model in Grasshopper, where variables such as openings size and floor area (35–45 m^2) were encoded. Optimisation was conducted using the Biomorpher tool, based on three performance criteria from the ACB: (i) ventilation (> 40% of floor area, 70/30 façade distribution); (ii) formal proportion (linear/rectangular); (iii) material use (minimised wall, roof, floor volume). Twelve families were generated within a population of 48 individuals (crossover 0.5, mutation 0.01). After calibration, the best result came from family 7, achieving 45.852% ventilation, 52.221 m^3 of material, and 36.566 m^2 of floor area. The findings indicated that the images generated by DALL·E 2 could meaningfully contribute to the development of climate-responsive architecture, particularly in the early conceptualization and parti definition phases [40].

In a subsequent follow-up study, Chornobai et al. [41] investigated the use of the GenAI text-to-image model DALL·E 3, integrated within ChatGPT, during the conceptual phase of the architectural design process to generate visual references that combined biological concepts with bioclimatic architectural principles. In this case, by drawing on qualitative parameters of the climatic archetypes, the analog tool supported both the contextualization of the ChatGPT model—guiding the generation of prompts—and in the evaluation of the images generated by DALL·E 3 [41]. The study concluded that the use of DALL·E 3 within ChatGPT demonstrated significant potential as a collaborative tool in the design process, supporting outcomes more closely aligned with the concept of environmental sustainability [41].

Building on these initial investigations, the current developments—using the latest text-to-image capabilities embedded in ChatGPT, namely model 4o—suggest that when used in conjunction with digital GenAI tools that integrate text and image generation, the analog tool can effectively support the three stages of the design process identified by Lawson [1], in a framework described as such:

I. Analysis: to contextualize the problem, by providing the guidelines and parameters to be considered;
II. Synthesis: to structure the prompt and generate solutions, based on the selection of design guidelines and parameters;
III. Evaluation: to verify the generated solutions, comparing the reference parameters with the characteristics presented in the results.

In this process (Fig. 2), the analysis and synthesis phases involve the combined use of the analog tool and the AI tool—first to generate text and then to generate images. In the evaluation stage, the verification of the AI-generated result is carried out manually by the designer, who can choose to adjust the final outcome by repeating one or all stages of the process. It is understood that the three applications of the analog tool can be used independently or in combination. In this case, a hybrid procedural framework may be considered (Fig. 2). It is important to highlight that, when used in combination, the process remains non-linear and iterative, allowing for sequences of stages to be repeated until a suitable design solution is found. Future research includes testing the framework with other text-to-image digital tools in order to explore alternative design outcomes.

As visual demonstrations, the figures below underscore how different levels of prompt articulation can support architecture students in translating abstract bioclimatic strategies into more defined spatial and formal ideas. Figures 3 and 4 illustrate the synthesis and evaluation phases of the Hybrid Design Process. Figure 3 demonstrates a basic prompt structure built from a straightforward selection of guidelines and parameters. By contrast, Fig. 4 shows a more advanced prompt structure that not only applies those same guidelines and parameters but also weaves in specific conceptual references—such as the biomimicry metaphor of snowflake geometry—to enrich the design output.

6 Considerations

In light of the development of this work, it is possible to discuss some opportunities, limitations, and challenges encountered regarding the proposal and applications of the analog tool, the AI tool, and the hybrid design process.

In order to support analysis, synthesis, and evaluation in the early stages of the design process, the analog tool is presented as a reference framework or informational guide on general guidelines and parameters of bioclimatic architecture. Therefore, although the specialized literature identifies four main climatic archetypes, the general nature of these guidelines and parameters can be seen as a limitation regarding the regional specificities found in the diversity of environmental contexts around the world. This limitation also extends to its restriction to the conceptual phase of sustainable architectural design, as more precise and locally representative data are required in later stages of the process in order to develop more accurate computational models and simulations concerning the building's environmental performance. Moreover, it is important to highlight that due to its referential/informational nature, the analog tool is not interactive or generative, meaning it cannot be considered a creative tool on its own.

Regarding the AI tool for image generation, it is important to consider that, although the generated content may present graphical and aesthetic quality, this does not ensure the functional or performance quality of sustainable architectural design. Challenges include technical issues—such as data inconsistency and the ongoing improvement of systems—as well as operational issues—such as data security policies and limited access to certain features of the tools. Another key point concerns the responsible use of these tools, which should go beyond instrumental learning

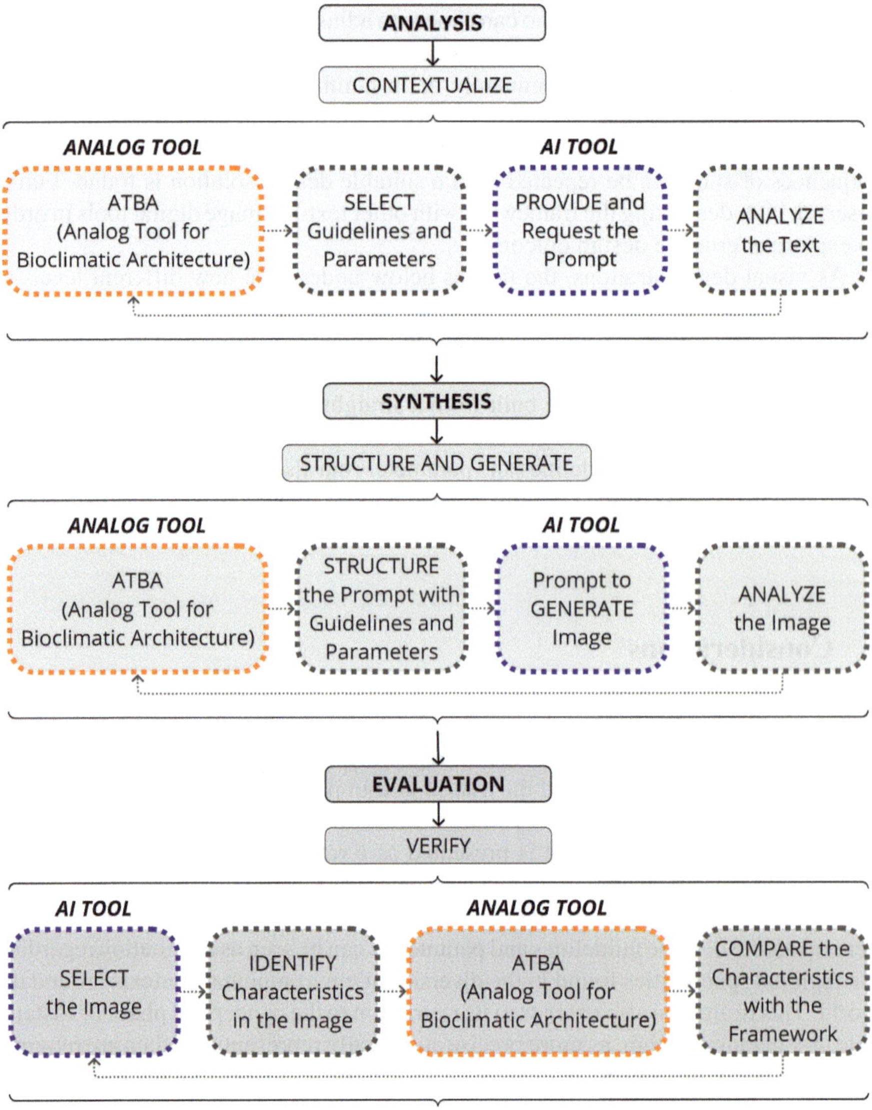

Fig. 2 Hybrid design process

and involve collective awareness and professional training, fostering the development of critical and reflective thinking. In the academic context, constant faculty development and the encouragement of integrating new tools and methodologies are necessary to support the advancement of knowledge and disciplinary innovation.

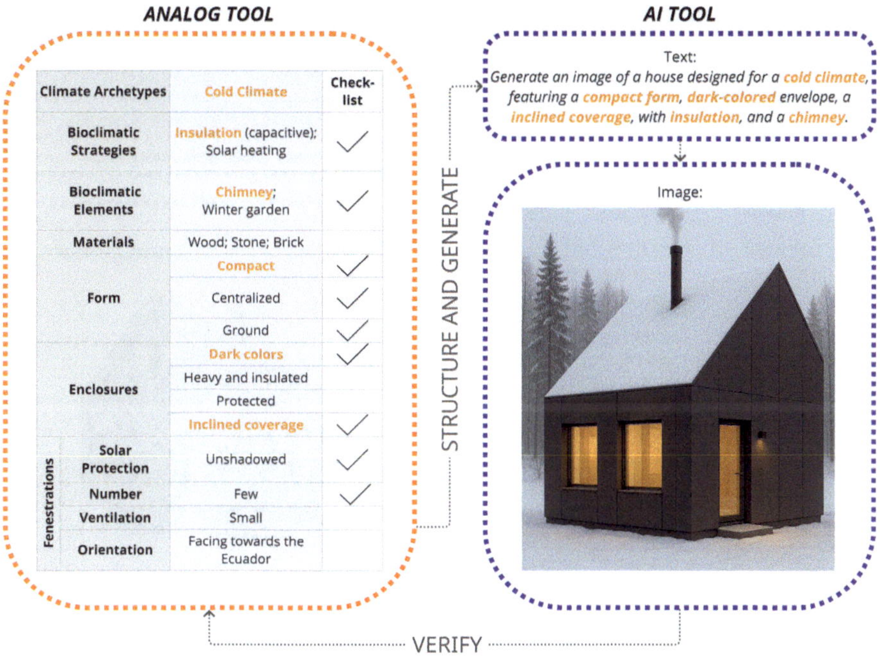

Fig. 3 Illustration of the hybrid design process—basic prompt structure. AI Tool (bottom right image): ChatGPT 4o

Regarding the hybrid design process, it is possible to understand that the technique can support the creative process of architectural design by offering opportunities for conceptual exploration and visual reference generation for the designer. However, the creative stimulus lies in the user's interpretative ability, in how the analog and intelligent interfaces are connected, and in how the references are incorporated into the design development. Therefore, design skills such as abstraction, reflection, and critical thinking—inherent to human creativity—remain essential. It is also important to consider that the proposed hybrid design process functions as a trial-and-error creative technique, which may affect the efficiency of its application and lead to frustration among students or even professionals unfamiliar with the proposed tools. Furthermore, this research envisions future experimentation within real academic contexts to identify convergences and divergences from its originally intended applications.

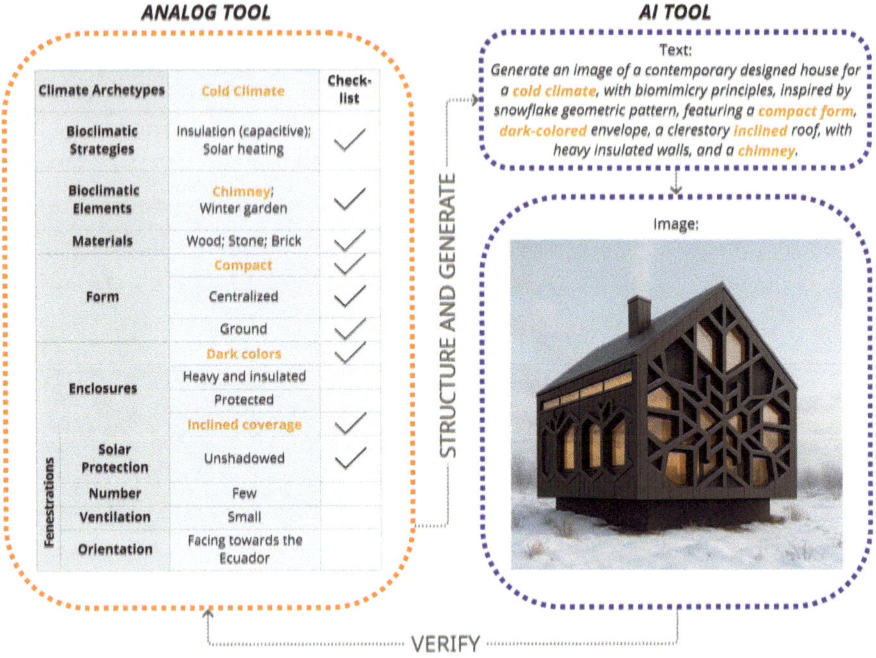

Fig. 4 Illustration of the hybrid design process—advanced prompt structure. AI Tool (bottom right image): ChatGPT 4o

7 Conclusion

Based on questions regarding the integration of GenAI models into the architectural design process in academic contexts, this research aimed to develop an analog tool that could serve as a reference support for environmentally responsible designs. This tool presented clear and easily understandable content on the guidelines and parameters of climatic archetypes, assisting students in the analysis, synthesis, and evaluation stages during the creative phases of the design process.

During the investigations, it was possible to understand that the developed analog tool should not be characterized as creative in itself. This is because the tool has a more informational role in verifying and comparing architectural guidelines and parameters across different climatic contexts. However, when applying the analog tool in combination with GenAI models, the process forms a hybrid structure with a creative nature, as it can be used to generate innovative alternatives based on fundamental architectural characteristics. Thus, it is concluded that the proposed approach does not replace design thinking, but instead, it can contribute to creative exploration within the architectural discipline when employed in a responsible, reflective, and critical manner.

References

1. Lawson B (2011) *Como arquitetos e designers pensam.* (Trad.: M.B. Medina). 1ª ed. Oficina de Textos, São Paulo
2. Brundtland GH (1991) Nosso futuro comum: comissão mundial sobre meio ambiente e desenvolvimento, 2ª. Fundação Getúlio Vargas, Rio de Janeiro
3. IPCC (Intergovernmental Panel on Climate Change) (2022) *Sixth Assessment Report.* https://www.ipcc.ch/report/ar6/wg2/. Accessed 23 Nov 2023.
4. McDonough W, Braungart M (2013) Dos princípios às práticas: Criando uma arquitetura sustentável para o século XXI. In: Sykes A (ed) O campo ampliado da arquitetura. Cosac Naify, São Paulo, pp 165–172
5. Heywood H (2015) 101 regras básicas para uma arquitetura de baixo consumo energético. (Trad.: A. Salvaterra). Gustavo Gili, São Paulo
6. Gurgel M (2012) Design Passivo: Baixo consumo energético. Guia para conhecer, entender e aplicar princípios do Design Passivo em residências. Editora Senac, São Paulo
7. Lamberts R, Dutra L, Pereira FOR (2014) Eficiência Energética na Arquitetura. 3ª ed. ELETROBRAS/PROCEL, Rio de Janeiro. https://labeee.ufsc.br/sites/default/files/apostilas/eficiencia_energetica_na_arquitetura.pdf. Accessed 15 Sep 2022
8. Szokolay SV (2019) Introdução à ciência arquitetônica: a base do projeto sustentável. (Trad.: M.C. Cescato). Editora Perspectiva, São Paulo
9. Roth LM (2017) Entender a Arquitetura: seus elementos, história e significado. (Trad.: J. Canêdo). Gustavo Gili, São Paulo
10. Creswell JW, Creswell JD (2018) Research design: qualitative, quantitative, and mixed methods approaches, 4th edn. Sage, Los Angeles
11. Turing A (1950) Computing machinery and intelligence. Mind 59:433–460. https://courses.cs.umbc.edu/471/papers/turing.pdf. Accessed 17 Feb 2025
12. McCarthy J, Minsky ML, Rochester N, Shannon CE (2006) A proposal for the Dartmouth Summer Research Project on Artificial Intelligence (original work published 1955). AI Mag 27(4). https://ojs.aaai.org/aimagazine/index.php/aimagazine/article/view/1904. Accessed 16 Feb 2025
13. Moor J (2006) The Dartmouth college artificial intelligence conference: the next fifty years. AI Mag 27(4). https://doi.org/10.1609/aimag.v27i4.1911. Accessed 16 Feb 2025
14. Krizhevsky A, Sutskever I, Hinton GE (2012) ImageNet classification with deep convolutional neural networks. In F Pereira, CJC Burges, L Bottou, KQ Weinberger (eds), Advances in neural information processing systems (NeurIPS) 25, pp 1097–1105. https://papers.nips.cc/paper_files/paper/2012/hash/c399862d3b9d6b76c8436e924a68c45b-Abstract.html. Accessed 17 Feb 2025
15. LeCun Y, Bengio Y, Hinton G (2015) Deep learning. Nature 521:436–444. https://doi.org/10.1038/nature14539. Accessed 17 Feb 2025
16. Bommasani R, Hudson DA, Adeli E, Altman R, Arora S, von Arx S et al. (2021) On the opportunities and risks of foundation models. arXiv:2108.07258. https://arxiv.org/abs/2108.07258
17. OpenAI (2023) DALL·E 3. https://openai.com/dall-e-3. Accessed 16 Feb 2025
18. OpenAI (2023) Improving image generation with better captions. Retrieved from https://cdn.openai.com/papers/dall-e-3.pdf
19. Rombach R, Blattmann A, Lorenz D, Esser P, Ommer B (2022) High-resolution image synthesis with latent diffusion models. arXiv:2112.10752. https://arxiv.org/abs/2112.10752. Accessed 17 Feb 2025
20. Saharia C, Chan W, Saxena S, Li L, Whang J, Denton E, Seyed Ghasemipour SK, Karagol Ayan B, Mahdavi SS, Gontijo Lopes R, Salimans T, Ho J, Fleet DJ, Norouzi M (2022) Photorealistic text-to-image diffusion models with deep language understanding. arXiv:2205.11487. https://arxiv.org/abs/2205.11487. Accessed 17 Feb 2025

21. Brown TB, Mann B, Ryder N, Subbiah M, Kaplan J, Dhariwal P, Neelakantan A, Shyam P, Sastry G, Askell A et al. (2020) Language models are few-shot learners. arXiv:2005.14165. https://arxiv.org/abs/2005.14165. Accessed 17 Feb 2025
22. Kingma DP, Welling M (2013) Auto-encoding variational Bayes. arXiv:1312.6114. https://arxiv.org/abs/1312.6114. Accessed 17 Feb 2025
23. Goodfellow I, Pouget-Abadie J, Mirza M, Xu B, Warde-Farley D, Ozair S, Courville A, Bengio Y (2014) Generative adversarial networks. arXiv:1406.2661. https://arxiv.org/abs/1406.2661. Accessed 17 Feb 2025
24. Eroglu R, Gül LF (2022) Architectural form explorations through generative adversarial networks—Predicting the potentials of StyleGAN. In Pak B, Wurzer G, Stouffs R (eds) Co-creating the future: inclusion in and through design—Proceedings of the 40th Conference on Education and Research in Computer Aided Architectural Design in Europe (eCAADe 2022)—Volume 2, Ghent, 13–16 September 2022, pp 575–582. https://doi.org/10.52842/conf.ecaade.2022.2.575
25. Sohl-Dickstein J, Weiss E, Maheswaranathan N, Ganguli S (2015) Deep unsupervised learning using nonequilibrium thermodynamics. arXiv:1503.03585. https://arxiv.org/abs/1503.03585. Accessed 17 Feb 2025
26. Ho J, Jain A, Abbeel P (2020) Denoising diffusion probabilistic models. arXiv:2006.11239. https://arxiv.org/abs/2006.11239
27. Vaswani A, Shazeer N, Parmar N, Uszkoreit J, Jones L, Gomez AN, Kaiser L, Polosukhin I (2017) Attention is all you need. arXiv:1706.03762. https://arxiv.org/abs/1706.03762. Accessed 17 Feb 2025
28. Ronneberger O, Fischer P, Brox T (2015) U-Net: convolutional networks for biomedical image segmentation. arXiv:1505.04597. https://arxiv.org/abs/1505.04597. Accessed 17 Feb 2025
29. LeCun Y, Bottou L, Bengio Y, Haffner P (1998) Gradient-based learning applied to document recognition. Proc IEEE 86(11):2278–2324. https://doi.org/10.1109/5.726791
30. Tang Y, Zhang N, Ciancia M, Wang Z (2024) Exploring the impact of AI-generated image tools on professional and non-professional users in the art and design fields. arXiv preprint arXiv:2406.10640. https://arxiv.org/abs/2406.10640. Accessed 17 Feb 2025
31. Po R, Wang Y, Golyanik V, Aberman K, Barron JT, Bermano AH, Chan ER, Dekel T, Holynski A, Kanazawa A, Liu CK, Liu L, Mildenhall B, Nießner M, Ommer B, Theobalt C, Wonka P, Wetzstein G (2023) State of the art on diffusion models for visual computing. arXiv:2310.07204. https://arxiv.org/abs/2310.07204. Accessed 17 Feb 2025
32. Ko J, Ajibefun J, Yan W (2023) Experiments on generative AI-powered parametric modeling and BIM for architectural design. arXiv:2308.00227. https://arxiv.org/abs/2308.00227. Accessed 17 Feb 2025
33. Li C, Zhang T, Du X, Zhang Y, Xie H (2024) Generative AI models for different steps in architectural design: a literature review. arXiv:2404.01335. https://arxiv.org/abs/2404.01335. Accessed 17 Feb 2025
34. OpenAI (2023) GPT-4o Technical Report. https://arxiv.org/abs/2303.08774v1 Accessed 12 May 2025
35. Chen S, Bai J, Zhao Z, Ye T et al. (2025) An empirical study of GPT-4o image generation capabilities. arXiv:2504.05979. https://arxiv.org/abs/2504.05979. Accessed 12 May 2025
36. Kowaltowski DCCK, Bianchi G, Petreche JRD (2011) A criatividade no processo de projeto. In: Kowaltowski DCCK et al (eds) O processo de projeto em arquitetura. Oficina de Textos, São Paulo, pp 21–56
37. Boden MA (1998) Creativity and artificial intelligence. Artif Intell 103(1–2):347–356. https://doi.org/10.1016/s0004-3702(98)00055-1. Accessed 30 Nov 2023
38. Tschimmel KC (2010) Sapiens e Demens no pensamento criativo do design. PhD thesis, Universidade de Aveiro, Departamento de Comunicação e Arte, Aveiro
39. Fabricio MM, Melhado SB (2011) O processo cognitivo e social de projeto. In: Kowaltowski DCCK et al (eds) O processo de projeto em arquitetura. Oficina de Textos, São Paulo, pp 57–63
40. Ponzio AP, Chornobai SR, Fagundes CVM, Rodrigues RC, José GCH (2023) Exploring creative AI thinking in the design process: the design intelligence strategy. In Koh I, Reinhardt D,

Makki M, Khakhar M, Bao N (eds) HUMAN-CENTRIC—Proceedings of the 28th CAADRIA Conference, Ahmedabad, 18–24 Mar*ch 2023*, pp 49–58. https://doi.org/10.52842/conf.caadria.2023.2.049

41. Chornobai SR, Ponzio AP, Miotto Bruscato LM, Fagundes CVM (2023) Fostering sustainability in the early stages of the architectural design process: a creative exploration of AI generative models. In García Amen F, Goni Fitipaldo AL, Armagno Gentile Á (eds) Accelerated Landscapes—Proceedings of the XXVII International Conference of the Ibero-American Society of Digital Graphics (SIGraDi 2023), Punta del Este, Maldonado, Uruguay, 29 November–1 December 2023, pp 831–842. https://papers.cumincad.org/cgi-bin/works/Show?sigradi2023_439

Erasing the Human Hand: AI Anxiety in Architecture

Maycon Sedrez and Gregory Pitts

Abstract One of the primary concerns of Artificial Intelligence (AI) in architecture revolves around the automation of creative tasks, a fear that resembles the anxiety over replacement of jobs that architects faced in the 1980s with the introduction of Computer-Aided Design (CAD) tools. This chapter discusses critical challenges that AI presents to architecture, specifically when it comes to replacing human tasks. In addition, it provides a narrative framework for understanding the dual nature of AI's potential in shaping the future of the architectural profession: replacing or enhancing. We analyse sixty architectural tasks (performance criteria) defined by the National Standard of Competency for Architects produced by the Architects Accreditation Council of Australia, which are directly embedded in architecture curriculum, by examining the extent to which they are susceptible to automation and discussing the potential transformation of these tasks considering AI's capabilities. Our findings indicate that automation of tasks by AI will affect architects in Australia requiring a review of the profession role and skills.

Keywords Artificial intelligence · Architecture · Design · Automation · Employment

1 Computerization of Architecture

AI has experienced rapid development over the past years, particularly due to advancements in accessibility of resources such as OpenAI. These innovations are enabling a release of many interfaces that facilitate generation of highly complex text, images and 3D shapes as this chapter is being written (first months of 2025). It

M. Sedrez (✉) · G. Pitts
School of Architecture and Built Environment, Deakin University, Geelong, VIC, Australia
e-mail: m.sedrez@deakin.edu.au

G. Pitts
e-mail: gregory.pitts@deakin.edu.au

M. Sedrez and G. Pitts (eds.), *Artificial Intelligence-Aided Design for Sustainability*, Urban Sustainability, https://doi.org/10.1007/978-981-95-1349-9_8

has been challenging to keep up with the evolution of these interfaces as their iterations for generating designs get progressively more sophisticated. An analysis of this extremely fast change is demonstrated in the Gartner Hype Cycle for Emerging Technologies that between 2020 and 2022 mapped respectively *AI-Assisted Design* [1], *AI-Augmented Design* [2] and *Generative AI Design* [3] at the beginning of innovation trigger. The Gartner's Hype Cycle plots innovations on a graph that represents the development, expectations and time to be commercialized in large scale—the plateau of productivity. It is predicted that these AI-Aided tools will be fully integrated to production in 5 to 10 years. However, recently, Perri [4] and STAMFORD [5] mapped generative AI in the peak of inflated expectations in the Gartner's Hype Cycle, moving faster to reach productivity in just 2 to 5 years. In addition, there are many limitations of current AI models and needs for hardware advancements. This fast-paced development and quick practical implementation has been creating anxiety in all professional fields [6, 7].

Despite the current AI frenzy, employing AI in architectural design is not something entirely new (see Chapter 2 for a thorough review of AI and design history). In 1986, Celestino Soddu developed an AI-generative system capable of generating 3D forms [8]. Other machine learning and AI applications became popular through programming interfaces in the 1990's and 2000's (evolutionary and generative algorithms). These were largely black-box systems that gave users little understanding of the computing processes but significantly impacted the efficiency of input vs output.

Over a decade prior to these changes in software capability, Cooley [9] predicted rising tensions between standard routines, optimization of tasks and productivity expectations. These were identified as emerging threats to designers' creativity, born from computational capabilities. Along with the dissemination of Computer-Aided Design (CAD) tools in architecture companies through this period, came a profession-wide anxiety about the automation of creative tasks and the potential replacement of jobs that architects could face. At that time, Löwstedt [10] reported lower levels in job satisfaction, challenges in adapting to new tools and extra workload in Swedish companies. Maver [11] also echoed these issues through the recognition that the introduction of Computer-Aided Architectural Design (CAAD) software would traumatically require re-skilling architects, and that medium-size practices could disappear.

These uncertainties led to a transformation in the field, such as changes in the role of architectural drafter with optimization of architectural representation through CAD/CAAD interfaces. Today, we are already seeing a change in the role of architectural visualization with integration of AI rendering models capable of creating many iterations and with faster speed. Current trends indicate that computers are being used to accelerate tasks and deal with complex problems while AI is being gradually integrated to daily work routines [12]. This emphasis on productivity and speed, comes with a promise of freeing more time for architects to spend on the design conceptualization and creative tasks. But much like the apprehensions surrounding job challenges and displacement in the 1970's and 80's, AI's increasing capabilities have sparked widespread anxiety about the potential replacement of human roles in the creative process. A fear of architects becoming obsolete has been expressed by

Leach [7], Erickson [8], Trabucco [13], Kołata and Zierke [14], and Hoque [15], to name a few. In addition, the notion that AI is dangerous to people if not used with caution has been presented by Brundage et al. [16] raising critical ethical questions regarding its use.

AI advances have already disrupted architecture discourse and practice even before these tools have reached an advanced level of implementation. For example, Trabucco [13] believes that the replacement of architects by AI might be experienced on lower levels, starting from the premise that computers are capable of replacing humans, particularly for repetitive tasks. We have seen computers replacing humans in many professions in the past such as, interactions with automated machines in customer service for instance. In construction, robots are being implemented to fabricate, demolish and perform repetitive or labour-intensive tasks such as painting, plastering walls, assembling bricks, inspection, etc.

Stewart et al. [17] suggests that architects can benefit more from adopting AI as the profession shows evidence of resistance to job replacement. Despite some positive prospects, there is a general fear that AI represents extreme risks to the profession [6, 7] and that will, at best, begin increasing efficiency in production and removing the need for so many human designers involved in a project and at worst, completely erase the human hand in the architectural design process. This might lead to a forced transition to highly automated digital design processes. Meanwhile, architects largely base design in analogic processes, systematic documentation and critical thinking, which cannot be performed by AI. For instance, the post-digital representation phenomena [18], that emerged a few years before AI integration in architectural visualization, prompted less realistic renders in favour of more subjective concepts, which demonstrates the profession's predilection to analogic even when using digital tools.

Undoubtedly, AI is being incorporated at large by companies across all professional fields. But, creative industries, amid views of fascination and rejections, have focused on how AI innovation can assist, augment or generate designs. In the architecture field, AI is enhancing architects' capacity to solve routine problems while potentially supporting creative processes. A large focus of architectural research shifted to AI and design in recent years, with some looking into which architect's tasks might be replaced [12]. Neil Leach [7] notes one of the most important things for architects to do "*is to be aware of the problem. For once a problem has been recognised, it becomes a different kind of problem—not one by which we are trapped, but one we can address.*"

There are already AI-driven solutions in the market supporting a range of design tasks, such as: project management, decision making, the evaluation of site options, space planning, feasibility study, structural design, complex modelling, design documentation, sustainable design, real-time rendering, building information modelling (BIM), modelling, urban design, sketching/ideation, environmental design, presentation, planning, concept generation, and generate design options [19–21]. However, due to the complexity of tasks involved in designing a building and setting AI algorithms, these tools are limited to certain aspects of the design process. For instance, Bölek et al. [22] listed practical applications already implemented in the field that optimize building performance (energy efficiency, passive design), support design (form-finding, modularity), provide spatial distribution (plan solutions, massing solutions), enhance restoration processes (complete missing parts, digitalization), and solve specific design problems (targeted tools).

With a diverse range of AI integration in architecture, the future of our profession lies in proactively engaging with digital tools in order to implement them towards shaping a successful future. However, if designers are still learning skills that can be easily replaced by AI without critically reflecting on its impact, then the profession is at risk of stumbling blindly into a future that erases the human hand in the design process. Since computerization of architecture might lead to a progress in digital design development and construction, architects will need to direct which type of design will be created. In this sense, architects can benefit from such computerization expanding design capabilities through AI instead of being replaced.

This chapter provides a narrative for understanding the dual nature of AI's potential in shaping the future of the architectural profession: replacing or enhancing, as introduced in chapter 1. In this sense, the chapter assess the impact of AI in tasks performed by architects, particularly in the context of Australia. The aim is to gauge industry expectations and skill sets that future architects will need and understand possible changes in the role of architects.

2 Automation of Jobs

Automation and computerization involve the employment of machines or computers to perform tasks previously done by people. Despite the hype surrounding job replacement, both recently and historically, these technological innovations have often led to the emergence of novel industries and new jobs. To date, this has frequently resulted in the reshaping of labour markets as opposed to the shrinking of job opportunities [23]. In the Australian context, a recent government study [24] projects a moderate growth in design professions over the next 10 years (see Table 1), with 'Construction' being the third industry with major employment growth behind 'Health Care and Social Assistance' and 'Professional, Scientific and Technical Services'. Alternatively, the World Economic Forum [25] predicts that 35% of Australian jobs will suffer some kind of disruption in the next 5 years, including issues caused by impacts of AI.

Despite these growth projections, the scope and limitations of automation as well as the future job impact remain topics of critical inquiry. This is particularly difficult to define in disciplines that rely heavily on human cognition, creativity and interpersonal skills. Frey and Osborne [26], argue that while automation is well-suited to tasks governed by clear rules and structured inputs, it is less capable of replicating the subjective and context-specific functions that define creative and interpersonal relations.

Architecture is an exemplar of a human-centered, creative profession. It is conventionally (and somewhat reductively) defined as the art and science of building,

Table 1 Employment projections by occupation in Australia. *Data* Jobs and Skills Australia [24]

Occupation	5-year change (2029)	10-year change (2034)
Architects, designers, planners and surveyors nfd*	+11.2%	+20.3%
Architects and landscape architects	+10.5%	+19.0%
Interior designers	+10.9%	+19.4%
Urban and regional planners	+8.2%	+15.8%

* *nfd: not fully described*

combining aesthetic creativity with technical expertise. In practice, this combines aspects of an architects intuition, creative aspirations and critical reflection to create functional and meaningful spaces. Architects must interpret client needs, coordinate with stakeholders, and respond to social, cultural and environmental contexts, all of which require empathy and interpersonal intelligence. When considering all these factors, architecture is an inherently complex profession, intertwined with a mix of subjective and empirical tasks.

Architecture requires a higher level of cognitive and abstract thinking in combination with analytical and problem-solving skills, thus understanding which tasks can be automated is beneficial to the advancement of the profession. In a study aimed at better defining the categories of cognition in architectural design, Celani et al. [27] revealed that 60% of architects' tasks, as defined by the American Institute of Architects, fall into a category that make them more susceptible to automation in the future. Despite early studies such as this, there is generally still very little reflection on the number of architecture jobs that might be affected due to AI replacement or which areas might be more at risk than others.

This returns us to the idea described by Leach [7] of 'recognising the problem' and the requirement to assess our profession for current weaknesses and opportunities that position the capabilities of AI alongside the core value of the Architectural profession and help us consider a more robust, AI inclusive future.

3 The Architecture Profession in Australia

The architecture profession in Australia is recognised for its high value, but it is not fully leveraged or utilized to its full potential. Housing currently makes up a large proportion of the Australian construction sector and is only expected to grow over the coming decades. This recognition of growing demand/housing shortage is largely driving the above job growth numbers presented in Table 1. Despite this demand, it is also recognised that Australian housing projects are mostly developed without the engagement of architects or designers. Data varies on the actual percentage but generally shows less than 10%, with sources like the Architecture Registration Board of Victoria (ARBV) stating only a 2% involvement of architects in the residential building market [28]. This factor is largely due to high fees charged by architects, relative to alternative paths to a building permit such as architectural drafters or builders producing 'in house' drawings. The higher fee comparison largely results from the increased scope of design consideration and processes, but is compounded by highly manual and repetitive tasks that consume a significant percentage of an architect's time.

This data highlights a current problem in the architectural profession in Australia, marginalising the profession into highly bespoke work. This, when combined with the intense competition for construction design and delivery in the housing market, leads to bleak projections of decline in architectural demand over the next 5 years [28]. There are, however, also more positive future opportunities hidden in these trends that hint at a need for greater competitive fee structures and efficiency in work processes. Greater efficiency in highly manual tasks has the potential to both, reduce the number of architects that need to have a hand in producing each individual project, while simultaneously opening up new avenues into the existing housing market and allowing a greater coverage through the housing sector.

Reflecting on this opportunity for computationally driven efficiency, we are reminded that digital tools have historically tended to augment rather than replace architects' work. Parametric design systems, generative modelling and Building Information Modelling (BIM) are now commonplace in the field largely due to their facility as 'enablers'. Architects can now rapidly explore complex geometries, analyse building performance and manage project data. As Cao [29] observes, parametric and BIM tools have "*been the driving force behind the most advanced architectural projects and the most innovative and creative professionals in this field*". Far from making architects obsolete, such tools expand design possibilities: architects use them to test hundreds of options for energy efficiency, daylight, structure, etc., far beyond what manual drafting could allow. In practice, these systems manage highly manual and repetitive tasks, calculations and coordination (e.g. automatic detailing or clash detection), freeing architects to focus on high-level design and client communication—the higher order functions more closely aligned with the true professional value of an architects contribution.

4 Assessing the Cognitive and Manual Aspects of Architectural Design

This chapter adapts a method described by Celani et al. [27], to help predict the impact of an increased AI presence in the Australian architecture profession. This approach produces an overview of automation of tasks, with architecture as case study. The method starts with the categorization of tasks in a matrix with two opposite axis: manual vs cognitive, routine vs non-routine, see Fig. 1.

Based on Celani et al. [27], non-routine tasks can be divided into: complex communication, for example interactions with humans and explain things; or expert thinking

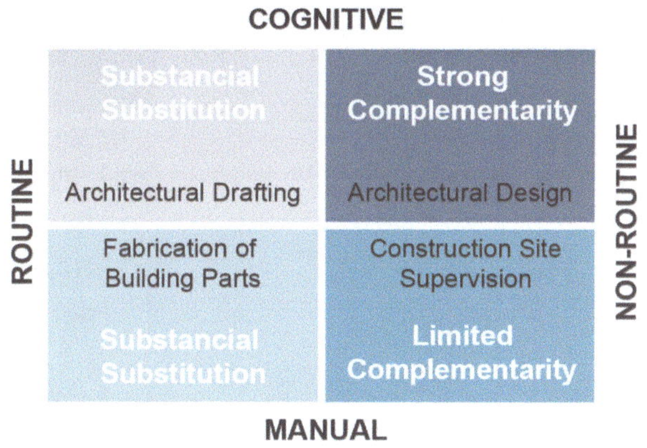

Fig. 1 Tasks and likelihood of computerization Adapted from Celani et al. [27]

(true pattern recognition including creative intelligence) which requires high skilled workers, and solving new problems where there is no ruled-based solution. Routine tasks can be divided into: inductive rules involving complex and sophisticated pattern recognition which can be provided by AI-trained models; or deductive rules that follow ruled-based logic (yes or no answer) thus automation is straightforward [27].

Starting from the premise that computers can replace humans in performing certain tasks [30], it is possible to analyse which tasks are more susceptible to automation. Human work is composed of cognitive activities that lead to decision-making, while computers execute rules that are linked in a particular way to generate an outcome. If a human task consists of a set of ruled-base solutions and the information can be translated to computer languages, then the task can be replaced by a computer. This is more complex when it comes to AI, especially because it is possible to observe multiple emerging applications in the architecture field. Additionally, architectural design has a creative aspect that is highly subjective and also linked with physical forms and tectonics. Since AI—particularly Machine Learning—performs on trained data, a "machines creativity" will be limited to a specific data set, while human brain can associate multiple subjective and objective ideas simultaneously.

In order to assess this in relation to the architectural profession in Australia, we analyse sixty architectural tasks defined by the Architects Accreditation Council of Australia (AACA) in the National Standard of Competency for Architects (NSCA) [31]. These tasks are directly embedded into architecture curriculum as requirement for architects knowledge and practice. The study examines the extent to which these tasks are susceptible to automation and discusses the potential transformation of these tasks considering AI's capabilities. The purpose of this chapter is not to offer definitive solutions to the complex issues surrounding AI-Aided Design. The issues at hand are still unfolding, with many of the most pressing concerns being shaped now as technology continues to advance. Instead, our goal is to discuss the replacement and enhancement of tasks to consider how and where the profession needs to integrate knowledge that requires human skills as a core professional value. By raising these critical questions, we hope to encourage ongoing reflection and discourse on the

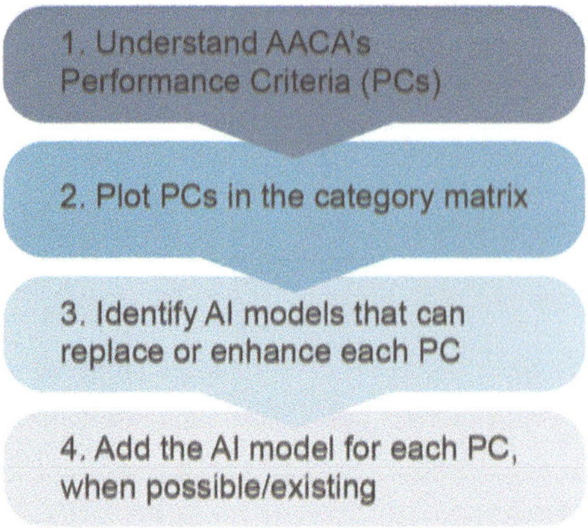

Fig. 2 Computerization skills method

intersection of AI and human creativity in architecture, emphasizing the importance of understanding the broader implications for the future of the profession.

This method employs Levy and Murnane [30] analysis of computerization of skills in its adapted version to architecture tasks [27]. In this chapter, we inform possible AI automations based on McKinsey and Company [32] descriptions of AI models usability for businesses, because of its existing practical implementation in several fields. In this sense, the brief explanation provided by McKinsey and Company [32] about the machine learning models and its applications in business will be used to confront: a. the Performance Criteria (PCs) informed by the NSCA , b. involved tasks, and, c. possibility of automation. Figure 2 illustrates the method steps, beginning with an analysis of NSCA's PCs and plotting each PC in the task category matrix (Fig. 1). Then, we proceed by identifying AI models for each PC where there is substantial substitution or complementation.

4.1 Performance Criteria

The Royal Institute of British Architects (RIBA) and the American Institute of Architects (AIA) deployed policies and guidelines after major concerns about the risks posed by AI emerged. Their attention focused on the efficiency of performing certain tasks, new workflow models and calculations, but expressed which architects' characteristics cannot be replaced by machines: reliable, stable and functional [6]. The RIBA [33] report presented that 37% of interviewed architects consider AI as a threat to the profession, in addition, ethical challenges such as intellectual property (designs being imitated) and lowering fees were also mentioned. However, only 7% of the architects interviewed by the RIBA report agreed that AI led to staff cuts. Meanwhile, the AIA [34] guide for specification revealed the expectation to use AI for product research and specification. Alternatively, the Royal Australian Institute of Architects

(RAIA) identified possible benefits of AI: automatizing manual tasks and reviewing of regulatory code for design irregularities [6].

The AACA issued the latest version of the NSCA in 2021 - a document that "*identifies the skills, knowledge and capabilities required for the general practice of architecture in Australia*" [31]. It defines the criteria to assess architecture programs and registration in Australia, and therefore, it is integrated in the curriculum of architecture schools. In Australia, according to the NSCA [31], architects' Professional Capabilities encompass Professionalism, Communication and Environmental Practice; these three contain skills and knowledge required for professional architects and support architectural education. The Professional Capabilities are evaluated through Performance Criteria (PCs) which "*describe discrete aspects of architectural practice*" [31]. PCs intrinsically orient what is the work of architects, including skills, business framework, their responsibilities (with the environment, indigenous communities, and clients), and ethics. The NSCA distributes sixty PCs into four main units of competency without any prioritization:

1. *Practice management and professional conduct*: corresponds to the architecture business administration, provision of quality services and ethical and legal aspects of the building industry. PCs 1 to 16.
2. *Project initiation and conceptual design*: corresponds to the design concept and ideation (research, analysis, design methods, conceptual design). PCs 17 to 35.
3. *Detailed design and construction documentation:* corresponds to produce design details, select materials and technical solutions. PCs 36 to 47.
4. *Design delivery and construction phase services*: corresponds to project execution and construction. PCs 48 to 60.

Table 2 presents all the sixty PCs distributed in the task category matrix adapted from Celani et al. [27]. This assignment was judged by reviewing each PC description and assigning the type of task. PC47 informs several different tasks including 'producing models', thus, it is the only duplicated PC in the table, because model-making and drawing are considered a manual task. The second part of the method is identifying AI models that can replace or enhance the task. We adopted McKinsey and Company [32] evidence of AI in connection with business and examples of technologies already being used. McKinsey and Company [32] defines four major types of machine learning applications according to its training procedure: 1. Supervised Learning (SL) uses labelled data to learn relationships and make predictions, 2. Unsupervised Learning (UL) uses unlabelled data to identify pattern(s) and recognize it, 3. Reinforcement Learning (RL) uses digital incentives to optimize data, and 4. Deep Learning (DL) consists of a neural network of information.

McKinsey and Company [32] machine learning classifications and its applications in businesses are then attributed to each PC in Table 2, where we suggest possible AI tools that can replace or enhance the required skill. This is a speculative attribution, since some of these AI tools are not ready for application in the architecture field or have not found its purpose in architectural businesses so far. In addition, AI systems are limited to deal with complex ideas and externalities of its own system, such as abstract, symbolic or subjective interpretations.

Table 2 Task category matrix for the 2021 NSCA performance criteria (DL: Deep Learning; SL: Supervised Learning; UL: Unsupervised Learning; RL: Reinforcement Learning)

Type	Cognitive tasks			
Tasks	Non-routine: 34 (~57%)		Routine: 26 (~43%)	
NSCA PCs	Complex communication *(Interaction with humans, explain things)*	Expert thinking (True Pattern Recognition) *(High skilled workers, solving new problems, no ruled-based solution)*	Inductive rules *(Complex/ sophisticated pattern recognition)*	Deductive rules *(Ruled-based logic)*
Practice Management,Professional Conduct (PC1-16)	PC5, PC7, PC8	PC6:SL, PC9:SL, PC15	PC3:UL, PC4:UL, PC11:UL, PC12:SL	PC1:SL, PC2:SL+UL, PC10:RL, PC13:UL, PC14:SL, PC16: SL
Project Initiation, Conceptual Design (PC17-35)	PC23, PC34	PC17, PC19:SL, PC22, PC24:SL, PC25, PC26:DL, PC27, PC28, PC29:DL, PC30:DL, PC32	PC18: DL	PC20:SL, PC21:SL, PC31:SL+RL, PC33:RL+DL, PC35:SL
Detailed Design, Construction Documentation (PC36-47)	PC43, PC44	PC36:DL, PC40, PC41	PC38:SL, PC42:SL	PC37:DL, PC39:SL, PC45:SL, PC46:SL, PC47:SL+DL
Design Delivery, Construction Phases (PC48-60)	PC50, PC51, PC53	PC48:SL, PC49:UL, PC 57:RL	PC59	PC52:RL, PC60:RL
Impact	Complementation		Substantial substitution	
Type	Manual tasks			
Tasks	Non-routine		Routine	
Detailed Design, Construction Documentation (PC36-47)	PC47:RL		PC47:RL	
Design Delivery, Construction Phases (PC48-60)	PC54, PC55, PC56, PC58:DL			
Impact	Complementation		Substantial substitution	

Categorizing NSCA tasks resulted in 37 PCs (61%) that can potentially employ AI. In addition, 43% of the PCs are assigned as routine tasks and might be impacted by a substantial substitution, which corroborates with results presented by Celani et al. [27], where 40% of tasks (as per the American Institute of Architects definitions) were considered routine. This consistency demonstrates that automation will transform the way architects perform tasks and that the power of AI might speed the automation of tasks. In this scenario, we call universities and professional organizations to reevaluate skills required to become an architect. We selected a sample of tasks informed by PCs to illustrate skills that can be automated or supported by AI, and tasks that are already being replaced by AI tools.

- Practice management and professional conduct

PCs potentially impacted by AI: 1, 2, 3, 4, 6, 9, 10, 11, 12, 13, 14, 16

PC 12: "*Provide independent, culturally responsive and objective advice in accordance with relevant building codes, standards, technical specifications and guidelines, and planning regulations, including climate change implications, across all aspects of architectural practice.*" [31]

PC 13: "*Be able to identify and apply strategies, programming and processes for documentation through all project stages to facilitate project delivery, as appropriate to selected procurement processes.*" [31]

AI is capable of identifying patterns and recognizing strategies with greater efficiency than humans. We expect that AI applications will analyse project data and identify recurring design solutions (patterns) in design documentation. Additionally, AI could be used to verify designs against building codes and other quantifiable regulatory information.

- Project initiation and conceptual design

PCs impacted by AI: 18, 19, 20, 21, 24, 26, 29, 30, 31, 33, 35

PC 20: "*Be able to assess project budget and timeframe against project requirements and objectives, relevant legislation, statutory planning requirements, building codes and standards.*" [31]

PC21: "*Be able to apply project budgets, or work with quantity surveyor to establish project budgets, based upon understanding of cost planning, value management and factors influencing.*" [31]

AI can analyse data and requirements to support budget estimations and define timeframes based on recurring patterns. Although no tool exists that comprehensively considers costing in relation to regulatory and budget considerations, it is possible to imagine that the quantifiable aspects of these PC's could be highly automated to predict both the potential cost of a design and the time required for its implementation.

- Detailed Design and construction documentation

PCs impacted by AI: 36, 37, 38, 39, 42, 45, 46, 47

PC 37: "*Be able to produce timely, accurate, complete and comprehensible documentation of the design so that it can be constructed.*" [31]

PC 39: "*Be able to integrate the material selection, structural and construction systems established in the conceptual design into the detailed design and documentation.*" [31]

Although there is currently no tool available to generate architectural drawings as vectors (so far), emerging tools are capable of creating 3D forms. It is reasonable to expect that in the coming years, AI plugins will be able to suggest materials and structural elements based on design concepts. This highlights the innovative aspects of the field, where the creative component will be influenced by the capabilities of the AI tool and its dataset. Therefore, collaborative approaches will be essential.

- Design delivery and construction phases

PCs impacted by AI: 48, 49, 52, 57, 58, 60.

PC 52: "*Be able to apply the principles and mechanisms implicit in the selected procurement method and associated construction contract(s), based on an understanding of the implications of differing contractual relationships.*" [31]

PC 60: "*Apply appropriate methodologies for undertaking post occupancy evaluations and life cycle assessment where required under terms of engagement.*" [31]

Administrative tasks will be heavily impacted by AI, particularly due to exceptional efficiency in textual analysis and interpretation. AI systems will streamline the management of documentation, contracts, and communication by automatically processing vast amounts of text and identifying key information quickly. In addition, AI systems will become pervasive in building management and assessment of performance. By utilizing advanced sensors and data analytics, AI can provide real-time insights into building performance, such as energy efficiency, occupant comfort, and provide alerts for maintenance needs.

Despite these and many other predictions about the future integration of AI [12], there are still limitations regarding development of new systems and data availability. These critical challenges to automate tasks might slow down the transfer of technology to practice but won't reduce its feasibility. With this in mind, as tasks are gradually automated, we already see new professional profiles emerging in architecture, for example, 'AI Prompt Curator' and 'Creative AI Storyteller'. Thus, sophisticated AI systems will assist architects in several stages of the design and ultimately will need human control, input, feedback, provision of datasets, and a cognitive alignment between the system, the design and the architect vision.

The World Economic Forum [25] identified the core skills for 2025, with 'analytical thinking' topping the list as a key cognitive skill. In second place is 'resilience, flexibility, and agility,' which correlates with self-efficacy. 'Leadership and social influence' comes in third, reflecting the ability to collaborate effectively with others, while 'creative thinking' ranks fourth, another essential cognitive skill to architects. Together, these top four skills highlight the profile of a professional adept at effectively solving problems in a collaborative and creative manner. These skills align closely with the practice of architecture, where the ability to think critically and creatively, and to collaborate is essential. These competencies are independent of current AI development, suggesting that while automation will transform routine tasks, creative architects will prevail. This encourages architecture schools in Australia to emphasize the development of creative and critical thinking skills. However, the Association of Consulting Architects (ACA) in Australia, an entity "*representing architectural practices in business and employment matters*" [35] issued a paper reflecting on current issues over AI use. Their main concern is the replacement of architects' tasks by AI disguised as a cost-saving action. In addition, the lack of regulations in Australia combined with undefined transparency requirements might create several safety and ethical challenges to the profession.

5 Conclusion

In this chapter, we have demonstrated that the AI will impact heavily in architecture tasks and architects' skills. Today, architectural visualization, particularly the creation of renders, is increasingly reliant on AI-powered image-generation tools, making these technologies more accessible to a wider audience and, therefore, less specialized. One of the key challenges we faced in this research was identifying the practical attributes of NSCA PCs and assessing their potential connection to AI solutions. Many AI applications remain speculative or are tailored to non-design professionals, making this alignment complex.

Furthermore, the fact that many buildings in Australia are not designed by architects means that the broader public is often unaware of AI's impact on design, exacerbating concerns within the profession. Architects increasingly fear being replaced, a concern heightened by the rapid evolution of AI and the slow pace of adaptation in both architectural education and professional practice. While architecture students may encounter AI tools through personal initiative or tutor expectations, the absence of formal curriculum integration represents a significant gap. As Bernstein [12] suggests, architectural schools and professional associations must urgently reconsider how architects are trained to ensure they remain relevant in an AI-driven landscape.

The accelerating computerization of architectural practice over the past three to four decades has both challenged and changed the role of the human hand in the design process. At this transformation moment, it is crucial to articulate how design should evolve in response to these technological shifts. Our method is inherently limited as it examines AI applications solely from the perspective of the authors. A more comprehensive analysis would require input from multiple stakeholders in the design field, along with collaboration from computer scientists, to refine the classifications outlined in Table 2. Nevertheless, since Celani et al. [27] first explored this topic in 2015, there has been a clear trajectory of increasing automation. In just the past five years, AI has become a central topic of discussion across all disciplines, profoundly impacting professionals.

This chapter does not investigate broader implications of AI, such as ethical concerns, transparency, biases, computational and energy demands, political ramifications, safety, or digital security. Moreover, it focuses purely on architecture as a business that performs PCs, without addressing human interactions and relationships that are fundamental to the profession. Looking ahead, we anticipate the emergence of new architectural roles dedicated to integrating AI into professional practice. These facilitators will require skill sets that are not currently embedded within architectural education, creating a growing competency gap. Rethinking architecture education and AI needs to align to the Beijing Consensus [36] and therefore contemplate multidisciplinarity, digital skills and sustainability to prepare the profession to AI advancements.

References

1. Panetta K (2020) 5 Trends Drive the Gartner Hype Cycle for Emerging Technologies, 2020. In Gartner. https://www.gartner.com.au/en/articles/5-trends-drive-the-gartner-hype-cycle-for-emerging-technologies-2020. Accessed 30.Oct.2024
2. Panetta K (2021) 3 themes surface in the 2021 hype cycle for emerging technologies. In Gartner. https://www.gartner.com/smarterwithgartner/3-themes-surface-in-the-2021-hype-cycle-for-emerging-technologies. Accessed 30.Oct.2024
3. Perri L (2022) What is new in the 2022 gartner hype cycle for emerging technologies. In Gartner. https://www.gartner.com.au/en/articles/what-s-new-in-the-2022-gartner-hype-cycle-for-emerging-technologies. Accessed 30.Oct.2024
4. Perri L (2023) What's new in the 2023 gartner hype cycle for emerging technologies. In Gartner. https://www.gartner.com/en/articles/what-s-new-in-the-2023-gartner-hype-cycle-for-emerging-technologies. Accessed 30.Oct.2024
5. STAMFORD, Conn (2024) Gartner 2024 hype cycle for emerging technologies highlights developer productivity, total experience, AI and security. In Gartner. https://www.gartner.com/en/newsroom/press-releases/2024-08-21-gartner-2024-hype-cycle-for-emerging-technologies-highlights-developer-productivity-total-experience-ai-and-security. Accessed 30.Oct.2024
6. Barker N (2023) Architecture bodies respond to "real risks" of AI. In Dezeen's AItopia Series. https://www.dezeen.com/2023/08/15/architecture-bodies-ai-issues-riba-aia-aitopia/. Accessed 30.Oct.2024
7. Leach N (2023) AI is putting our jobs as architects unquestionably at risk. In Dezeen's AItopia Series. https://www.dezeen.com/2023/02/13/ai-architecture-jobs-risk-neil-leach-opinion/. Accessed 30 Oct 2024
8. Erickson K (2024) AI and work in the creative industries: digital continuity or discontinuity? Creat Ind J. https://doi.org/10.1080/17510694.2024.2421135
9. Cooley M (1977) Impact of CAD on the designer and the design function. Comp Aided Des 9(4):238–242. https://doi.org/10.1016/0010-4485(77)90004-5
10. Löwstedt J (1988) Prejudices and wishful thinking about computer aided design. N Technol Work Employ 3:30–37. https://doi.org/10.1111/j.1468-005X.1988.tb00086.x
11. Maver TW (1896) Social impacts of computer-aided architectural design. Des Stud 7(4):178–184. https://doi.org/10.1016/0142-694X(86)90034-7
12. Bernstein P (2022) Machine learning: architecture in the age of artificial intelligence. RIBA Publishing
13. Trabucco D (2021) Will artificial intelligence kill architects? An insight on the architect job in the AI future. TECHNE J Technol Archit Environ 2:128–132. https://doi.org/10.13128/techne-10696
14. Kołata J, Zierke P (2021) The decline of architects: can a computer design fine architecture without human input? Buildings 11. https://doi.org/10.3390/buildings11080338
15. Hoque F (2024) Does artificial intelligence have the possibility of taking over designers' jobs in the future? Intern J Sci Bus 31:1:26–35. https://doi.org/10.58970/IJSB.2273

16. Brundage M et al. (2018) The malicious use of artificial intelligence: forecasting, prevention, and mitigation. Future of Humanity Institute, University of Oxford, Centre for the Study of Existential Risk, University of Cambridge, Center for a New American Security, Electronic Frontier Foundation, OpenAI
17. Stewart C, Caldwell GA, Teixeira MBF, Roberts J (2020) Robotics and artificial intelligence in architecture: what skills will architects need in 2050? In Hearn G (ed) The future of creative work. Edward Elgar Publishing Limited, Cheltenham, pp 108–122. https://doi.org/10.4337/9781839101106.00015
18. Jacob S (2017) Architecture enters the age of post-digital drawing. Metropolis. https://metropolismag.com/projects/architecture-enters-age-post-digital-drawing/. Accessed 07.Dec.2024
19. Frew S (n.d.) 15 top AI tools for architects and designers. Architizer. https://architizer.com/blog/practice/tools/top-ai-tools-for-architects-and-designers/. Accessed 30.Dec.2024
20. Archgyan (2024) Top AI tools for architects in 2024. Archgyan. https://archgyan.com/top-ai-tools-for-architects-in-2023/. Accessed 30.Dec.2024
21. Nasir O (2024) 15 trendy AI tools to enhance architectural practices. https://parametric-architecture.com/15-trendy-ai-tools-to-enhance-architectural-practices/. Accessed 30.Dec.2024
22. Bölek B, Tutal O, Özbaşaran H (2023) A systematic review on artificial intelligence applications in architecture. J Des Resil Archit Plan 4:1:91–104. https://doi.org/10.47818/DRArch.2023.v4i1085
23. Acemoglu D, Restrepo P (2019) Automation and new tasks: how technology displaces and reinstates labour. J Econ Persp 33(2):3–30
24. Jobs and Skills Australia (2024) Employment projections. https://www.jobsandskills.gov.au/data/employment-projections. Accessed 09.Feb.2025
25. World Economic Forum (2025) Future of jobs report 2025: insight report. Switzerland. https://www.weforum.org/reports/the-future-ofjobs-report-2025/. Accessed 15.Apr.2025
26. Frey CB, Osborne MA (2017) The future of employment: how susceptible are jobs to computerisation? Technol Forecast Soc Chang 114:254–280. https://doi.org/10.1016/j.techfore.2016.08.019
27. Celani G, Sedrez M, Lenz D, Macedo A (2015) The future of the architect's employment. In Celani G, Sperling D, Franco J (eds) Computer-aided architectural design futures. The Next City - New Technologies and the Future of the Built Environment. CAAD Futures 2015. Communications in Computer and Information Science, 527. Springer, Berlin, Heidelberg. https://doi.org/10.1007/978-3-662-47386-3_11
28. The Market for Architectural Services. ARBV Industry resources. https://www.arbv.vic.gov.au/systemic-risks-australian-architecture-sector/market-architectural-services. Accessed 13.May.2025
29. Cao H (2024) Innovative applications of parametric design and digital tools in architecture: exploring the integration of generative design and BIM technology. Appl Comput Eng 98(1):79–84
30. Levy F, Murnane RJ (2005) How computerized work and globalization shape human skill demands. MIT Industrial Performance Center working paper series, Cambridge
31. AACA (2021) National standard of competency for architects - NSCA. Architects Accreditation Council of Australia. https://aaca.org.au/national-standard-of-competency-for-architects/2021nsca/. Accessed 06.Oct.2024
32. McKinsey and company (n.d.) An executive guide to AI. Quantum Black: AI by Mckinsey. https://www.mckinsey.com/capabilities/quantumblack/our-insights/an-executives-guide-to-ai?cid=othereml-alt-mip-mck&hdpid=d36c6b61313b-431c-b3d4-141bc805e7e28&hctky=116253808&hlkid=36ab108977a849a. Accessed 30.Dec.2024
33. RIBA (2024) RIBA AI report. Royal Institute of British Architects. https://www.architecture.com/knowledge-and-resources/resources-landing-page/riba-ai-report-2024?srsltid=AfmBOoofEdbm8SWTOMmUuQJUvjrULvifUvto3FEQjOENft3DfAObvs72. Accessed 19.Jul.2024
34. AIA (2023) The architects journey to specification. American Institute of Architects. https://www.aia.org/resource-center/the-architects-journey-to-specification. Accessed 22.Sep.2024

35. ACA (2023) Responsible AI in Australia: have your say. Association of Consulting Architects Australia. Melbourne. https://aca.org.au/responsible-ai-in-australia/. Accessed 30.Aug.2024
36. UNESCO (2019) Beijing Consensus: on artificial intelligence and education. Outcome document of the International Conference on Artificial Intelligence and Education. Planning Education in the AI era: Leap the leap. United Nations Education, Scientific and Cultural Organization, Paris. https://unesdoc.unesco.org/ark:/48223/. Accessed 13.Oct.2024

Mapping Emerging Machine Learning Methodologies in Architectural Design: A Pathway for Present Integration and Future Research

Victor Calixto, Camilo Cruz Gambardella, Darcy Zelenko, Wanchao Su, and Duncan Maxwell

Abstract The architectural design process spans a complex, iterative value chain, requiring diverse expertise across conceptualisation, technical development, and building use. While recent advances in machine learning (ML) have sparked growing interest in its application to architectural design, most research remains narrowly focused on task-specific automation in early-stage design. This paper addresses a critical gap by systematically mapping ML methodologies—alongside computational design techniques—onto architectural processes and principles, with a focus on the RIBA Plan of Work. Emphasising the conceptual design phase (Stage 2), the study reveals both a disproportionate research emphasis on early-stage applications and a lack of integration across later lifecycle stages such as construction and post-occupancy. Through a synthesis of quantitative and qualitative data, the paper identifies methodological imbalances, sustainability blind spots, and risks posed by opaque ML systems. It argues for a more holistic, transparent, and lifecycle-oriented approach to ML in architecture—one that supports creativity, decision-making, and design agency throughout the built environment's full trajectory.

Keywords Machine learning · Computational design · Design value chain · Architectural design process · Spatial planning

These authors contributed equally to this work

V. Calixto (✉) · C. C. Gambardella · D. Zelenko · D. Maxwell
Future Building Initiative, Monash University, Melbourne, VIC, Australia
e-mail: victor.calixto@monash.edu

C. C. Gambardella
e-mail: camilo.cruzgambardella@monash.edu

D. Zelenko
e-mail: darcy.zelenko@monash.edu

D. Maxwell
e-mail: duncan.maxwell@monash.edu

C. C. Gambardella · W. Su
SensiLab, Monash University, Melbourne, VIC, Australia
e-mail: wanchao.su@monash.edu

M. Sedrez and G. Pitts (eds.), *Artificial Intelligence-Aided Design for Sustainability*, Urban Sustainability, https://doi.org/10.1007/978-981-95-1349-9_9

1 Introduction

Architecture is an inherently multidisciplinary practice, encompassing a spectrum of activities that span the entire lifecycle of the built environment, from strategic definition to design, construction, and post-occupancy evaluation. Each stage within this process demands collaboration and specialised expertise, in order to produce the distinct deliverables required. However, the increasing complexity and fluidity of project workflows make it difficult to demarcate these stages clearly. As it is well known in the field, the design process does not operate as a linear sequence; instead, it emerges iteratively, shaped by feedback loops that blur the boundaries between ideation, evaluation, and execution.

Recent advancements in Machine Learning (ML) have introduced new possibilities for engaging with this complexity. Within architecture and related design fields, ML has been used to automate tasks, optimise design outcomes, and generate novel formal solutions. However, much of the current research remains narrowly focused—tending to address isolated design problems or stage-specific applications without critically interrogating how ML can be holistically integrated into architectural practice. As a result, a key question remains underexplored: how can ML reshape not just the output of design processes, but their structure, rationale, and impact across the lifecycle of buildings, from design to construction?

This study addresses that question by systematically mapping the application of ML across the stages different stages of the design process, using the Royal Institute of British Architects (RIBA) Plan of Work [1] as a guiding framework. We pay particular attention to the conceptual design phase (Stage 2), where experimentation with generative, predictive, and optimisation models is most prevalent. By analysing how ML techniques intersect with architectural tasks, design objectives, and sustainability goals, this paper identifies not only technical opportunities but also methodological gaps and epistemic risks. The aim is to move beyond a tool-centric view of ML and toward a critical framework that supports the development of transparent, accountable, and creative computational systems to support architectural design.

2 Background and Related Work

2.1 The Architectural Design Process and the RIBA Plan of Work

The RIBA Plan of Work provides a widely accepted framework for structuring architectural design practice across eight stages [1]. While variations exist, the core stages include (0) Strategic Definition, (1) Preparation and Brief, (2) Concept Design, (3) Spatial Coordination, (4) Technical Design, (5) Manufacturing and Construction,

(6) Handover, and (7) Use. Each stage has distinct tasks and deliverables, requiring specialised expertise and tools.

Although the RIBA Plan of Work is presented through a linear progression of design steps, the architectural design process is inherently complex, involving feedback loops between stages. The Concept Design stage, which includes spatial planning, is where design teams typically concentrate their efforts, as its outcomes significantly shape subsequent phases. However, if developed in isolation, later stages can strongly impact the initial concept. Reconciliation between technical requirements—which ensure the building's structural integrity and functionality—and spatial planning, defined in the Concept Design stage, is crucial for a cohesive design process. Similarly, spatial planning, which includes site analysis, massing studies, and initial spatial organisation, plays a fundamental role in shaping subsequent design phases. Effective spatial planning requires a nuanced understanding of building codes, environmental factors, user needs, and construction feasibility. Given this complexity, it presents a compelling opportunity to explore both the potential benefits and challenges of integrating ML into the architectural design process.

2.2 *Machine Learning Methods in Architectural Design*

The integration of ML techniques and methods into architectural design can be effectively understood through the lens of the four learning approaches: supervised learning, unsupervised learning, semi-supervised learning, and reinforcement learning. In Supervised Learning [2] models are trained on large amounts of labelled data, where inputs are mapped to expected outputs. This technique is used in the development of predictive models—e.g., classifiers and regression models, which are well suited for tasks like energy consumption prediction and structural performance analysis. Unsupervised Learning, on the other hand, focuses on the identification of patterns in unlabelled data. This technique is helpful in the exploration of large datasets, as it helps uncover hidden relationships, such as grouping similar design elements based on thermal efficiency or material usage, or reducing the complexity of high-dimensional data with techniques like Principal Component Analysis (PCA) and t-Stochastic Neighbour Embedding (t-SNE). Semi-supervised learning [3], which combines both labelled and unlabelled data to improve learning efficiency, is especially useful when labelled data is scarce, allowing for more robust models without needing extensive datasets. This is particularly beneficial for predicting building performance based on limited data. Lastly, Reinforcement Learning (RL) [4], introduced earlier in control theory, where agents learn through interactions and rewards offers an innovative approach by training agents through trial-and-error as they interact with an environment, which has proven effective for optimising design parameters such as spatial configurations or structural integrity. By applying RL, architects can explore and refine timber structures or automate space planning, learning from feedback to maximise desired outcomes. These paradigms, collectively, provide powerful tools for enhancing architectural design, pushing the boundaries of creativity, efficiency, and sustainability in the built environment.

2.3 Related Work

The integration of ML and AI into architectural design has concentrated significant research interest over the decades, with scholars and practitioners investigating its potential to enhance various aspects of the design process. Early studies laid foundational frameworks and identified challenges, while recent reviews have shifted toward a more systematic exploration of ML applications in architecture, focusing on emerging methodologies and their practical implications.

Zwierzycki [5] analysed nearly 200 publications on AI in architectural design since the 1980s, building on Gero's seminal "Ten Problems for AI in Design" [6]. The study categorised these publications by problem, tool, solution, stage, and future work, identifying persistent barriers to AI adoption. These include methodological inconsistencies, such as a lack of standardised evaluation metrics and subjective property assessments, as well as misaligned research priorities, with a bias toward layout generation and optimisation over foundational issues. Additional challenges stem from overly ambitious projects, inappropriate applications of existing AI tools, inadequate representation methods, and limited innovation in translating research into practical tools. External barriers, such as interface limitations, insufficient computational power, and systemic issues like funding and education, further hinder adoption. Zwierzycki updates Gero's framework by emphasising unresolved challenges in inference, generalisation, and the need for strategic, foundational research before pursuing complex applications.

Topuz and Çakici Alp [7] conducted a systematic review of ML applications in architecture from 2007 to 2022, analysing 60 articles from eight academic databases. Their study categorised applications across Computer-Aided Design (CAD), Computer-Aided Engineering (CAE), and Computer-Aided Manufacturing (CAM), identifying 21 architectural subfields where ML techniques have been applied, such as building information modelling (BIM), sustainability, urban design, and historical building preservation. They also examined the programming languages, software, ML platforms, libraries, and algorithms used, noting the prevalence of Python, Rhino-Grasshopper, TensorFlow, and neural networks (e.g., ANN, CNN, SVM). The study highlights the increasing adoption of ML across the architectural lifecycle and suggests expanding the scope of future research to include broader datasets and search parameters for a more comprehensive analysis.

Rhee et al. [8] performed a systematic review of neural network applications in Computer-Aided Architectural Design (CAAD) from 1990 to 2021, analysing 187 articles from CumInCAD. They identified six primary problem categories—generation, recognition, evaluation, analysis, fabrication, and theory—tracking trends over two significant periods of research activity: 1995–2005 and 2015–2021. The latter period coincides with the rise of deep learning. Their analysis explored the types of neural networks used (e.g., CNNs, GANs), data formats (matrices, images, point clouds), and the reproducibility of research outputs. A critical finding was the widespread lack of data and model accessibility, which hinders reproducibility and impedes the field's progress. The authors proposed guidelines to improve reproducibility in future CAAD research, including detailed dataset descriptions, parameter settings, and model specifications. Rhee et al. [8] provide a focused historical analysis of neural networks, Topuz and Çakici Alp [7] address a broader spectrum of ML techniques. Together, these studies underscore the growing importance of ML in architectural design but highlight gaps in reproducibility, methodology, and integration across different design stages.

The studies listed above provide valuable insights into the challenges and opportunities of ML in architecture. However, they remain limited in scope, as they are either historically focused, methodologically narrow, or too broad to capture the nuanced interplay between ML approaches and specific design tasks. This paper builds on these contributions by offering a stage-based mapping of ML methodologies onto the RIBA Plan of Work. In doing so, it seeks to fill the critical gap between theoretical exploration and practical integration, providing a higher-level synthesis of how ML can augment design processes—particularly during spatial planning and early-stage decision-making—while identifying areas where current applications fall short of architectural practice's broader epistemic and ethical demands.

3 Methods

3.1 Search Strategy and Data Collection

The search strategy was carefully designed to ensure comprehensive coverage of publications addressing the intersection of ML and architectural design. The literature search was conducted across two leading academic databases—Scopus and Web of Science—as well as the *Cumulative Index on Publications in Computer Aided Architectural Design (CUMINCAD)*. The latter is a specialised digital archive supported by prominent conferences and journals within the CAAD research community, offering access to targeted resources in computational design and architecture.

A detailed search was executed using the *ALL* fields function in combination with Boolean operators (AND, OR). The strategy involved pairing AI-related keywords (e.g., *"Artificial Intelligence"*, *"Machine Learning"*, *"Deep Learning"*, *"Neural Networks"*, *"Reinforcement Learning"*, *"Natural Language Processing"*, *"Supervised"*, *"Unsupervised"*, *"Semi-Supervised"*, *"Transformer"*, *"Recurrent Neural Network"*, *"Convolutional Neural Network"*) with architecture-related terms (e.g., *"Architecture Design"*, *"Spatial Planning"*, *"Spatial Coordination"*, *"Architectural Concept"*, *"Early Stages of Design"*, *"Space Planning"*).

The structured search query was as follows:

ALL = ("Artificial Intelligence" OR ... OR "Convolutional Neural Network") AND

ALL = ("Architecture Design" OR ... OR "Space Planning")

This search retrieved 144 papers from Scopus and 149 from Web of Science. Additionally, 990 papers were scraped from CUMINCAD using a custom Python script. After removing duplicates, a total of 1,267 unique papers were identified. These papers passed through a classification process, in which each paper received a label according to its objectives, artefact type (model, approach, or tool), machine learning approach, rule-based approach, RIBA Stage, Sustainable Goals. This process was conducted with a mix approach of classification based on Large Language Model (LLM) prompts with Llama 3, and a rule-based inclusion/exclusion script focusing on the *Title* and *Abstract* and a comprehensive and exhaustive keywords list, including acronyms and regular expressions. Then, the paper selection was further refined with an inclusion/exclusion automated script. Figure 1 provides an overview of the paper selection process.

3.2 Inclusion/Exclusion Criteria

The literature review was conducted using a rigorous set of inclusion and exclusion criteria to ensure relevance and methodological integrity. Studies were included if they specifically addressed architectural design processes, tasks, elements, and principles outlined in the RIBA Plan of Work. Additionally, included studies focused on the application of ML in architectural design, examined computational design techniques, and explored the integration of ML within architectural practice. Research investigating emerging ML approaches and published in peer-reviewed sources in English was also considered.

Conversely, studies falling outside the RIBA framework, such as those related to arts, urban design, or planning, were excluded. Non-peer-reviewed studies were omitted unless deemed highly relevant. Additionally, studies solely focused on AI methodologies that do not involve ML—including Expert Systems and Symbolic Reasoning—were excluded, as were those centred exclusively on rule-based systems, parametric methods, and generative design approaches. Studies published in languages other than English were also excluded. Furthermore, studies where machine learning and artificial intelligence were vaguely described in the abstract, without clear evidence of the learning approach or algorithm used, were not considered.

3.3 RIBA's Value Chain Classification Methodology

To systematically process and standardise the 1,266 unique papers collected from CUMINCAD, Scopus, and Web of Science, a structured classification and refinement

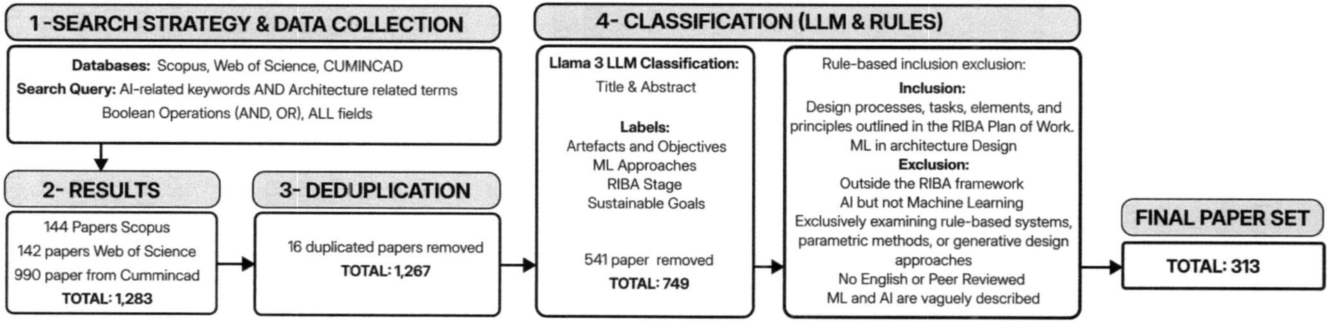

Fig. 1 Research methods diagram

pipeline was implemented. This methodology combined LLM classification, rule-based filtering, and targeted exclusion criteria to ensure a transparent and systematic selection process.

The initial phase involved cleaning and standardising the dataset to ensure consistency across all columns in the *.csv* file containing the paper metadata. This step eliminated inconsistencies in formatting, missing values, and duplicate records, creating a uniform dataset for classification. Once the data was pre-processed, the Llama 3 model was employed to classify the primary objectives of each paper. The LLM was used to infer the thematic focus of each study, aligning papers with predefined categories to support subsequent rule-based filtering.

To further refine the classification, a rule-based script was implemented, leveraging an exhaustive keyword-matching system for each column of interest. These included *RIBA Stage*, *Learning Approach*, *Rule-Based Approach*, *Algorithm*, *Artefacts*, *Objectives*, and *Sustainable Goal*. Each category was assigned a curated list of keywords to ensure consistency and improve classification accuracy. For example, papers classified under *Supervised Learning* were identified through keywords such as:

> Linear Regression, OLS, Ridge, Lasso, Elastic Net, Bayesian Regression, Polynomial Regression, and Support Vector Regression (SVR)

This method provided fine-grain control over classifications, improving precision and minimising ambiguity in how papers were categorised.

Following the rule-based classification, a secondary script was applied to refine the dataset based on inclusion and exclusion criteria, focusing on the paper abstracts. This phase ensured that review papers, studies outside the RIBA framework, and non-ML AI methodologies (e.g., Expert Systems, Symbolic Reasoning) were removed. Additionally, predefined keyword filters were applied to exclude irrelevant studies and ensure that only papers containing machine learning-related experimental indicators were retained. This filtering process resulted in the exclusion of 541 papers, leaving 749 papers for further analysis.

The classification of RIBA Stages was further refined, particularly for papers labelled as *Unsigned* by the LLM. To mitigate uncertainty, an additional keyword-matching approach was applied, aligning textual references within abstracts with corresponding *RIBA Stage* definitions. Papers that lacked a determined *RIBA Stage* classification were removed from the dataset, leading to the exclusion of an additional 436 papers. After this final refinement, a total of 313 papers remained for detailed analysis.

While this methodology provides systematic filtering and classification, it has certain limitations. The reliance on LLM classification introduces potential biases, as the model's predictions are influenced by the training data and context-specific interpretations. Also, rule-based filtering depends on predefined keywords, which could result in the exclusion of relevant papers if alternative terminology is used. The strict filtering criteria, although designed to reduce noise, may inadvertently exclude studies that require manual verification.

Despite these challenges, this approach enhances transparency, replicability, and systematic selection, ensuring that the dataset is both structured and relevant to the research focus. By combining LLM-driven classification with structured rule-based refinement, this methodology effectively identifies AI-driven design methodologies within the RIBA framework. Future refinements could involve hybrid approaches, integrating human-in-the-loop validation to improve classification accuracy and address edge cases more effectively.

4 Analysis

4.1 RIBA's Plan of Work and Machine Learning

The mapping of existing ML-related research on the RIBA stages over time reveals a significant increase in academic interest, particularly after 2010, with a peak around 2020–2022, as shown in Fig. 2. Before 2000, publications were sparse, typically fewer than ten papers per year, but this number grew substantially in the following decades, reaching over fifty at its highest point. Notably, Stage 3 (Concept Design) dominates the research landscape, reflecting its critical role in shaping architectural projects through early spatial planning, generative design, and layout optimisation. In contrast, Stages 4 (Technical Design) and 5 (Manufacturing and Construction), while historically underrepresented, have seen a marked increase post-2015, aligning with the rise of digital fabrication, BIM integration, and AI-driven automation in construction processes. On the other end of the spectrum, Stages 0 (Strategic Definition), 1 (Preparation and Brief), and 7 (Use) remain less explored, despite their potential for machine learning applications in feasibility analysis, ideation and synthesis, and post-occupancy evaluation. The sharp decline in publications after 2022 may indicate a shift in research priorities, external constraints, or dataset limitations. Overall, Fig. 2 highlights that while ML research has largely concentrated on early-stage design decision-making, there is growing recognition of its relevance in later project phases, suggesting an opportunity for further exploration of automation in construction and long-term building performance optimisation.

4.2 (Machine) Learning Approaches

The distribution of (ML) learning approaches across RIBA stages, shown in Fig. 3, reveals a strong concentration in the Concept Design stage (Stage 2), where approaches, particularly Supervised and Reinforcement Learning, are widely applied to tasks such as form generation, spatial planning, and optimisation processes. This dominance suggests that early-stage decision-making in architecture benefits significantly from data-driven methodologies. Stages 4 (Technical Design) and

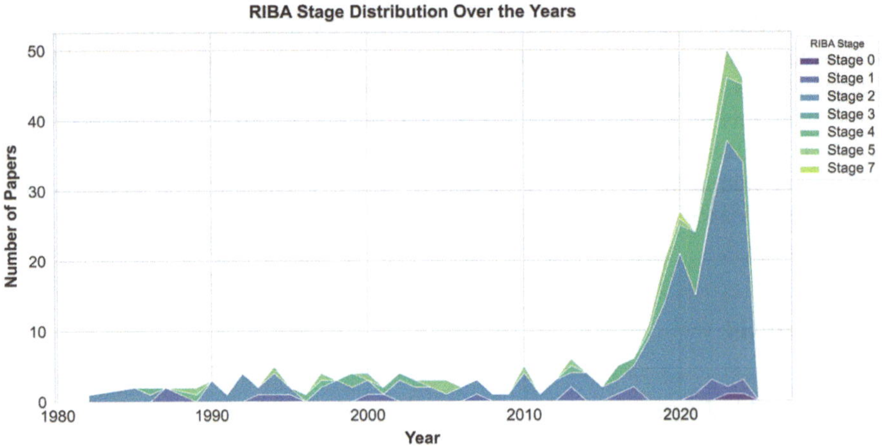

Fig. 2 Mapping of ML-focused research papers on the RIBA plan of work over time

5 (Manufacturing and Construction) also show moderate ML adoption, reflecting the increasing role of computational tools in material selection, structural analysis, and AI-driven automation in construction workflows. However, early-stage strategic planning (Stages 0 and 1) and post-occupancy analysis (Stage 7) remain largely underexplored, likely due to the complexity of integrating ML into feasibility studies and long-term building performance assessment. Additionally, Unsupervised Learning is minimally represented, indicating that clustering and pattern detection techniques are underutilised in architectural design. Figure 3 highlights the concentration of ML research in generative and optimisation-driven processes, while pointing to future opportunities for broader integration in areas such as adaptive reuse, and operational performance monitoring.

From the papers where the algorithm could be inferred, the distribution of machine learning (ML) algorithms across RIBA stages as shown in Fig. 4 reveals a strong concentration in Stage 2 (Concept Design), where Generative Adversarial Networks (GANs) is the most frequently utilised algorithm, appearing 48 times.

This dominance suggests a heavy reliance on generative modelling and data-driven spatial exploration early in the design process. Additionally, Genetic Algorithms (GA) appear 14 times in Stage 2 and are also present in Stage 5 (Manufacturing and Construction), indicates that there is also an interest in optimisation tasks, presumably for spatial planning—in Stage 2—and material selection—later in the process. Convolutional Neural Networks (CNNs), with 12 occurrences in Stage 2, further highlight the use of deep learning for image-based decision-making and automated spatial analysis. In contrast, later design stages, including Stage 4 (Technical Design) and Stage 5 (Manufacturing and Construction), show limited ML applications, primarily leveraging evolutionary algorithms such as Genetic Algorithms and Particle Swarm Optimisation for structural optimisation. Notably, Stages 0 (Strategic Definition) and 7 (Use) exhibit minimal ML adoption, suggesting an underexplored

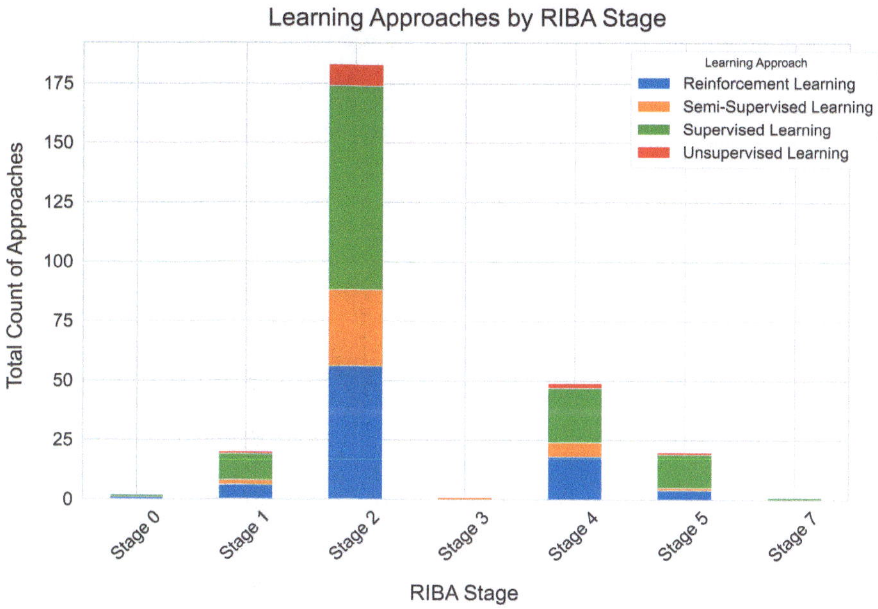

Fig. 3 Distribution of ML approaches for each of the RIBA stages

opportunity for integrating predictive modelling in early feasibility studies and post-occupancy performance assessment, indicating that ML applications in architecture remain heavily concentrated in generative and optimisation-driven processes during conceptual design, with emerging yet relatively sparse applications in technical detailing and construction workflows. This result suggests a growing potential for expanding ML research beyond early-stage generative design into adaptive, lifecycle-based applications that support sustainable and performance-driven architectural decision-making.

4.3 RIBA Stages, Learning Approaches, Artefacts, and Objectives

Figure 5 presents an alluvial diagram illustrating the relationships between RIBA Stages, Learning Approaches, Artefacts, and Objectives in the context of machine learning applications within architectural workflows. The diagram is structured with categories arranged as vertical columns, while the relationships between them are represented by flowing bands across the horizontal (XY) plane. These bands visually trace how machine learning methods are applied across different stages of the architectural process, connecting each RIBA Stage to specific learning approaches, their resulting artefacts, and the corresponding design or operational objectives.

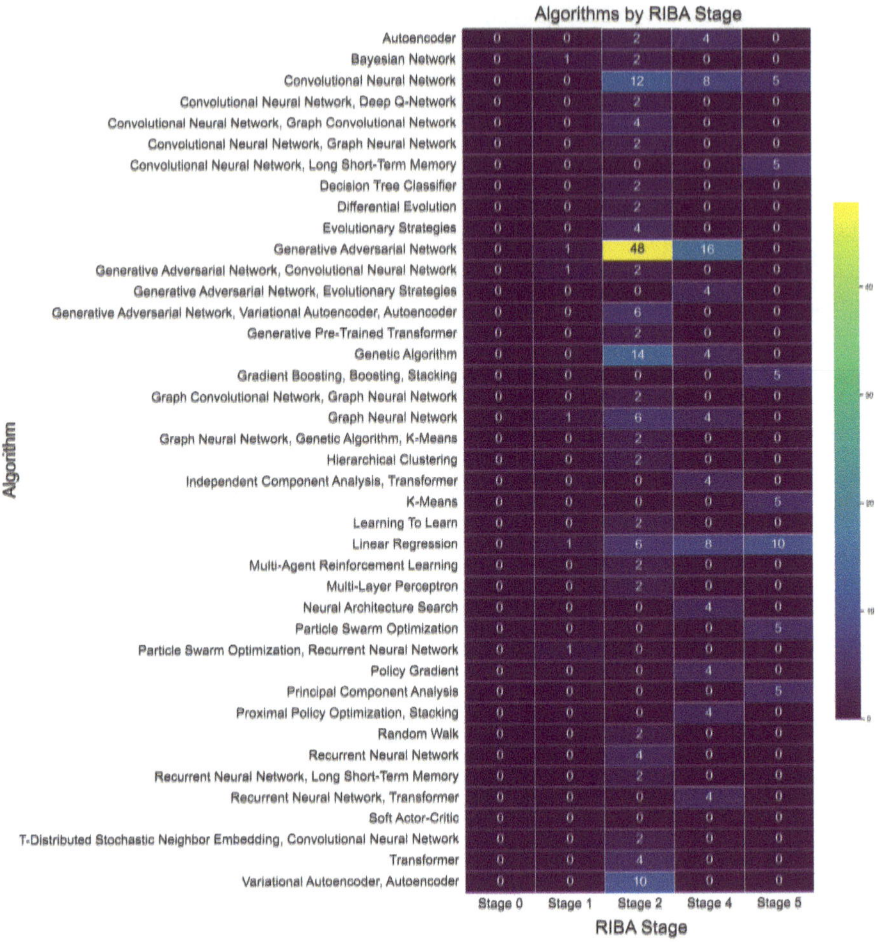

Fig. 4 Use of specific ML algorithms by RIBA stage

The diagram reveals a strong concentration of ML usage during early-stage design, particularly in Stage 2 (Concept Design), where Supervised Learning (136 occurrences) dominates. This suggests that ML is primarily leveraged for predictive modelling, performance evaluation, and spatial planning tools rather than as a generative design methodology. Reinforcement Learning (85 occurrences) and Semi-Supervised Learning (42 occurrences) appear to play a smaller yet significant role, particularly in design exploration and optimisation tasks, while Unsupervised Learning (13 occurrences) remains marginal, indicating limited application of clustering or pattern-recognition methods in architectural workflows. The most prevalent Artefact category is Tool (272 occurrences) highlighting that ML is largely employed as a computational support mechanism, rather than an independent driver of design approaches, with only a minor representation of ML-driven conceptual models. Objectives category emphasises Concept (123 occurrences), Compliance (42 occurrences), and Education (62 occurrences), reinforcing ML's use in spatial planning, regulatory validation, and knowledge dissemination rather than in construction and post-occupancy evaluation. Despite its strong presence in conceptual stages, ML adoption declines in later RIBA stages, with Stage 4 (Technical Design, 57 occurrences) and Stage 5 (Manufacturing & Construction, 23 occurrences) receiving significantly less attention. This suggests that ML-based generative approaches and optimisations are not yet fully integrated into fabrication and execution workflows. Moreover, ML applications in Stage 7 (Use) are nearly absent, pointing to an underexplored opportunity for post-occupancy analysis, predictive maintenance, and adaptive building performance modelling. The findings indicate that current research primarily views ML as a computational augmentation tool rather than a transformative design methodology, underscoring the need for further investigation into its role in construction automation, real-time design optimisation, and lifecycle-based architectural applications. Addressing these gaps could facilitate a more holistic integration of machine learning into architectural workflows, extending beyond early-stage design assistance to construction execution and long-term building performance monitoring.

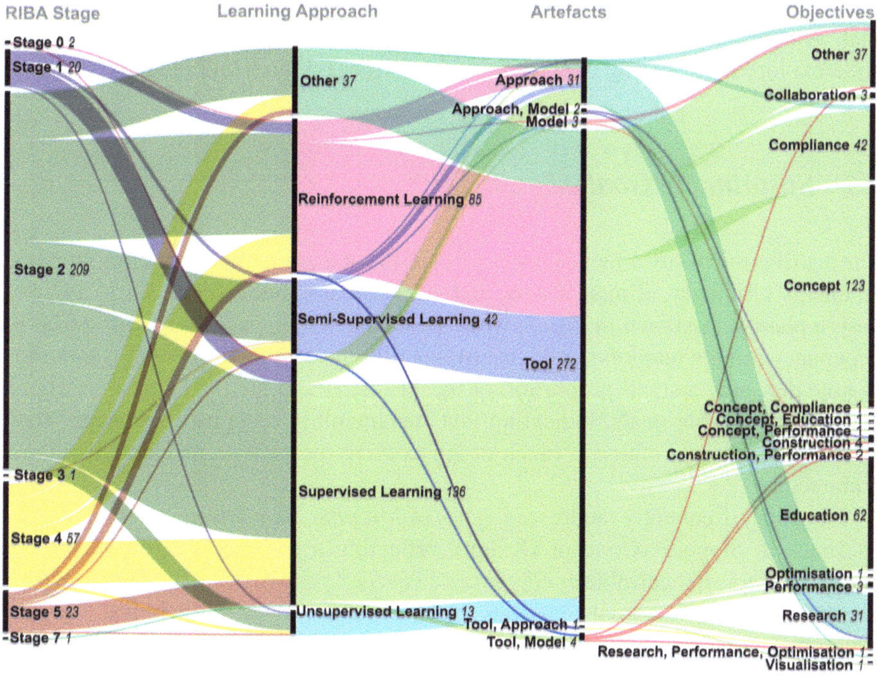

Fig. 5 RIBA stages, learning approaches and objectives

4.4 RIBA Stages, Rule-Based and Machine Learning

The integration of ML with rule-based design approaches is most prevalent in Stage 2 (Concept Design), as shown in Fig. 6, where Parametric Design combined with Supervised Learning emerges as the dominant approach, appearing 26 times. This highlights the widespread use of data-driven parametric modelling for spatial optimisation and form generation, with additional applications in Stage 4 (Technical Design) (8 occurrences), indicating its role in structural analysis and refinement. Similarly, Generative Design leveraging both Supervised and Reinforcement Learning is also concentrated in Stage 2, with 8 and 10 occurrences, respectively, demonstrating the application of ML-driven generative modelling techniques in architectural design. Expert Systems (ES) appear across multiple stages, particularly in Stages 1, 2, and 3, with Reinforcement and Supervised Learning approaches, underscoring their role in rule-based decision-making and compliance verification. Reinforcement Learning also appears in Parametric and Generative Design, reflecting its potential in adaptive, iterative optimisation workflows. However, the integration of ML and rule-based approaches remains largely absent in Stages 5, 6, and 7, indicating a significant research gap in construction automation, robotic fabrication, and post-occupancy performance evaluation, suggesting that while ML-rule-based approaches are well-established in early and mid-design phases, their application in later-stage processes such as material selection, assembly automation, and lifecycle adaptation remains underexplored, presenting opportunities for further research and industry adoption.

4.5 Sustainable Goals

The distribution of sustainable goals across RIBA stages highlights a strong emphasis on energy efficiency, climate action, and infrastructure innovation in early and mid-design phases, as shown in Fig. 7, while later stages such as construction and post-occupancy remain underexplored in sustainability research. However, it is important to note that only a small proportion of the 313 papers in the final dataset explicitly address sustainable goals, indicating that sustainability remains a secondary focus or is not clearly articulated in much of the existing research on machine learning in architecture.

The most frequently addressed goal, *Affordable and Clean Energy*, appears 14 times in Stage 2 (Concept Design), reflecting a growing interest in net-zero energy strategies, renewable energy integration, and passive design approaches at the conceptual stage. This goal continues to be relevant in Stage 4 (Technical Design, 4 occurrences) and Stage 5 (Manufacturing & Construction, 5 occurrences), indicating its importance in both technical detailing and implementation. Similarly, *Climate Action* (8 occurrences in Stage 2 and 4 occurrences in Stage 4) is a key consideration in early design and technical coordination, yet its absence in Stage 5 suggests a research gap in applying climate-responsive strategies during construction.

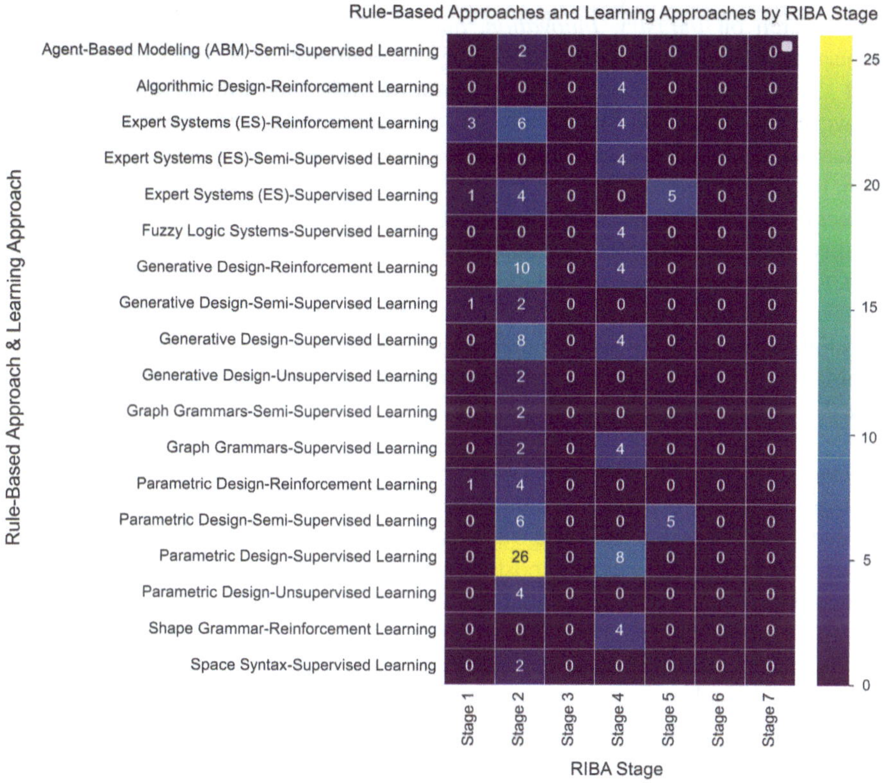

Fig. 6 Mixing of learning approaches and rule-based approaches by RIBA stage

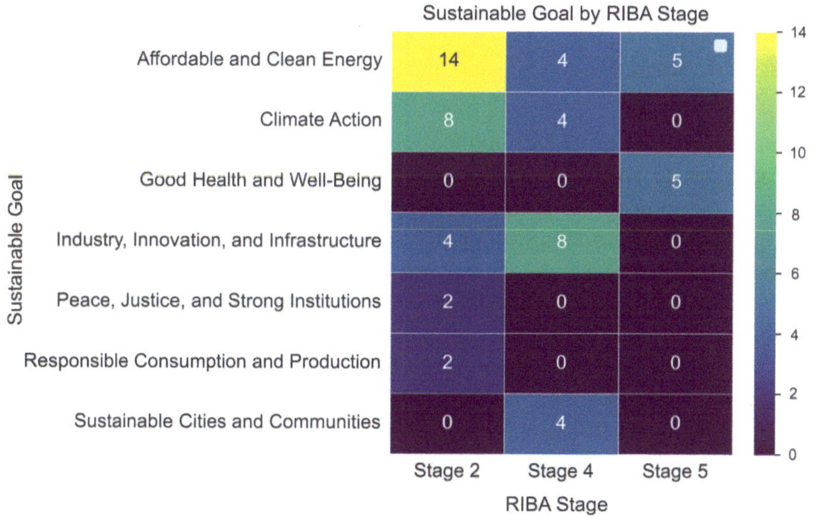

Fig. 7 Sustainable goals by RIBA stage

Research on *Industry, Innovation, and Infrastructure* is also prominent, with 4 occurrences in Stage 2 and 8 in Stage 4, emphasising a focus on technological advancements, innovative construction methods, and infrastructure planning. However, its absence in Stage 5 suggests that while these innovations are explored conceptually and technically, fewer studies focus on their execution during construction. In contrast, *Good Health and Well-Being* is only addressed in Stage 5 (5 occurrences), indicating a focus on post-construction aspects such as indoor environmental quality, occupant comfort, and well-being. Other sustainability goals, including *Peace, Justice, and Strong Institutions* (2 occurrences in Stage 2) and *Responsible Consumption and Production* (2 occurrences in Stage 2), are minimally represented, highlighting a limited focus on social equity, governance, and material sustainability. Similarly, *Sustainable Cities and Communities* is only considered in Stage 4 (4 occurrences), suggesting its relevance in technical coordination and regulatory compliance at the urban scale.

Overall, besides the lack of papers explicit addressing the subject, suggesting that sustainability research in architecture is primarily driven by energy and climate-related concerns during early-stage and technical design phases, while social sustainability, material circularity, and post-construction sustainability remain underrepresented. The absence of sustainable construction research in Stage 5 points to an opportunity for further exploration of life-cycle assessment, circular construction practices, and post-occupancy sustainability monitoring. Expanding research beyond conceptual and technical phases to encompass implementation and long-term performance evaluation could bridge critical gaps in sustainable architectural practice.

5 Discussion

5.1 RIBA Stages

The systematic analysis of ML applications across the RIBA Plan of Work reveals a disproportionate emphasis on Stage 2 (Concept Design). This dominance is not accidental: the conceptual stage invites creative exploration, benefits from fewer regulatory constraints, and is structurally well-suited to the integration of generative and predictive algorithms. As shown in Fig. 2, scholarly output has grown exponentially since 2010, peaking around 2020–2022, with GANs, CNNs, reinforcement learning, and Bayesian neural networks used extensively for early-phase tasks such as spatial generation, classification, and embodied carbon estimation [9–11].

While this focus is understandable, it exposes a problematic imbalance. RIBA Stage 2 has become a sandbox for ML experimentation, but the methods developed rarely mature into tools applicable to later project phases. This limits the transformative potential of ML across the building lifecycle. Sebestyen et al. [12], for example, highlight the capacity of VAEs to support abductive reasoning in early design, but such insights remain siloed in preliminary phases.

Encouragingly, there is evidence of expansion into later stages. Post-2015, research has grown in Stage 4 (Technical Design) and Stage 5 (Manufacturing and Construction), driven by increased interest in digital fabrication, automation, and model-based delivery. Gara et al. [13] employ swarm intelligence for FE model calibration in heritage buildings, while He et al. [14] apply Bayesian neural networks to structural updating—signifying a shift toward real-world implementation.

Nonetheless, Stages 0 and 1 (Strategic Definition and Briefing) and Stage 7 (Use) remain conspicuously underexplored. These stages are essential for performance benchmarking, user feedback, and long-term sustainability, yet receive minimal attention in current ML research. Higgs and Doherty [15] provide a rare exception, examining privacy-preserving ML techniques for sanitary facility usage. However, the scarcity of studies in operational and post-occupancy contexts reveals a critical research gap: the absence of closed-loop feedback mechanisms that integrate building use data into continuous design learning.

This reveals a broader issue—the architectural ML research community is largely oriented toward design ideation and optimisation, often overlooking the full temporal scope of architectural practice. To realise ML's potential, future efforts must prioritise lifecycle integration, long-term performance monitoring, and interoperability across digital tools and project phases. Without this, ML risks remaining a proof-of-concept technology, rather than an embedded architectural methodology.

The observed overrepresentation of RIBA Stage 2 in ML research not only highlights the fascination of conceptual design for algorithmic exploration but also reflects a deeper tension within architectural practice one that Bernstein [16] frames as the evolving interplay between human creativity, procedural automation, and the need for integrative thinking across the project lifecycle. While Stage 2 offers a low-barrier context for testing generative and predictive systems, such as GANs or reinforcement models, their confinement to early-stage ideation risks marginalising the broader transformative potential of AI in architecture. Bernstein's notion of "projective responsibilities" underscores this gap: the role of the architect is not merely to generate options, but to foresee, integrate, and coordinate across disciplinary systems and temporal scales. ML, if siloed in Stage 2, remains a sandbox tool; to become what Bernstein calls the "glue code" of architectural knowledge, it must evolve into a connective tissue that spans design, delivery, and use. This requires a strategic rethinking of data infrastructure, as well as the interoperability of digital platforms to support AI-enabled feedback loops—especially in underexplored phases like briefing and post-occupancy. As empiricist AI systems inch toward cognitive capabilities, the architectural profession faces a pivotal moment: either embrace ML as a lifecycle-integrated collaborator that augments human foresight, or risk relegating it to a domain of isolated optimisation divorced from the architect's ethical and systemic responsibilities. Bernstein's provocation is clear—AI must be guided not by what is computationally convenient, but by what is architecturally necessary [16].

5.2 *Machine Learning Approaches*

Supervised learning continues to dominate architectural ML research due to its compatibility with labelled datasets and its capacity to deliver predictable, measurable outcomes. Applications range from spatial classification [17] to carbon estimation [11] and perceptual modelling using omnidirectional imagery [18]. However, the prevalence of supervised methods often reflects a dependency on data-rich environments, potentially excluding design challenges where data are sparse or highly variable.

Reinforcement learning introduces a more interactive and adaptive paradigm, enabling real-time design feedback and iterative improvement. Akizuki et al. [10] and Wu et al. [19] show how RL can be embedded into generative design workflows and performance-driven optimisation. Yet, these efforts remain limited to early design stages, and rarely extend into dynamic decision-making environments such as construction coordination or building operations.

More critically, the underutilisation of unsupervised and semi-supervised methods underscores a reluctance to embrace ambiguity and discovery in architectural ML. Tools such as VAEs, as deployed by Sebestyen et al. [12], offer pathways to abductive reasoning and conceptual exploration, yet these approaches are often sidelined in favour of more deterministic frameworks. Bielski et al. [20] argue for greater transparency and explainability, positioning semi-supervised learning as a middle ground between designer agency and algorithmic inference.

This imbalance has methodological consequences. Overreliance on supervised learning can lead to reductive design solutions, where ML models reinforce known patterns rather than provoke new ones. Unsupervised approaches—though harder to validate—offer opportunities for unexpected insight and generative creativity, especially when combined with rule-based or parametric frameworks. A more balanced research agenda would embrace hybrid models that combine the predictability of supervision with the exploratory power of unsupervised methods.

Ultimately, the choice of learning approach should reflect not only technical feasibility but also design intent, ethical transparency, and the pedagogical need to cultivate designers who can navigate, critique, and creatively engage with intelligent systems.

5.3 *Rule-Based Approaches*

The integration of ML with rule-based design systems represents a critical intersection between automation and domain expertise in architectural workflows. This hybridisation is most prevalent in RIBA Stage 2 (Concept Design), where parametric and generative design methods provide a fertile ground for data-driven exploration. As shown in Fig. 6, Supervised and Reinforcement Learning are frequently coupled with parametric frameworks to enable adaptive form generation, spatial optimisation, and constraint-driven iteration.

This dominance reflects a broader epistemological trend: the conceptual stage is where design rules—whether geometric, programmatic, or environmental—are most fluid, and where ML can enhance rather than replace creative agency. However, the reliance on predefined rule sets also limits the generalisability of these approaches. Many applications assume static constraint environments, reducing their utility in later stages where rules are dynamic, regulatory, or construction-dependent.

Stages 4 (Technical Design) and 5 (Manufacturing and Construction) show early signs of hybrid adoption, particularly through expert systems and evolutionary algorithms embedded in structural analysis and fabrication logic. Yet the coupling of ML with rule-based systems in these contexts remains rudimentary, often lacking interoperability with BIM standards or real-time construction data. This disconnect inhibits lifecycle integration and reinforces silos between design ideation and execution.

Critically, the near absence of hybrid models in Stages 6 and 7 (Handover and Use) represents a missed opportunity. These stages are rich with data—from as-built models to occupancy behaviour—that could inform adaptive rule systems. For example, integrating post-occupancy feedback into generative models could close the loop between design intent and building performance, turning static rules into evolving heuristics.

The challenge moving forward lies not only in technical implementation, but in rethinking the nature of rules themselves. Rather than fixed parameters, rules in hybrid ML systems should be treated as soft constraints—responsive to context, updated by feedback, and negotiated by both human and artificial agents. Achieving these demands more than parametric scripting; it requires robust data infrastructures, transparent inference mechanisms, and a redefinition of authorship in computational design. The fusion of ML and rule-based design is not merely a technical convenience but a methodological imperative. It offers a path toward intelligent systems that are not only generative but contextually aware, traceable, and ethically grounded across all RIBA stages.

5.4 ML Algorithms

The algorithmic landscape in architectural ML research is dominated by GANs, CNNs, and GA's, particularly within Stage 2 (Concept Design). GANs are widely adopted for spatial form generation and topological variation [9, 21], enabling rapid prototyping of diverse architectural layouts with minimal human input. CNNs, by contrast, support automated pattern recognition and classification, as seen in facade segmentation. These applications reflect a trend toward leveraging high-capacity models to automate visual and spatial reasoning tasks once considered subjective or manual.

Evolutionary algorithms, including GA's and swarm intelligence, are increasingly applied in later stages of the RIBA workflow, such as structural optimisation and material calibration [13]. Their strength lies in navigating large, multi-objective search spaces where traditional optimisation methods fall short. However, their computational expense and reliance on predefined fitness functions often limit adaptability to real-time design workflows or changing project constraints.

A more recent and significant development is the emergence of Bayesian Neural Networks (BNNs), which offer a probabilistic framework for dealing with uncertainty in architectural decision-making. Applications such as early-stage embodied carbon estimation [11] and structural model updating [14] exemplify the potential of BNNs to encode uncertainty, rather than obscure it. This represents a methodological shift: from deterministic optimisation toward risk-informed design, where ML models assist in quantifying confidence levels and supporting informed trade-offs.

Despite these advancements, algorithm deployment remains uneven across the RIBA lifecycle. Most applications are confined to design ideation, with limited translation into construction, post-occupancy evaluation, or adaptive reuse scenarios. Moreover, the black-box nature of many advanced models—especially GANs and deep CNNs—raises concerns about transparency and traceability, particularly in regulatory or collaborative environments.

Future research must go beyond showcasing algorithmic novelty to critically assess implementation viability. This includes improving model explainability, integrating with BIM and simulation tools, and aligning algorithmic outcomes with stakeholder needs across disciplines. Only then can ML algorithms mature from experimental prototypes into reliable, accountable components of the architectural design process.

5.5 Sustainable Goals

Despite the urgency of environmental challenges, only a limited subset of the reviewed literature explicitly aligns with the United Nations Sustainable Development Goals (SDGs). While themes like Affordable and Clean Energy, Climate Action, and Industry Innovation dominate the sustainability discourse—particularly in Stages 2 and 4 [11, 19, 22]—this focus reveals both a thematic narrowness and a missed opportunity for comprehensive impact.

What exists is promising but fragmented. Turhan [22] exemplifies how integrating life cycle assessment (LCA) with large language models can evaluate unconventional materials. Similarly, Akbar et al. [23] and Yousif et al. [24] demonstrate how machine learning can decode the behaviour of adaptive materials like wood and mycelium. These hybrid workflows point to a future where ML enables deeper material intelligence, yet their implementation remains niche.

The prevailing issue is not the absence of sustainability intent but its implicitness. Many studies reference energy or carbon metrics, yet rarely frame their contributions in terms of SDG alignment. Moreover, few projects track sustainability objectives across multiple RIBA stages, leaving lifecycle assessments disjointed. This undermines the potential of ML to act as a continuous evaluative and generative tool for sustainable design.

Equally, lacking is a commitment to social sustainability. Goals such as Responsible Consumption, Equity, or Well-being are rarely addressed through ML applications. This suggests a technocratic framing of sustainability, overly focused on quantifiable outputs while neglecting user-centred and systemic metrics. Such omissions are not merely academic—they risk perpetuating unsustainable practices under the guise of efficiency.

Future research should move beyond treating sustainability as a technical constraint and instead embed it as a multidimensional design driver. This requires formalising SDG metrics in ML workflows, expanding focus to underrepresented goals, and embracing qualitative measures such as occupant experience and material provenance. Only then can machine learning meaningfully contribute to holistic and equitable sustainable architecture.

5.6 Opportunities, Gaps, and Risks

The current landscape of ML in architecture is marked by a paradox: while early-stage design enjoys extensive ML integration, the remainder of the architectural lifecycle is comparatively neglected. This imbalance presents both a challenge and an opportunity. Underrepresented stages—such as Stage 0 (Strategic Definition), Stage 6 (Handover), and Stage 7 (Use)—are critical for aligning design intent with user experience, operational performance, and post-occupancy adaptation. The absence of ML tools in these phases undermines efforts to create intelligent, responsive, and resilient buildings.

Reinforcement learning and agent-based models offer promising avenues to address this gap. Zhang et al. [25] illustrate how multi-agent reinforcement learning can optimise spatial configurations in renovation contexts. However, such methods remain largely experimental, with limited uptake in mainstream workflows. This highlights a persistent barrier: the translation of innovative algorithms into practical, interoperable tools for architectural practice.

Equally, pressing are the epistemic risks posed by black-box models. Calixto and Croffi [26] warn that the increasing opacity of AI-driven tools can erode designer agency, obscuring the rationale behind computational decisions. This concern is particularly salient in educational settings, where intuitive interfaces may inadvertently reduce critical engagement with algorithmic logic. Without adequate transparency, ML risks becoming an authoritarian tool rather than a collaborative partner.

Addressing these concerns requires a multipronged strategy. First, explainable AI must be prioritised, especially in design contexts that value traceability and interpretability. Second, pedagogical frameworks should empower architects to interrogate, adapt, and co-create with intelligent systems. Third, ethical integration of ML should extend beyond bias mitigation to include questions of authorship, accountability, and design autonomy. Finally, ML research must move toward lifecycle implementation—embedding adaptive intelligence from ideation through construction, handover, and building use.

In reframing ML not as a surrogate for design intelligence but as an augmentative agent, the architectural discipline can better harness its creative and operational potential. Realising this vision demands critical reflection, institutional support, and a commitment to designing with, rather than merely through, intelligent machines.

6 Conclusion

This paper set out to address a critical gap in architectural research: the lack of a comprehensive understanding of how ML integrates across the full architectural design process. While ML has rapidly permeated the conceptual design phase—driven by the availability of data and the generative flexibility of early-stage design—it remains largely siloed within Stage 2 of the RIBA Plan of Work. This concentrated focus has fostered innovation in generative modelling, form exploration, and spatial classification, yet it has also reinforced a narrow conception of ML's role in architecture: one that privileges ideation over implementation, and optimisation over integration.

Through a systematic review, we have revealed significant disparities in how ML is applied across RIBA stages, with late-stage processes such as technical detailing, construction, handover, and post-occupancy use receiving minimal attention. These gaps undermine the potential for ML to contribute to adaptive building performance, lifecycle sustainability, and user-centred feedback loops. Despite emerging research in structural updating, fabrication, and agent-based renovation systems, these innovations remain isolated, lacking systemic integration into architectural workflows.

Moreover, the dominance of supervised learning reflects a methodological bias toward measurable, labelled data at the expense of exploratory, abductive reasoning. Unsupervised and semi-supervised methods—though better suited to ambiguous, open-ended design challenges—are notably underrepresented. This imbalance not only restricts innovation, but also risks reducing architectural creativity to pattern recognition rather than critical generation.

Sustainability, while a recurring theme, is inconsistently operationalised. Few studies frame their contributions within the UN SDGs, and even fewer apply these goals across multiple project phases. This reflects a technocratic view of sustainability, focused on energy and materials, rather than a holistic framework encompassing social equity, ethical material sourcing, and long-term well-being.

The risks posed by opaque, black-box systems further compound these issues. Without explainability and critical pedagogical grounding, ML can displace rather than augment design agency. As intelligent systems become more embedded in architectural education and practice, transparency, traceability, and ethical reflexivity must become core design values.

To fulfil ML's transformative potential in architecture, future research must move beyond technical novelty and toward systemic, lifecycle-oriented integration. This means embedding ML into feedback-rich design processes, enhancing interoperability with BIM and fabrication systems, and aligning computational tools with sustainability metrics and design ethics. Machine learning must evolve from a task-specific optimiser into a design collaborator—one that amplifies creativity, supports informed decision-making, and reinforces the social and environmental responsibilities of architecture.

In short, ML should not be confined to generating architectural form; it should help architects rethink form, meaning, performance, and impact—across time, scale, and stakeholder. This demands a recalibration of the discipline: one that positions machine learning not as a final answer, but as a critical question within the broader project of architectural intelligence.

Acknowledgements This research is supported by Building 4.0 CRC. The support of the Commonwealth of Australia through the Cooperative Research Centre Programme is acknowledged. The authors would also like to acknowledge Lendlease Digital.

References

1. Royal Institute of British Architects (2020) Riba plan of work 2020. https://www.architecture.com/knowledge-and-resources/resources-landing-page/riba-plan-of-work
2. Mitchell TM (1997) Machine learning, 1st edn. McGraw-Hill, New York, URL https://www.cs.cmu.edu/~tom/mlbook.html
3. Chapelle O, Sch¨olkopf B, Zien A (2006) Semi-supervised learning, 1st edn. MIT Press, Cambridge, MA. https://doi.org/10.7551/mitpress/9780262033589.001.0001
4. Sutton RS, Barto AG (1998) Reinforcement learning: an introduction, 1st edn. MIT Press, Cambridge, MA. https://mitpress.mit.edu/9780262193986/reinforcement-learning/
5. Zwierzycki M (2020) On AI adoption issues in architectural design, pp 515–524. https://doi.org/10.52842/conf.ecaade.2020.1.515
6. Gero JS (1991) Ten problems for AI in design. In: Amarel S, Despain A, Nii P, et al. (eds) Proceedings of the Twelfth International Joint Conference on Artificial Intelligence (IJCAI-91), Workshop on AI in Design. Morgan Kaufmann, San Mateo, CA, p unpaginated. https://papers.cumincad.org/data/works/att/46ce.content.04041.pdf
7. Topuz B, Çakici Alp N (2023) Machine learning in architecture. Autom Constr 154:105012. https://doi.org/10.1016/j.autcon.2023.105012. https://linkinghub.elsevier.com/retrieve/pii/S0926580523002728
8. Rhee J, Veloso P, Krishnamurti R (2023) Three decades of machine learning with neural networks in computer-aided architectural design (1990–2021). Des Sci 9:e25. https://www.cambridge.org/core/product/identifier/S2053470123000215/type/journalarticle
9. Zheng H, An K, Wei J et al. (2020) Apartment floor plans generation via generative adversarial networks. In: Proceedings of the 25th International Conference on Computer-Aided Architectural Design Research in Asia (CAADRIA 2020), Bangkok, Thailand, pp 599–608. https://doi.org/10.52842/conf.caadria.2020.2.599

10. Akizuki Y, Bernhard M, Kakooee R, et al. (2020) Generative modelling with design constraints—reinforcement learning for object generation. In: Proceedings of the 25th International Conference of the Association for Computer-Aided Architectural Design Research in Asia (CAADRIA). CAADRIA, Bangkok, Thailand, pp 445–454, https://doi.org/10.52842/conf.caadria.2020.1.445.. http://papers.cumincad.org/cgi-bin/works/paper/caadria2020306
11. Budig M, Heckmann O, Boon ANQ, et al. (2020) Data-driven embodied carbon evaluation of early building design iterations. In Holzer D, Nakapan W, Globa A et al. (eds) RE: anthropocene, design in the age of humans—proceedings of the 25th international conference on computer-aided architectural design research in Asia (CAADRIA 2020), vol 2. The Association for ComputerAided Architectural Design Research in Asia (CAADRIA), pp 303–312. https://doi.org/10.52842/conf.caadria.2020.2.303
12. Sebestyen A, Rock J, Hirschberg UL (2021) Towards abductive reasoning-based computational design tools—using machine learning as a way to explore the combined design spaces of multiple parametric models. In: Proceedings of the eCAADe 2021 Conference, Novi Sad, Serbia, pp 141–150. 10.52842/conf.ecaade.2021.1.141
13. Gara F, Nicoletti V, Arezzo D, et al. (2025) Model updating of cultural heritage buildings through swarm intelligence algorithms. Intern J Archit Herit 19(2):259–275. https://doi.org/10.1080/15583058.2023.2277324
14. He Y, Yin T, Wang X (2025) Design of multi-hidden-layer bayesian neural networks for model updating. Adv Struct Eng 28(5):952–972. https://journals.sagepub.com/doi/https://doi.org/10.1177/13694332241298020
15. Higgs B, Doherty B (2020) Sanitary sanity: Evaluating privacy preserving machine learning methods for post-occupancy evaluation. In: Proceedings of the 25th International Conference on Computer-Aided Architectural Design Research in Asia (CAADRIA 2020), Bangkok, Thailand, pp 697–706. https://doi.org/10.52842/conf.caadria.2020.2.697
16. Bernstein P (2022) Machine learning: architecture in the age of artificial intelligence. Riba Publishing
17. Eisenstadt V, Arora H, Ziegler C, et al. (2021) Exploring optimal ways to represent topological and spatial features of building designs in deep learning methods and applications for architecture. In: Proceedings of the 26th International Conference on Computer-Aided Architectural Design Research in Asia (CAADRIA 2021), Hong Kong, pp 191–200. https://doi.org/10.52842/conf.caadria.2021.1.191
18. Kinugawa H, Takizawa A (2019) Deep learning model for predicting preference of space by estimating the depth information of space using omnidirectional images. Proceedings of the 37th eCAADe and 23rd SIGraDi Joint Conference, pp 61–68. https://doi.org/10.5151/proceedings-ecaadesigradi2019339
19. Wu Z, Li M, Liu W, et al (2024) A Data-Driven Model for Sustainable Performance Prediction of Residential Block Layout Design Using Graph Neural Network. In: Proceedings of the eCAADe 2024 Conference, Nicosia, pp 575–584. https://doi.org/10.52842/conf.ecaade.2024.1.575
20. Bielski J, Langenhan C, Ziegler C, et al. (2022) The what, why, what-if and how-to for designing architecture, explainability for auto-completion of computer-aided architectural design of floor plan layouting during the early design stages. In: Proceedings of the 27th International Conference on Computer-Aided Architectural Design Research in Asia (CAADRIA 2022), Sydney, Australia, pp 435–444. https://doi.org/10.52842/conf.caadria.2022.2.435
21. Yang X, Bao DW, Yan X, et al. (2022) Optigan: topological optimization in design form-finding with conditional gans. In: Proceedings of the 27th International Conference on Computer-Aided Architectural Design Research in Asia (CAADRIA 2022), Bangkok, Thailand, pp 121–130. http://papers.cumincad.org/cgi-bin/works/paper/caadria2022121
22. Turhan GD (2023) Life cycle assessment for the unconventional construction materials in collaboration with a large language model. In Proceedings of the eCAADe 2023 Conference, Graz, Austria, pp 39–48. https://doi.org/10.52842/conf.ecaade.2023.2.039
23. Akbar Z, Wood D, Kiesewetter L, et al. (2022) A data-driven workflow for modelling self-shaping wood bilayer, utilizing natural material variations with machine vision and machine

learning. In: Proceedings of the 27th International Conference on Computer-Aided Architectural Design Research in Asia (CAADRIA 2022), Sydney, Australia, pp 393–402. https://doi.org/10.52842/conf.caadria.2022.1.393
24. Yousif S, Akyuz Y, Giffoni L, et al. (2023) Incorporating physical experimentation into creative dl-driven design space exploration. In: Proceedings of the 28th International Conference of the Association for Computer-Aided Architectural Design Research in Asia (CAADRIA), Ahmedabad, India, pp 99–108. https://doi.org/10.52842/conf.caadria.2023.1.099. http://papers.cumincad.org/cgi-bin/works/paper/caadria2023197
25. Zhang Z, Guo Z, Zheng H, et al. (2024) Automated architectural spatial composition via multi-agent deep reinforcement learning for building renovation. Autom Constr 167:105702. https://doi.org/10.1016/j.autcon.2024.105702. https://linkinghub.elsevier.com/retrieve/pii/S0926580524004382
26. Calixto V, Croffi J (2024) Back to black boxes? An urgent call for discussing the impacts of the emergent ai-driven tools in the architecture design education. In: Proceedings of the 29th CAADRIA Conference, Singapore, pp 39–48. https://doi.org/10.52842/conf.caadria.2024.3.039

Democratizing Advanced AI Methods in Post-participation: From Citizen Data to Decision-Making

Cem Ataman, Simon Perrault, and Bige Tunçer

Abstract Digital participation has transformed how communities engage in urban decision-making, generating vast amounts of citizen-generated data. Despite the potential for inclusive governance, the scale and unstructured nature of these inputs pose substantial analytical challenges. Traditional qualitative and quantitative methods often prove inadequate for managing large textual datasets, leading to incomplete or biased interpretations of public discourse. Emerging AI-driven techniques, such as topic modeling, natural language processing, and large language models, address this complexity by preserving contextual depth. However, interdisciplinary barriers, concerns about transparency, and oversimplification limit the effective incorporation of these techniques into post-participation. This chapter introduces a framework for democratizing advanced AI methods, bridging the divide between computational efficiency and qualitative depth. Emphasizing the user-centric design of a digital participation data analysis toolkit, the chapter demonstrates how AI-driven insights can be made accessible to stakeholders with diverse disciplinary backgrounds. The framework supports structured data preprocessing, multi-layered visualization, and iterative loops, ensuring that citizen inputs inform evidence-based decision-making. Drawing on three contrasting urban contexts, Singapore, Madrid, and Hamburg, this chapter illustrates the transformative impact of inclusive AI analytics on real-world governance and presents how responsible integration of AI methods can reinforce digital participation, linking complex data analysis with practical urban applications.

C. Ataman (✉)
Urbanism and Urban Architecture, Eindhoven University of Technology, Eindhoven, Netherlands
e-mail: c.ataman@tue.nl

S. Perrault
Information Systems Technology and Design, Singapore University of Technology and Design, Singapore, Singapore
e-mail: simon_perrault@sutd.edu.sg

B. Tunçer
Information Systems in the Built Environment, Eindhoven University of Technology, Eindhoven, Netherlands
e-mail: b.tuncer@tue.nl

M. Sedrez and G. Pitts (eds.), *Artificial Intelligence-Aided Design for Sustainability*, Urban Sustainability, https://doi.org/10.1007/978-981-95-1349-9_10

Keywords Citizen engagement · Digital tools · Post-participation · Data analytics · Urban governance · Explainable AI (XAI)

1 Introduction

The rapid expansion of digital participation has generated an unprecedented volume of citizen-generated data in urban decision-making processes [1]. Cities worldwide increasingly rely on online platforms to engage communities, host public debates, and gather feedback on policy proposals. In this chapter, we focus on digital participation processes in which citizens contribute feedback, ideas, or concerns through structured online platforms, open forums, or consultation tools. These inputs are typically collected in natural language form and are shaped by local contexts, institutional frameworks, and the design of the participation mechanism. While we do not describe the participation process in full detail, we treat these citizen-generated inputs as a form of distributed contextual knowledge. The expert's role in this setting is not to replace citizen insight, but to help synthesize, interpret, and operationalize it, translating distributed public knowledge into data-informed strategies that support more inclusive urban decision-making.

However, large-scale discussions often unfold in unstructured textual formats, making it difficult to derive insights systematically [2]. In response, scholars and practitioners have turned to advanced AI-driven methodologies and computational techniques, such as topic modeling, Natural Language Processing (NLP), and Large Language Models (LLMs), to manage this complexity [3–5]. Yet, despite the promise of enhanced analytical rigor, the integration of AI methods into post-participation stages is fraught with challenges related to data quality, interpretative nuance, interdisciplinary collaboration, and the sheer scale of textual content, all of which can undermine democratic access to these tools.

Existing research highlights two broad concerns. First, conventional methods, ranging from manual discourse analysis to word-frequency and co-occurrence mapping, often struggle to handle the volume and intricacy of online public discourse [6]. While these approaches offer valuable qualitative perspectives, they become prohibitively time-consuming in cases involving thousands of contributions. Second, although emerging AI-driven tools address issues of scale, they risk reducing interpretative depth when deployed as "black-box" systems [5], whose internal logic is not easily understood or questioned by users, limiting transparency and critical engagement with the results. In such cases, even when algorithms generate plausible outputs, their lack of transparency can hinder critical interpretation and create misplaced trust in automated results. Researchers have thus emphasized the importance of integrating human-centered approaches that retain contextual depth, while ensuring transparency and accountability in algorithmic outcomes. In this context, the quest to democratize advanced AI methods remains vital for realizing inclusive, evidence-based urban interventions.

The approach presented in this chapter rests on the premise that bridging computational efficiency with contextual depth is essential to fully leverage citizen input in urban design and planning. By examining current shortcomings in post-participation, the stage that transforms raw citizen input into actionable insights, this work demonstrates how AI-driven techniques can be employed responsibly, enhancing both the quality and applicability of participation data in decision-making. In particular, this chapter focuses on empowering stakeholders who lack specialized domain knowledge, thereby maintaining a balance between quantitative rigor and qualitative interpretation. Such a balance supports more transparent and equitable urban governance. Nonetheless, a key question persists: How can AI be leveraged to handle large-scale textual data without sacrificing the inclusion of all stakeholders? In this context, inclusion refers not only to access but also to the meaningful representation of diverse voices and perspectives, which is an aim closely aligned with ongoing global commitments and the goals of SGD 11, which advocates for inclusive, participatory, and sustainable urban development.

This chapter addresses that question by introducing the Digital Participation Data Analysis Toolkit (DIP-DAT) that unifies AI-driven methodologies and data visualization techniques within a single web-based platform. Drawing on multiple urban contexts, Singapore, Madrid, and Hamburg, DIP-DAT underscores the importance of interdisciplinary collaboration and stakeholder inclusivity. It seeks to reduce subjectivity through criteria-based selections and data-driven benchmarks, thereby supporting more robust AI-driven post-participation workflows.

The remainder of this chapter is organized as follows. The next section reviews the limitations of existing analytical methods in large-scale citizen engagement and explains the post-participation dilemma. It then examines the evolving role of AI, highlighting key methodological advancements and barriers to real-world implementation. Subsequently, the chapter proposes a conceptual framework that links complexity to usability, emphasizing visualization as a vital communicative bridge. Afterward, DIP-DAT is introduced, with an overview of its objectives and applications. Finally, the chapter concludes with lessons learned and practical recommendations aimed at guiding future efforts to democratize AI-driven data analysis in post-participation.

2 Understanding the Gap in Citizen Engagement Analysis

2.1 Key Techniques in Large-Scale Discussion Analysis

Citizen participation is increasingly recognized as a foundational element of sustainable urbanism. Beyond their normative value in promoting democratic governance, participatory processes enable the co-creation of contextually grounded solutions that reflect the lived experiences and priorities of local communities. Accordingly, the analytical methods used to interpret citizen input directly influence how effectively participatory data can inform resilient, inclusive, and sustainable urban strategies.

With the expansion of digital participation, various data collection methods have been developed to extract meaningful insights from citizen engagement. Structured interfaces such as polls, proposal feedback mechanisms, and systems enabling public support through likes and dislikes offer straightforward interaction modes. However, the most prevalent and dynamic medium remains online debate platforms, where participants engage in asynchronous and often anonymous discussions in natural language [2]. Analyzing these discussions requires techniques such as thematic analysis, which identifies recurring topics by categorizing discussions into dominant themes, and discourse analysis, which examines argument structures and the interplay of perspectives [7, 8]. While these methods provide valuable insights into overarching concerns, power dynamics, and narrative construction, they are highly labor-intensive and difficult to scale for large datasets. Their reliance on close reading renders them impractical for efficiently processing thousands of contributions.

To address the scalability limitations of qualitative methods, quantitative approaches such as word frequency analysis and co-occurrence mapping offer alternative strategies. These methods identify frequently used terms and their interrelationships, providing a macroscopic view of discussion patterns [9]. However, they fail to capture the semantic depth and contextual nuances of citizen discourse. Similarly, network analysis maps connections between ideas and participants, revealing influential voices and emergent discourse clusters [10]. While effective in uncovering relational structures, this approach remains limited in extracting deeper, context-dependent insights from textual data.

Consequently, while online discussion platforms and digital participation tools generate rich public discourse, they also introduce significant analytical challenges. The sheer volume, complexity, and fluidity of textual data exceed the capabilities of traditional methods, necessitating more advanced computational approaches. Conventional discourse analysis techniques, though well-established, were not originally designed to process the scale, dynamism, and contextual variability of digital participation data. As participatory urban design and planning increasingly rely on large-scale citizen input, refining these methodologies is crucial in the post-participation phase. AI-driven processing and analysis techniques offer promising advancements, enhancing analytical capabilities and ensuring that citizen discourse can be systematically integrated into evidence-based urban decision-making.

2.2 *The Post-Participation Dilemma*

Since traditional analytical techniques are often inadequate for processing citizen engagement data in digital mediums, the post-participation phase in urban decision-making faces a critical challenge: transforming vast volumes of citizen-generated discourse into meaningful and actionable insights. Digital participation platforms generate extensive textual datasets, encapsulating a wide range of public concerns, interests, and perspectives. However, the sheer scale and complexity of this data pose substantial barriers to its effective interpretation and integration into urban planning and design processes [3]. The inability of conventional methods to capture intricate semantic relationships within large-scale textual data restricts their capacity to extract meaningful insights that inform evidence-based decision-making. Consequently, valuable citizen contributions risk being underutilized, creating a disconnect between participatory engagement and subsequent policy implementation.

Compounding this challenge is the diversity of stakeholders involved in digital participation, each approaching the data from distinct disciplinary perspectives. Architects, urban designers, planners, and policymakers engage with participatory data differently; architects may focus on spatial design implications, while policymakers prioritize governance strategies [1]. This variation underscores the need for analytical frameworks capable of accommodating multiple conceptual perspectives while maintaining the interpretability and practical applicability of citizen inputs. Without a structured post-participation process, participatory data remains fragmented, hindering its capacity to inform comprehensive urban interventions.

Thus, aligning citizen-derived knowledge with systematic decision-making processes has become a central objective of the post-participation phase. Ensuring data integrity and analytical rigor is essential for fostering inclusive and responsive urban governance. However, even with robust data management, a fundamental dilemma persists: while large-scale citizen input holds immense potential for shaping planning and design interventions, ensuring its interpretability and cross-disciplinary applicability remains an enduring challenge. Addressing this issue necessitates methodological advancements at the intersection of AI and urbanism, facilitating a more effective bridge between participatory discourse and evidence-based decision-making. Multi-layered information frameworks, designed as dynamic and data-driven mechanisms, provide a structured approach for translating complex participatory insights into actionable strategies, ultimately enhancing the responsiveness and inclusivity of urban design and planning. In this sense, participation extends beyond a procedural step; it contributes to the adaptive capacity and long-term resilience of urban systems. When citizen insights are systematically integrated into design and planning processes, they foster greater responsiveness to local needs and support more context-sensitive, sustainable urban development strategies.

3 Evolving Role of AI in Post-Participation

3.1 Topic Modeling, Natural Language Processing, and Large Language Models

To address the limitations of traditional methods, AI-driven methodologies have emerged as a pivotal force in post-participation, providing scalable and computationally efficient solutions for processing and analyzing large-scale textual data [11].

Among the most advanced techniques are Natural Language Processing (NLP), topic modeling, and Large Language Models (LLMs), each contributing distinct analytical capabilities that enhance both qualitative and quantitative interpretations of citizen discourse. These approaches collectively facilitate a deeper understanding of participation inputs, enabling more systematic and data-driven decision-making in urban design and planning.

NLP has evolved into a specialized subfield of AI, playing an essential role in digital participation within urban design and planning [12]. It enhances post-participation by extracting semantic structures, sentiment orientations, and interrelations among key concepts in citizen discussions. Deployed primarily for data preprocessing, cleaning, and substantive analysis, NLP leverages advanced computational linguistic techniques to conduct dependency parsing and sentiment detection. These capabilities allow stakeholders to discern nuanced variations in public sentiment, identify recurring policy concerns, and map key actors within participatory dialogues [13]. Furthermore, NLP facilitates multilingual processing, ensuring that diverse linguistic contributions are effectively analyzed, thereby promoting a more inclusive and representative participatory process.

Complementing NLP, topic modeling serves as a critical unsupervised machine learning technique for structuring large-scale discussions by identifying latent thematic patterns in unstructured textual data. Unlike traditional thematic coding, which relies on predefined categories, topic modeling dynamically groups textual data into coherent thematic clusters, making it particularly suited for exploratory analysis [14]. Methods such as Latent Dirichlet Allocation (LDA), Latent Semantic Indexing (LSI), and Hierarchical Dirichlet Process (HDP) have demonstrated their efficacy in uncovering citizen concerns, policy priorities, and emerging discourse trends [15]. By automating the identification of thematic structures within participatory discussions, topic modeling enables urban planners, designers, and policymakers to efficiently process vast amounts of text and extract actionable insights, thereby enhancing the responsiveness of urban governance [16].

Recent advancements in LLMs, including transformer-based architectures such as GPT and BERT, have further transformed the landscape of post-participation analysis. Unlike earlier approaches that primarily relied on statistical models and keyword-based analytics, LLMs employ deep learning techniques to capture intricate semantic relationships, rhetorical structures, and even latent biases within citizen discourse [4]. These models facilitate automated text summarization, question-answering, and dialogic interaction, thereby significantly enhancing the interpretation and synthesis of public input. The capacity of LLMs to process nuanced, large-scale textual data in real time enhances participatory governance, enabling decision-makers to engage with citizen feedback in a more context-sensitive manner.

Together, NLP, topic modeling, and LLMs represent a transformative shift in how citizen-generated data is processed, analyzed, and integrated into urban decision-making. By leveraging these AI-driven methodologies, stakeholders can move beyond conventional participation frameworks and develop more responsive and analytically rigorous approaches to urban design and planning.

3.2 AI as a Catalyst for Transforming Urban Design and Planning

Integrating the above-mentioned AI methodologies into post-participation has fundamentally reshaped the relationship between citizen engagement and urban decision-making. By automating the structuring and analysis of large-scale public discourse, AI systematically translates citizen feedback into evidence-based design and planning strategies for sustainable cities. This enhanced capacity for data processing enables urban stakeholders to identify emerging spatial challenges, assess public sentiment toward specific projects, and formulate policies that are more attuned to community needs. Furthermore, AI addresses critical limitations related to scale, linguistic diversity, and bias in participation processes, thereby fostering greater inclusivity and ensuring the representation of marginalized voices. As a result, previously excluded or overlooked perspectives are increasingly integrated, strengthening the democratic foundations of urbanism.

Beyond enhancing inclusivity, these methods further facilitate dynamic and iterative feedback loops in urban decision-making, transforming participation from a static, retrospective practice into a continuous, forward-looking process. Through predictive analytics, longitudinal sentiment analysis, and AI-assisted scenario modeling, decision-makers can proactively identify potential urban challenges and refine policy directions in real time [17]. By embedding citizen perspectives into planning through post-participation, these approaches help align urban strategies with lived realities, enhancing legitimacy, reinforcing the social dimension of sustainability, and supporting more responsive governance. By embedding citizen perspectives into design and planning through post-participation, these adaptive mechanisms help align urban strategies with lived realities and the evolving needs of communities, enhancing legitimacy, reinforcing the social dimension of sustainability, and supporting more responsive urban governance.

Nevertheless, realizing AI's potential in post-participation requires a commitment to democratic practices, emphasizing transparency, accountability, and stakeholder collaboration. Ensuring interpretability is crucial for legitimacy and societal acceptance, as AI increasingly shapes how citizen contributions inform urban decision-making. By embedding AI-driven insights, participatory governance can become more responsive and inclusive, ultimately fostering equitable and transformative urban development processes.

3.3 Interdisciplinary Obstacles to Implementation

Despite the advantages of AI-driven approaches, integrating AI into post-participation presents significant interdisciplinary challenges that extend beyond technical feasibility. While AI enhances the capacity to analyze large-scale citizen discourse, its effective implementation necessitates close collaboration across multiple domains, including urban design, planning, architecture, data science, public policy, and the social sciences. These disciplines operate within distinct epistemological paradigms, methodological traditions, and professional cultures, creating barriers to seamless adoption. Therefore, developing an integrative framework for the democratic use of AI in participation processes requires addressing several critical obstacles.

Epistemological and Methodological Divergences: The integration of AI in post-participation is hindered by fundamental disciplinary differences. Urban stakeholders prioritize qualitative, context-sensitive interpretations, while AI-driven automation, if not carefully designed, risks oversimplifying complex citizen discourse. Resolving this tension requires frameworks that balance computational efficiency with interpretative depth, integrating AI analytics with human-centered participation to ensure meaningful and actionable outcomes in urban decision-making.

Lack of Standardized Analytical Frameworks: Urbanism lacks standardized methodologies for integrating AI into post-participation workflows, unlike disciplines with well-established AI benchmarks. Existing models, often trained on generic datasets, struggle to adapt to specific sociopolitical and linguistic contexts, leading to fragmented analytical approaches and limiting comparability. Developing domain-specific AI frameworks for participatory discourse is essential for ensuring methodological coherence and applicability across diverse urban environments.

Ethical, Legal, and Governance Challenges: AI-driven discourse analysis must uphold transparency, accountability, and data protection. However, the opacity of many models raises concerns about algorithmic bias and the misrepresentation of citizen input. Ethical priorities also diverge across disciplines: urban scholars emphasize equity and inclusivity, while AI researchers focus on optimization. Addressing these requires explainable AI (XAI), regulatory oversight, and interdisciplinary governance to ensure democratic integrity in participation processes.

Institutional Resistance and Capacity Gaps: Urban governance institutions often lack the technical expertise and infrastructure for AI integration, leading to reliance on external providers. Bureaucratic inertia further hinders digital transformation, as traditional urban practices are slow to adapt. Strengthening institutional capacity through digital literacy initiatives and cross-sector collaboration is essential to fostering AI readiness and ensuring its effective integration into decision-making structures.

Overcoming these interdisciplinary challenges demands a holistic and integrative approach that fosters cross-disciplinary collaboration among diverse stakeholders. Key strategies include:

- Developing hybrid analytical frameworks that combine computational techniques with qualitative, human-centered methodologies.
- Establishing interdisciplinary research consortia to define standardized AI evaluation metrics and best practices in participatory urban design and planning.
- Enhancing the transparency and interpretability of AI-driven insights through XAI models, ensuring that decision-making processes remain comprehensible and accountable.

- Promoting institutional adaptability by investing in digital capacity-building initiatives that empower urban agencies to engage meaningfully with AI-driven tools.

By bridging disciplinary divides through collaborative approaches, AI can move beyond being a purely technical instrument to serving as a catalyst for inclusive, equitable, and context-aware urban decision-making.

4 Developing a Framework to Bridge Complexity and Usability

This chapter has thus far examined contemporary digital participation practices, the challenges of analyzing large-scale textual data in post-participation, and the role of AI-driven methodologies in extracting meaningful insights. It has also highlighted interdisciplinary obstacles in implementing these methods, particularly the difficulty of translating data across domains in urban decision-making, where stakeholders have diverse expertise and analytical needs. This complexity is especially evident when professionals in urbanism, design, and policymaking utilize AI-driven methodologies that require specialized technical knowledge. The following section explores strategies to enhance the accessibility and usability of these advanced techniques, ensuring their effective application across disciplines in post-participation.

4.1 Conceptual Underpinnings: From Complexity to Usability

The proliferation of citizen-generated data in post-participation, coupled with the increasing reliance on AI-driven techniques like NLP, topic modeling, LLMs, and big data analytics, presents challenges for non-technical stakeholders. While these methods facilitate data translation, their complexity often limits accessibility for decision-makers, making it essential to synthesize and structure information inclusively to support informed, democratic decision-making. An effective framework must align AI-generated insights with the diverse needs of urban stakeholders by ensuring interpretability and contextual relevance. Presenting findings through accessible formats, such as visualizations or data summaries, bridges the gap between computational complexity and practical application, enabling urban practices to better harness collective intelligence in shaping the built environment.

To operationalize this, a post-participation data analysis framework is introduced, as illustrated in Fig. 1. Following data collection through intelligent digital participation tools, the raw input is stored in a post-participation data space, where it undergoes processing, analysis, and transformation into meaningful insights. Yet, due to

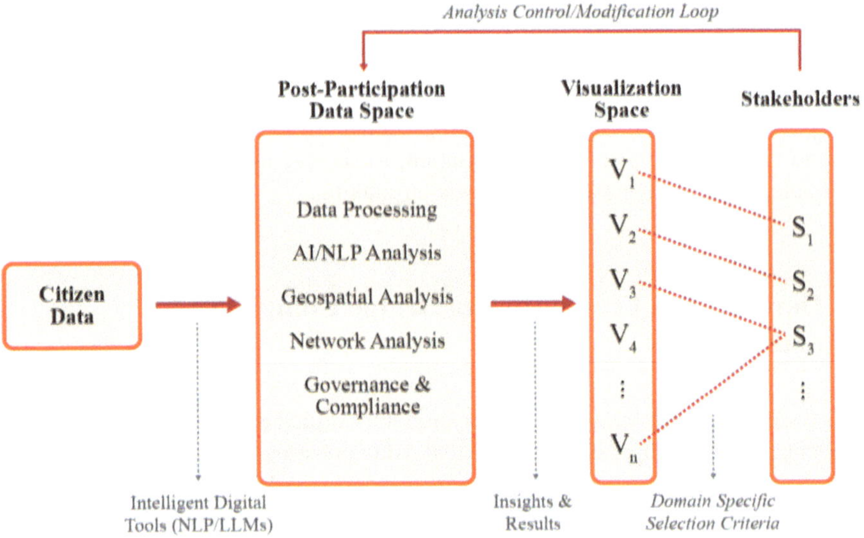

Fig. 1 Post-participation data analysis framework: Collected citizen data is processed within the data space, where analysis takes place, and then transferred to the visualization (V) space. In this space, stakeholders (S) can select relevant information to include in information packages based on their respective domains

the depth and complexity of the collected data, not all analyses are equally relevant or useful for every stakeholder group. To address this, the framework provides stakeholders with control over analytical processes, enabling them to select relevant analytical techniques based on their domain-specific requirements.

The framework further integrates multi-layered visualizations, allowing stakeholders to customize and combine different analytical outputs according to their selection criteria, thereby generating tailored information packages from unstructured citizen data. This modular approach enhances the efficiency and relevance of AI-driven analysis and significantly reduces computational costs, as it eliminates the need for exhaustive processing of all available data. Additionally, it mitigates the fragmentation of analytical tools, such as geospatial analysis, network analysis, and AI-based NLP models, by consolidating them within a user-friendly, interactive environment. Through dynamic control and iterative modification loops, advanced analytical techniques become democratized, ensuring that stakeholders, regardless of their technical or design expertise, can engage meaningfully with AI-driven insights.

4.2 Data Visualization as a Communicative Tool

To effectively integrate the proposed framework into post-participation processes, it is essential to recognize the role of data visualization (Fig. 1). As cities generate increasing volumes of citizen-generated content, urban stakeholders face the dual challenge of extracting meaningful insights from urban big data while ensuring accessibility for diverse audiences [18]. Raw analytical results alone are often difficult to interpret, and traditional reporting methods struggle to convey the nuances of open-ended textual data, spatial information, and emerging thematic patterns. While AI and data analytics efficiently detect trends [19], their outputs often require additional interpretation. Even advanced statistical methods can leave stakeholders with complex findings that hinder decision-making. Consequently, the importance of visualization in urban design and planning becomes evident for translating complex computational findings into clear, actionable representations that benefit both experts and non-experts [20].

A key advantage of data visualization is its ability to integrate multiple data types, including qualitative and quantitative inputs. Textual data from online debates, for instance, often contain semantic ambiguities that AI-driven techniques can identify but not fully contextualize [9]. Structured visual interfaces help communicate underlying patterns, allowing urban stakeholders to explore both broad trends and deeper contextual insights [21]. Furthermore, stakeholder diversity significantly influences visualization design, as domain-specific preferences, cultural and linguistic factors, and contextual considerations can shape perceptions of complexity or oversimplification. Achieving a balance between comprehensiveness and accessibility is therefore crucial to mitigating cognitive overload and fostering meaningful engagement.

Consequently, effectively leveraging visualization in large-scale participation processes requires careful consideration of both technical and human factors. From a technical perspective, data must be structured and processed systematically, with algorithms filtering noise and highlighting relevant insights. On the human side, variations in visualization literacy among stakeholders influence their ability to interpret data accurately. Moreover, given that urban projects often unfold over extended periods, sometimes spanning years or decades, visualization platforms must remain adaptable to evolving datasets and user requirements. Addressing these challenges strengthens the role of data visualization as a communicative tool, promoting inclusive and informed urban decision-making while ensuring that participation insights remain accessible, interpretable, and impactful in post-participation.

4.3 Ensuring Relevance to Diverse Stakeholders

A critical consideration in implementing a post-participation framework is ensuring that AI-driven methodologies for data collection and analysis remain relevant and accessible to diverse stakeholders. Given the varying expertise and priorities among end users, maintaining data quality, interpretive rigor, and meaningful engagement is essential for fostering stakeholder confidence and reinforcing the democratic accessibility of AI-driven methodologies in urban design and planning.

A key challenge lies in balancing data quantity with quality. While large datasets can enhance analytical robustness, volume alone does not guarantee representativeness [18]. Therefore, stakeholders must be able to navigate both structured and unstructured datasets effectively to identify and correct analytical outputs that may misalign with real-world conditions. Achieving this requires the integration of both quantitative metrics (e.g., demographics, mobility patterns) and qualitative insights (e.g., public forums, surveys) within the framework. A hybrid approach, combining algorithmic efficiency with domain expertise, enhances the depth and applicability of AI-driven findings in urban design and planning. AI can support this process by integrating machine learning techniques with qualitative coding methods [20]. Furthermore, iterative control and modification loops allow for the refinement of AI-driven interventions by incorporating evolving data and shifting stakeholder priorities (Fig. 1).

The effectiveness of this framework ultimately depends on the clarity and accessibility of AI-generated insights, ensuring that technical complexity does not alienate non-specialist stakeholders [21]. Continuous stakeholder engagement, transparent reporting, and user-friendly visualizations help align data interpretations with on-the-ground realities rather than reducing them to abstract academic exercises. To achieve this, challenges such as data overload, technological disparities, and interpretational biases must be addressed, promoting inclusive and reliable AI-driven processes. Only through such an approach can AI methodologies be aligned with stakeholder needs, balancing data integrity, analytical rigor, transparent communication, and adaptability.

5 DIP-DAT: The Digital Participation Data Analysis Toolkit

To facilitate communication between citizen-generated data and stakeholders across various domains, a web-based platform, Digital Participation Data Analysis Toolkit (DIP-DAT), has been developed. DIP-DAT enables the systematic analysis of large-scale digital participation data in post-participation through AI-driven methods for urban decision-making (Fig. 2). To assess its scalability and accessibility, DIP-DAT is implemented across three distinct case studies, each representing different governance structures and urban scales within the post-participation framework.

5.1 Insights from Singapore, Madrid, and Hamburg

DIP-DAT is implemented in three distinct urban contexts, Singapore, Madrid, and Hamburg, to illustrate contrasting models of digital participation in urban governance. These cities represent varying degrees of top-down and bottom-up governance structures, offering a diverse spectrum of participatory approaches.

Fig. 2 DIP-DAT homepage before data upload and analysis initiation

Singapore, governed primarily by the People's Action Party since its independence in 1965, exemplifies a predominantly top-down governance model that has gradually incorporated mechanisms for citizen input. While the introduction of digital platforms such as the National Conversation initiative and the REACH portal signals a shift toward broader civic engagement, the governance framework remains state-led and largely consultative rather than co-creative [22]. In contrast, Madrid adopts a bottom-up approach, empowering citizens to propose and vote on policy matters in real time. This model fosters a participatory democracy in which direct citizen involvement in decision-making processes is prioritized [23]. Hamburg, distinct from both Singapore and Madrid, operates under a decentralized, project-oriented model. Comprising seven boroughs and over a hundred neighborhoods, the city integrates localized strategies for civic engagement, combining digital tools, such as surveys and forums, with traditional participatory methods like community meetings [24]. This hybrid approach seeks to capture diverse viewpoints and tailor solutions to the specific conditions of each neighborhood.

Within these governance structures, DIP-DAT is applied to datasets collected from three different digital participation platforms, each corresponding to a different scale of civic engagement. The Consul Project in Singapore facilitates digital participation at the neighborhood scale, focusing on localized urban transformation projects [25]. Decide Madrid enables city-wide participation on various policy topics, fostering large-scale deliberative democracy [26]. In Hamburg, the Digital Participation System (DIPAS) supports project-level participation through more targeted data collection [27]. Together, these three datasets illustrate a broad spectrum of digital participation models, ranging from state-driven consultative mechanisms (Singapore) to large-scale direct democracy (Madrid) and decentralized, project-based engagement (Hamburg).

This diversity has informed the development of DIP-DAT as a flexible and adaptive toolkit, capable of addressing both the opportunities and constraints inherent in different governance structures, datasets, and stakeholder needs. By adhering to its core objectives and functionalities, DIP-DAT aims to provide an intuitive and meaningful medium for stakeholders to leverage advanced AI-driven analysis methods in digital participation processes.

5.2 Toolkit Objectives & Core Functionalities

DIP-DAT is designed to integrate advanced AI-driven workflows, topic modeling, content analysis, and data visualization into a unified system for analyzing digital participation data. Grounded in established research on digital participation and human–computer interaction, the framework incorporates multi-level analysis, iterative decision-making, and user-centric design principles [5]. By providing automated data preprocessing, structured filtering, and an interactive web-based interface, DIP-DAT enhances the integrity and clarity of information emerging from citizen engagement processes. This approach directly addresses the fragmentation often observed in post-participation workflows, where reliance on multiple disconnected tools and labor-intensive manual tasks can disrupt analytical continuity.

A primary objective of DIP-DAT is to optimize post-participation practices by streamlining the transition from raw data to actionable insights. This is achieved through a structured workflow that begins with data import and preprocessing, followed by content-driven operations such as hyperparameter tuning for topic modeling, sentiment analysis, and interactive visualization (Fig. 3). The system's user-friendly interface ensures that stakeholders with diverse technical backgrounds can navigate these processes efficiently, make informed methodological choices, and apply relevant analytical techniques suited to their specific domains.

A fundamental functionality of DIP-DAT is its automated and semi-automated data preprocessing capabilities, which include duplicate removal, normalization, and structured data transformation. These processes enhance data reliability by mitigating errors and inconsistencies, while intuitive filtering options allow stakeholders to exclude irrelevant segments and refine datasets for deeper analysis. Additionally, by integrating hyperparameter tuning within a simplified interface, DIP-DAT facilitates real-time iterative experimentation with different model configurations, allowing for more flexible and adaptive textual analysis.

Another key functionality centers on dynamic visualizations, where DIP-DAT consolidates multiple visualization techniques, such as heat maps, topic distributions, and sentiment graphs, into a single platform (Fig. 4). This integration minimizes disruptions caused by switching between different tools and platforms, enabling stakeholders to maintain analytical momentum. Interactive features, such as zooming, filtering, and variable highlighting, support an exploratory approach to analysis, allowing stakeholders to uncover detailed patterns in public discourse. By seamlessly

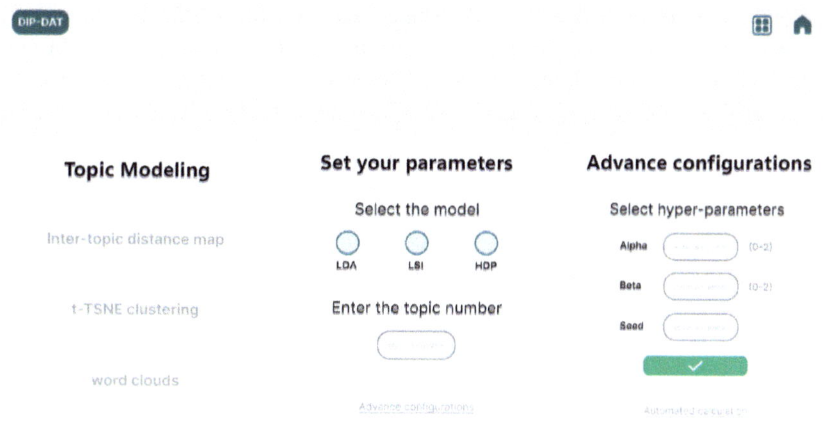

Fig. 3 Interfaces for topic model selection and hyperparameter tuning with automated calculation options for diverse stakeholders

merging analysis and visualization, DIP-DAT fosters a more comprehensive and interpretable examination of complex datasets.

Stakeholder accessibility is a central consideration in DIP-DAT's design. The web-based interface, accessible via any major browser, eliminates the need for specialized software installations. Clear navigation menus, guided prompts, and structured

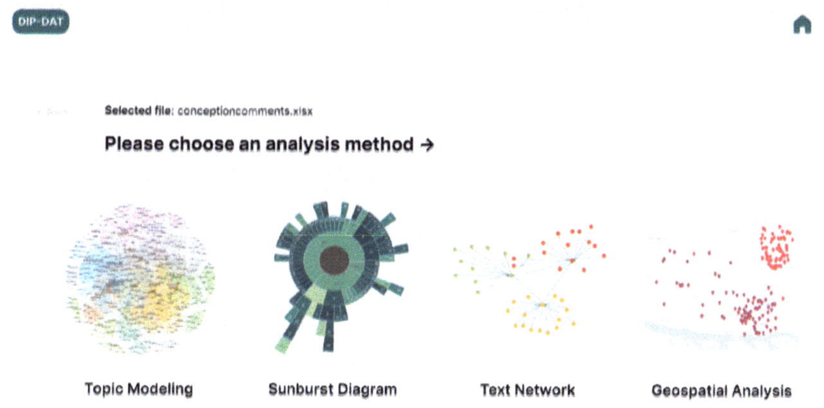

Fig. 4 Interface for selecting data analysis and visualization techniques in DIP-DAT

workflows support both novice and expert users, ensuring that rigorous analytical processes remain accessible across varying levels of expertise. Furthermore, DIP-DAT supports multiple data formats, including CSV, JSON, JPEG, and PDF, enhancing interoperability and facilitating seamless data exchange across analytical and reporting platforms. These features promote transparency, adaptability, and collaboration, allowing stakeholders to integrate participation outputs into broader decision-making frameworks or share findings across interdisciplinary teams.

Overall, DIP-DAT's objectives and core functionalities underscore a holistic approach to post-participation, integrating AI-driven techniques, topic modeling, intuitive data management, and dynamic visualization into a streamlined and cohesive platform. By ensuring seamless interaction between components, ranging from data preprocessing and filtering to modeling and visualization, DIP-DAT enhances both the analytical rigor and accessibility of insights derived from digital participation processes.

5.3 Adapting the Toolkit to Practical Constraints: Key Observations and Lessons Learned

DIP-DAT's deployment in real-world contexts revealed several pragmatic lessons regarding the adaptation of a data analysis toolkit to diverse stakeholder needs, technical settings, and methodological constraints. One recurring observation highlighted the system's capacity to streamline data analysis for users without specialized expertise. Most stakeholders indicated that logical organization and user-friendly interfaces substantially reduce complexity. Furthermore, the system's modular design allows stakeholders from policymaking, urban design and planning, and related fields to configure analytical pipelines for their specific domains. Whether a stakeholder conducts geospatial analysis, sentiment clustering, or focused policy investigations, DIP-DAT's architecture enables the selective use of the components most relevant to each task, underscoring the toolkit's interdisciplinary applicability.

Another crucial lesson emerged in the context of data preprocessing, where conventional manual filtering can be time-consuming and prone to errors. DIP-DAT's reliance on predefined filtering criteria and semi-automated workflows addresses this challenge by balancing user input with automated guidance. This dual approach not only accelerates data preparation but also increases confidence among those unaccustomed to advanced computational methods. Stakeholders found that the ability to engage thoughtfully with filtering decisions, while assigning routine tasks to the system, fosters both efficiency and rigor in data handling.

Advanced visualization methods further extend DIP-DAT's flexibility, enabling users to examine datasets from multiple perspectives and thereby enhance stakeholder engagement. The capacity to move seamlessly between high-level overviews and granular inspections of specific data subsets adds an additional layer of scalability. As a result, the toolkit meets broad and specialized analytical requirements in accordance with user preferences.

Minimizing cognitive load stands out as another significant consideration when adapting DIP-DAT to practical environments. A clear, structured interface allows stakeholders to concentrate on substantive tasks rather than navigating confusing layouts or ambiguous workflow steps, especially when employing advanced AI methodologies. By guiding users explicitly from data selection and preprocessing to visualization, the system promotes efficient progression and lowers the mental burden typically associated with complex analytics. Automated and semi-automated features further reduce strain on those lacking advanced technical backgrounds, an important factor in extending the reach of powerful data analysis tools. The option to preview data and visualize partial results before finalizing decisions additionally decreases error rates and supports reflective, yet expedited, exploration.

A final lesson involves the mitigation of subjectivity in post-participation data analysis. DIP-DAT addresses potential biases by encouraging explicit, criteria-based selections and offering quantitative metrics that ground decisions in observable evidence. This strategy leads to consistent and defensible outcomes, as it balances subjective choices with transparent, data-driven benchmarks. The toolkit's capacity for varied visual comparisons likewise enhances analytical reliability, permitting cross-validation of insights and prompting deeper investigation into anomalies or emergent trends. In practice, many users noted that the combination of sentiment clustering, geographic overlays, and content analysis metrics reduced arbitrary interpretations and helped preserve the data's fundamental message. Through structured decision-making, quantitative rigor, and multidimensional visualization, DIP-DAT demonstrated a notable ability to adapt to real-world constraints while sustaining analytical validity. Table 1 summarizes these observations and lessons learned under varied practical conditions.

6 Conclusion: Democratizing Advanced AI Methods in Post-Participation

This chapter has examined the challenge of democratizing advanced AI-driven methodologies for analyzing the escalating volume of citizen-generated data in urban decision-making. It has underscored the need for robust, scalable approaches that retain the contextual depth of public discourse while integrating the insights of all stakeholders. Traditional methods often prove inadequate for the scope of digital participation and frequently fall short of capturing essential citizen perspectives. Although AI-based techniques, including topic modeling, NLP, and LLMs, offer powerful ways to address the complexity and sheer volume of online debates, they introduce ethical, institutional, and methodological hurdles that complicate the responsible deployment of computational strategies in urban governance.

Table 1 Evaluation of five distinct aspects of DIP-DAT, focusing on challenges or concerns, proposed solutions, key features and strengths, and notable observations

Aspect	Challenges & concerns	Provided solution	Key features & strengths	Notable observations
Streamlined workflow for data analysis	Complexity in data analysis for non-experts	Streamlined, logical organization	Logical organization, User-friendly interface, Hyperparameter tuning	Suitable for those lacking specialized knowledge; praised for modularity and interdisciplinary relevance
Expedited decision-making in data processing	Time-consuming and error-prone manual filtering	Predefined filtering and preprocessing criteria	Predefined filtering criteria, Balance between manual and automated features	Streamlines decision-making, reducing time in data preprocessing
Navigation across analytical modalities	Complexity of analytical methods	Modular, customizable functionalities	Modular analysis features, Advanced visualizations	Accommodates various types of analytical tasks and needs; enhances stakeholder engagement
Minimization of cognitive load for post-participants	Cognitive overload due to interface complexity	Structured, intuitive design	Structured interface, Explicit steps and options, Automated features	Facilitates ease of use and focuses on analytical tasks, minimizing cognitive load
Mitigation of subjectivity in participation data analysis	Inherent subjectivity in data analysis	Clearly defined criteria and metrics	Clearly defined criteria, Quantitative metrics, Multiple visualization methods	Addresses issues of subjectivity, enhancing the integrity and reliability of the analysis

By reviewing current practices and highlighting ongoing limitations in large-scale citizen engagement, this chapter has demonstrated how AI can catalyze inclusive, data-driven, and context-sensitive urban design and planning when employed judiciously. Adopting novel tools in post-participation workflows enriches the ability to glean deeper insights from citizen input, provided these tools remain transparent, interpretable, and aligned with stakeholder needs. The DIP-DAT system illustrates one possible approach, showcasing how user-friendly interfaces, multi-layered data visualization, and (semi-)automated analysis mechanisms can simplify complex analytics without eroding analytical rigor. As demonstrated in diverse case contexts, DIP-DAT emphasizes interdisciplinary collaboration and interoperability, bridging gaps among urban planners, architects, data scientists, and social scientists.

Across its implementations in Singapore, Madrid, and Hamburg, DIP-DAT has also shown potential to support real-world urban design processes. In Singapore, it contributes to structuring state-led public engagement by translating citizen inputs into interpretable outputs for centralized decision-making. In Madrid, it facilitates participatory democracy at scale by organizing large volumes of public feedback into thematically coherent design priorities. In Hamburg, DIP-DAT supports decentralized and localized design interventions, helping designers and planners respond to borough-specific concerns through data-driven methods. These examples illustrate how AI-driven systems, when thoughtfully integrated, can serve not only as analytical engines but as enablers of more inclusive, transparent, and sustainable urban design practices.

Consequently, three core insights emerge from this chapter. First, AI-based methods are adept at organizing extensive textual data but require careful integration to preserve data integrity and uphold democratic accountability. Second, providing flexible control and iterative feedback loops ensures that stakeholders with varying degrees of technical proficiency remain active contributors in the analytical process. Third, comprehensive data visualization and unified, user-friendly interfaces help mitigate cognitive overload and foster shared understanding across different disciplines. Addressing these considerations can harmonize technical and human-focused perspectives in post-participation activities, ultimately amplifying the influence of citizen engagement on evidence-based urban decision-making.

References

1. Ataman C, Herthogs P, Tunçer B, Perrault S (2025) From insight to action: An integrated assessment framework for digital citizen participation in data-centric urban practices. Cities 156:105545. https://doi.org/10.1016/j.cities.2024.105545
2. Ataman C, Tunçer B, Perrault ST (2022) Asynchronous Digital Participation in Urban Design Processes: Qualitative Data Exploration and Analysis With Natural Language Processing. In: POST-CARBON - Proceedings of the 27th CAADRIA Conference. Sydney, Australia, pp 383–392
3. Cai M (2021) Natural language processing for urban research: a systematic review. Heliyon 7:e06322. https://doi.org/10.1016/j.heliyon.2021.e06322
4. Tupayachi J, Xu H, Omitaomu OA et al (2024) Towards next-generation urban decision support systems through AI-powered construction of scientific ontology using large language models—a case in optimizing intermodal freight transportation. Smart Cities 7:2392–2421. https://doi.org/10.3390/smartcities7050094
5. Yang Q, Steinfeld A, Rosé C, Zimmerman J (2020) Re-examining whether, why, and how human-AI interaction is uniquely difficult to design. In: Proceedings of the 2020 CHI Conference on Human Factors in Computing Systems. Association for Computing Machinery, New York, NY, USA, pp 1–13
6. Kinra A, Beheshti-Kashi S, Buch R et al (2020) Examining the potential of textual big data analytics for public policy decision-making: a case study with driverless cars in Denmark. Transp Policy 98:68–78. https://doi.org/10.1016/j.tranpol.2020.05.026
7. Guest G, MacQueen K, Namey E (2012) Applied thematic analysis. SAGE Publications, Inc., 2455 Teller Road, Thousand Oaks California 91320 United States

8. Serafis D, Greco S, Pollaroli C, Jermini-Martinez Soria C (2020) Towards an integrated argumentative approach to multimodal critical discourse analysis: evidence from the portrayal of refugees and immigrants in Greek newspapers. Crit Discourse Stud 17:545–565. https://doi.org/10.1080/17405904.2019.1701509
9. Ataman C, Tunçer B, Perrault S (2024) Digital participation in urban design and planning: addressing data translation challenges in urban policy- and decision-making through visualization techniques. In: Accelerated Design. Singapore, pp 201–210
10. Alvarez L, Borsi K, Rodrigues L (2017) The role of social network analysis on participation and placemaking. Sustain Cities Soc 28:118–126
11. Fu X (2024) Natural language processing in urban planning: a research agenda. J Plan Lit 08854122241229571. https://doi.org/10.1177/08854122241229571
12. Hirschberg J, Manning CD (2015) Advances in natural language processing. Science 349:261–266. https://doi.org/10.1126/science.aaa8685
13. Sayah I, Schnabel MA (2019) Amplifying citizens' voices in smart cities: an application of social media sentiment analysis in urban sciences. In Intelligent and Informed—Proceedings of the 24th International Conference on Computer-Aided Architectural Design Research in Asia, CAADRIA 2019. pp 773–782
14. Sbalchiero S, Eder M (2020) Topic modeling, long texts and the best number of topics. Some Problems and solutions. Qual Quant 54:1095–1108
15. Ataman C, Tunçer B, Perrault S (2023) Transforming large-scale participation data through topic modelling in urban design processes. In Turrin M, Andriotis C, Rafiee A (eds) Computer-Aided Architectural Design Futures (CAAD Futures). INTERCONNECTIONS: Co-computing Beyond Boundaries. Springer Nature Switzerland, Cham, pp 271–286
16. Hagen L, Uzuner O, Kotfila C, et al (2015) Understanding citizens' direct policy suggestions to the federal government: a natural language processing and topic modeling approach. In: 2015 48th Hawaii International Conference on System Sciences. IEEE, pp 2134–2143
17. Miller T, Durlik I, Kostecka E et al (2025) Integrating artificial intelligence agents with the internet of things for enhanced environmental monitoring: applications in water quality and climate data. Electronics 14:696. https://doi.org/10.3390/electronics14040696
18. Kitchin R (2014) Big Data, new epistemologies and paradigm shifts. Big Data Soc 1:205395171452848. https://doi.org/10.1177/2053951714528481
19. Batty M (2013) The new science of cities. MIT Press
20. Borner K (2015) Atlas of knowledge: anyone can map. MIT Press
21. Few S (2009) Now you see it: simple visualization techniques for quantitative analysis. Analyrics Press, Oakland, CA, USA
22. Soon C, Soh YD (2014) Engagement@web 2.0 between the government and citizens in Singapore: dialogic communication on Facebook? Asian J Commun 24:42–59. https://doi.org/10.1080/01292986.2013.851722
23. Susha I, Grönlund Å, Janssen M (2015) Driving factors of service innovation using open government data: an exploratory study of entrepreneurs in two countries. IP 20:19–34.
24. Simonofski A, Fink J, Burnay C (2021) Supporting policy-making with social media and e-participation platforms data: a policy analytics framework. Gov Inf Q 38:101590. https://doi.org/10.1016/j.giq.2021.101590
25. (2022) Consul Project. https://consuldemocracy.org/en/
26. (2023) Decide Madrid. https://decide.madrid.es
27. (2023) DIPAS. https://www.hamburg.de/dipas/.

Enhancing Building Performance Through AI: Techniques, Models, and Practical Challenges

Bahereh Vojdani Fakhr, Andressa Carmo Pena Martinez, and Deok-Oh Woo

Abstract This chapter thoroughly examines the rationale and impetus for incorporating high-performance and AI techniques in architectural design. It discusses the increasing significance of sustainable design in architecture and the role of AI in realizing these objectives, emphasizing its influence on energy usage and building efficiency. Artificial Intelligence has multiple benefits that improve high-performance buildings, rendering them more sustainable and efficient. We discuss three fundamental elements: (1) AI techniques: data-driven design, rule-based design, and generative design, (2) AI models such as machine learning, surrogate modeling, and hybrid methods, and (3) key implementation challenges such as data limitations, scalability, and ethical issues. We compare the constraints of conventional optimization techniques with the advantages of AI systems in addressing these issues. The results illustrate how these AI methodologies discern trends, forecast results, and propose optimization measures, all intended to improve building performance. In this sense, the chapter presents the practical implementation of AI in architecture through real-world examples, emphasizing its capacity to enhance building energy efficiency, daylighting, and other sustainability goals. As a result, this work highlights AI's significance in attaining sustainability and energy efficiency, provides perspectives on the future of sustainable construction methods, and contributes to the continuing discourse on AI's incorporation in architectural design.

Keyword Artificial intelligence (AI) · Building design optimization · High-performance buildings · Sustainable architecture

B. V. Fakhr · A. C. P. Martinez (✉) · D.-O. Woo
University of Maryland, College Park, MD, USA
e-mail: acpm@umd.edu

B. V. Fakhr
e-mail: vojdani@umd.edu

D.-O. Woo
e-mail: deokoh@umd.edu

M. Sedrez and G. Pitts (eds.), *Artificial Intelligence-Aided Design for Sustainability*, Urban Sustainability, https://doi.org/10.1007/978-981-95-1349-9_11

1 Introduction

The advanced capabilities of modern building technologies and their integration enable energy-efficient buildings and improvements to be financially viable. The utilization of AI across several domains is growing exponentially. Although this tendency has been shown in several industries before architecture and urban planning, recent decades have seen an increased interest in utilizing AI to improve sustainable building performance. Technological advancements have facilitated the convergence of AI, the Internet of Things, and big data in the Architecture, Engineering, and Construction (AEC) sector. The convergence has led to the emergence of intelligent and sustainable building concepts.

Since 1990, and particularly since 2000 to the present, a design process integrating building energy simulation with optimization has emerged. This process is referred to by various labels, including "computational optimization" [1], "simulation-based optimization" [2]," building performance optimization " [3], and "performance-driven design" [4]. The term "performance" encompasses a broader scope than merely energy efficiency. Energy-efficient design optimization for buildings is an emerging technology under active investigation.

In this sense, this chapter examines and critically assesses the implementation of AI methodologies, encompassing data-driven and rule-based design strategies, to improve sustainable building performance. AI models offer predictive analytics for design decision-making. Implementing AI in sustainable design encounters challenges concerning data accessibility, computational intricacy, and model interpretability. Figure 1 depicts the workflow of this chapter, these fundamental elements, and their significance in AI-facilitated sustainable design.

Currently, AI research has yielded a wide variety of procedures that can be directly applied or adapted to solve complex problems in the energy sector (such as in energy system planning, design, operation, investment, and dynamic control and stability), previously thought to be unsolvable without several simplified assumptions [5]. These methodologies utilize real-time data collection and analysis to guide decision-making, allowing stakeholders to act on efficient energy management designs and renovation initiatives [6].

This chapter explains how these techniques enhance energy efficiency, minimize environmental consequences, and augment building performance. It also emphasizes the correlation between advanced technologies in high-performance buildings (HPBs) and the Sustainable Development Goals (SDGs) established by the United Nations, considering the worldwide emphasis on sustainable development. The aim is to illustrate how the HPB industry can contribute to achieving SDGs 7 (Affordable and Clean Energy) and 11 (Sustainable Cities and Communities). Also, it demonstrates the revolutionary potential of AI in tackling significant problems in sustainable building design through the integration of advanced computational

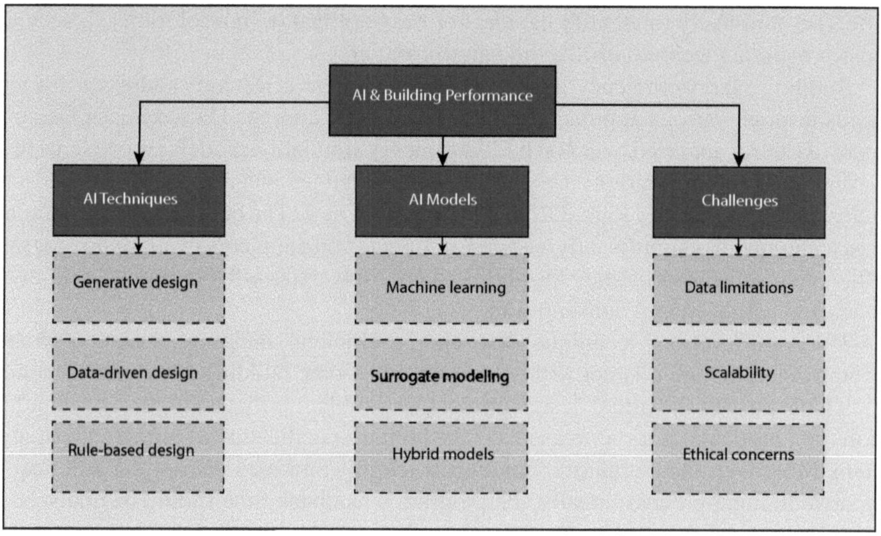

Fig. 1 Topics addressed by this chapter

approaches, including predictive analytics and real-time data processing. It analyses these methodologies' function in promoting innovation, facilitating evidence-based decision-making, and aligning construction practices with the international Sustainable Development Goals.

2 AI Techniques in High-Performance Buildings Design: From Traditional to Building Design Optimization (BDO)

The revolutionary capacity of AI in HPBs is in its ability to facilitate automation, improve accuracy, and support informed, data-driven decision-making. In contrast to traditional methods that frequently depend on human computations, heuristic estimations, and static modeling methods, AI offers dynamic functionalities, including real-time data analysis, predictive modeling, and optimization algorithms. These innovations optimize intricate operations and yield profound insights into energy efficiency, occupant comfort, and environmental impacts.

Some architects address energy-efficient design using traditional and, in some cases, ineffective methods. They lack proficiency in employing energy simulation software to help their designs, much less in utilizing design optimization techniques that integrate energy simulation and optimization algorithms. The methods use optimization algorithms to produce novel designs informed by simulation outcomes and specified design goals. In contrast to the traditional "trial-and-error" design methodology reliant on designers' expertise and experience, this strategy is more efficient, effective, and likely to identify the ideal or near-optimal design solution [7], such as post-processing techniques like the Pareto frontier.

Building energy efficiency, a longstanding discipline, effectively addresses diverse needs through various approaches. Almost a decade ago, the United States Department of Energy reported over 400 building energy simulation models and systems [8]. Several influential programs are globally acknowledged and utilized across various fields, including Energy Plus, TRNSYS [9], and DOE-2. The building energy simulation technique has significantly assisted architects and engineers in attaining energy-efficient designs by providing the ability to precisely and swiftly compute buildings' loads and actual energy consumption.

The optimization technique for energy-efficient building design appears promising, although it is not without constraints. Some building simulations cannot match the pace of the initial design phase. One problem is that configuring a simulation for a particular notion necessitates the human specification of numerous parameters. Moreover, the simulation duration is lengthy and may disrupt the architect's creative thought process; ideally, the program's feedback time should be under 10 s [10]. However, numerical optimization is an iterative process that requires numerous building simulation runs and can take hours or days to reach convergence. Consequently, requirements for preliminary design tools indicate that a tool for rapid global design space exploration is necessary to assess a wide range of early design concepts swiftly. In this case, only a few pertinent parameters should be considered to mitigate complexity.

Also, building simulations generally use intricate "black box" models featuring numerous design elements as variables. Consequently, users typically lack a comprehensive understanding of the fundamental equations and the interactions among the inputs. These models necessitate extensive input parameters, yield precise results, and are compatible with most architectural designs, as shown in Fig. 3. For example, the primary categories of physically based lighting modeling algorithms are ray tracing and radiosity, which involve extensive matrix operations contingent upon temporal and spatial precision. Numerous initiatives aim to expedite simulations by algorithmic simplicity and hardware acceleration, rendering them suitable for design assessment [11–13].

Nevertheless, building forms are iteratively adjusted and optimized throughout the initial design phase, necessitating corresponding adjustments to the simulation models, which might make the simulation process time-consuming. Furthermore, not all parameters relevant to the simulation are adjustable or require determination, such as materials, and precise calculations can be time-consuming and computationally intensive [14]. So, a more rapid prediction model with reduced input parameters is required.

One of the fastest-expanding areas of building simulation research is building design optimization (BDO) [15]. BDO using AI might be data-driven, rule-based, or even a hybrid model, depending on the strategies and tactics employed to optimize the building design. Data-driven strategies in building performance employ extensive datasets and sophisticated analytics to enhance energy efficiency, occupant comfort, and overall operational effectiveness. Data-driven technologies have initiated a new epoch in the construction sector, transitioning it from traditional methods to intelligent and sustainable practices.

Rule-based design strategies involve formulating and applying specific rules or guidelines to achieve the intended design objectives and enhance the overall performance of buildings. Rule-based optimization methods rely on predefined rules to guide the decision-making process, allowing designers to make well-informed decisions based on predetermined principles and criteria. Metaheuristic (near-optimal solutions to a broad range of problems) or heuristic (a quick solution for a specific situation) optimization is a subdivision of artificial intelligence that computes for global optimum solutions of non-deterministic polynomial time-hard (NP-hard) issues that are incompatible with exact mathematical optimization techniques [16].

Predefined rules, heuristics, or algorithms based on standards, building codes, domain knowledge, or expert input provide the foundation of a rule-based approach. The benefits are that they are straightforward and comprehensible, and do not require sophisticated computational capabilities or massive datasets. However, it has limited flexibility and scalability and could miss intricate, nonlinear linkages in building performance.

AI-driven design tools evaluate environmental data to model situations and provide energy-efficient, eco-friendly solutions. Incorporating these computer models in the AEC industry indicates a period of intelligent and sustainable practices, as they improve efficiency and reduce the building's environmental impact.

3 AI Models for Sustainable Building Performance

AI models such as machine learning, surrogate modeling, and hybrid algorithms are transforming multiple facets of the AEC industry. Machine learning proxy models analyze data to represent nonlinear correlations between input parameters and output metrics, aiding design review [17]. Instead of choosing all detailed parameters as input, making them particularly effective for iterative optimization [18], AI can also examine extensive data sets to identify patterns and trends, facilitating decision-making. Machine learning can enhance construction schedules by evaluating many elements and minimizing project durations and environmental consequences.

Figure 2 shows, as an example, a process linking building simulation outputs to neural network-based predictions to illustrate how simulation data and AI models can be integrated. In this case, black box models are trained using tabular findings from EnergyPlus simulations, such as temperature, material, and illumination measurements. The result is a heatmap showing each zone's energy use intensity (EUI) prediction. By connecting visual energy insights with raw simulation data, this representation facilitates well-informed design decisions.

Surrogate modeling is a nascent sub-discipline in building performance design that aims to develop streamlined, computationally efficient models to replicate intricate simulations [19]. Utilizing surrogate models enables designers and engineers to forecast building performance metrics [20], including energy consumption [21], thermal

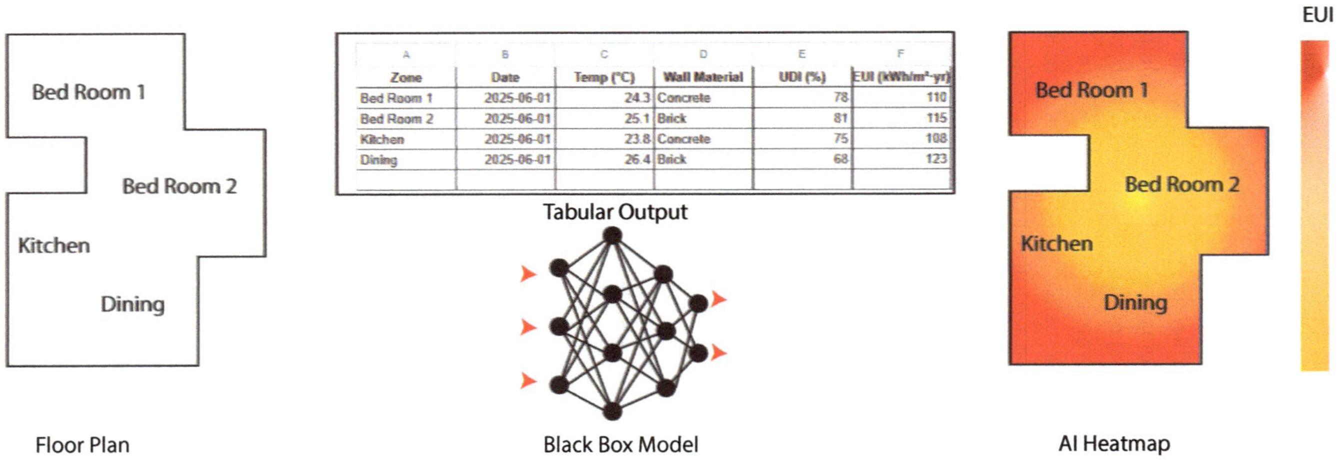

A	B	C	D	E	F
Zone	Date	Temp (°C)	Wall Material	UDI (%)	EUI (kWh/m²·yr)
Bed Room 1	2025-06-01	24.3	Concrete	78	110
Bed Room 2	2025-06-01	25.1	Brick	81	115
Kitchen	2025-06-01	23.8	Concrete	75	108
Dining	2025-06-01	26.4	Brick	68	123

Fig. 2 AI-driven zone-level prediction workflow: From building floor plan and tabular simulation

comfort, and daylighting, without relying on resource-intensive simulations. This methodology is especially beneficial in enhancing building design processes since it expedites decision-making and facilitates a more extensive examination of design alternatives.

Hybrid algorithms are optimization algorithms that combine the benefits of multiple techniques to enhance the overall performance of the optimization procedure [22]. They involve integrating various search strategies, optimization algorithms, or mathematical models to produce a more robust, efficient, and accurate optimization procedure. Hybrid algorithms can assist in overcoming the limitations and drawbacks of individual techniques by leveraging their complementary strengths and mitigating their shortcomings [23]. They can result in enhanced convergence, speedier convergence rates, enhanced exploration and exploitation of the search space, and higher-quality solutions.

The AEC industry is on a path of significant transformation driven by incorporating big data technologies [24]. In an age characterized by swift urbanization, heightened environmental awareness, and technological advancement, the AEC sector needs to pursue creative solutions for intelligent and sustainable development. Big data technology is a powerful accelerator, significantly transforming the industry's landscape [25]. By utilizing extensive data produced at each phase of construction, from planning and design to execution and maintenance, the AEC sector is facilitating a future in which buildings transform from simple constructions into intelligent, sustainable ecosystems.

The initial two categories are especially beneficial for optimizing building operations, devising building-specific upgrading alternatives, analyzing building sensor data, or doing energy mapping and building performance benchmarking within a specific geographic region included in the building stock data. Both categories of data consist of historical information about pre-existing buildings. Statistical prediction models developed from that data may not accurately apply to novel building technology or distinctive design concepts. Consequently, constructing simulations based on physical principles is essential for the design of novel buildings. The validity is not contingent upon observations; any novel design, retrofit alternative, or construction technology can be modeled.

Nonetheless, contemporary building simulation software incurs substantial computing expenses, and establishing a building model is time-consuming. Architects and designers do not adequately incorporate it into their daily practices. In contrast, AI models, or metamodels, are poised to deliver building performance assessments grounded in physical knowledge yet far more rapid than simulation-based design analysis [10].

A metamodel closely approximates the original simulation model with considerable accuracy through a simplified mathematical representation. A sequence of simulations subsequently requires only a fraction of the initial simulation duration, facilitating substantial computational savings. A metamodel, or surrogate model, serves as an approximation of a more intricate simulation model. It is developed using high-fidelity simulations or experimental data to analyze the correlation between inputs (e.g. building materials, geometry, occupancy) and outputs (e.g., energy consumption, daylighting, comfort). Metamodels facilitate expedited evaluations, rendering them suitable for real-time feedback, optimization tasks, and sensitivity analysis to discern crucial aspects influencing performance. The concept of surrogate modeling involves replicating an expensive, high-fidelity model, specifically a construction simulation model, using a statistical model. The surrogate is trained on a limited dataset of simulation inputs and outputs. Upon validation of its adequacy in approximating the detailed simulation model, it can be employed to rapidly forecast outcomes of the high-fidelity simulation based on a suitable set of building design parameters. The sophisticated surrogate models can effectively capture variable interactions and elucidate nonlinear, multi-modal behaviour [10].

Surrogate models are especially advantageous for expediting the development of building performance distributions using the external technique, which necessitates a substantial quantity of simulation samples. The external approach uses external sampling techniques to produce input data for simulations during sensitivity analysis, uncertainty analysis, or building performance optimization. These methods entail systematically sampling the input parameter space (e.g., building design variables, material attributes, ambient conditions) external to the simulation tool to provide a varied and thorough array of input possibilities.

Design objectives, or the surrogate model's inputs (sometimes called "features"), are defined in the first phase. Choosing the inputs and outputs is crucial since altering them later requires more high-fidelity model simulations. The design objective informs the selection of outputs. Like optimization techniques, a surrogate facilitates the study of a particular building design aspect, such as energy efficiency, encoded in the surrogate outputs. To avoid the curse of dimensionality, which states that the number of simulation samples required to produce an accurate surrogate of the design space increases exponentially with the number of parameters, the number of design parameters should be kept to a minimum [26]. In addition, the design team must evaluate many building performance indicators, including energy efficiency, thermal comfort, and daylight availability. The complexity and the form of the output distribution may vary significantly. Therefore, the metamodeling technique must exhibit robustness to the quantity of inputs and the nature of the output.

Figure 3 illustrates the integrated approach suggested in this chapter for using AI in high-performance building design. Numerous databases, such as those about climate and weather, building and material specifications, occupancy patterns, sensor data, geometric inputs like 3D models linked to BIM, and benchmark performance records, are used at the start of the process. These inputs feed into modeling tools such as EnergyPlus and OpenStudio, and simulation-ready real models derived from LiDAR or BIM scans. Tools like Honeybee and Dragonfly replicate the buildings' energy and daylighting performance. Data export tools prepare the tabular and structured simulation outputs for later use. For example, hourly EnergyPlus statistics can be exported in tabular CSV format. These findings are then utilized to train a neural network that predicts thermal comfort across various zones. Numerous black box models, including neural networks, support vector machines, random forests, surrogate models, and reinforcement learning algorithms, are subsequently informed by their outputs. Lastly, AI applications are built on this foundation to support real-time building operation, generative design, space planning, rule-based decision-making, and predictive energy forecasting. In addition to bridging data, simulation, and decision support, this all-inclusive and adaptable workflow structure supports the chapter's focus on switching from conventional energy modeling to AI-enhanced optimization for sustainable design.

4 Methodology and Data

AI algorithms analyze data from various sources, such as weather patterns, occupancy rates, and energy consumption trends, to enhance building energy efficiency. Predictive analytics facilitates intelligent control systems that dynamically regulate heating, cooling, and lighting, reducing energy waste and operational expenses. Machine learning algorithms consistently enhance energy-saving measures by analyzing past data.

The metamodeling types discussed in this chapter are AI techniques for various datasets. As a result, the metamodels are developed from a collection of input and output data to facilitate predictions of future outputs. This stage needs multiple building simulations to establish the metamodel for expedited forecasts of building performance.

5 Results

Combining AI with traditional methods like modeling, simulation, and analytics can transform several phases of building design, construction, and operations [27]. AI developments at the building design stage have previously unheard-of potential to maximize sustainability objectives across various engineering and architectural factors [28]. For example, AI-driven generative design may quickly evaluate several design alternatives while considering various characteristics to reduce embodied carbon and improve sustainability. AI-driven procedures can find the best solutions with a significantly lower environmental impact than traditional approaches by iteratively creating and assessing concepts. Predictive models for building energy consumption are used with optimization algorithms since they are one of the most important instruments for evaluating various energy-saving tactics. These models may be data-driven or physical. Software for building performance simulation (BPS) uses physical models. However, physical models take a long time to produce reliable results [29]. Because optimization techniques need many iterations to obtain convergence, their application is impractical in some situations. To get around this, prediction models for building energy usage based on data and machine learning were created and used in conjunction with optimization techniques [30].

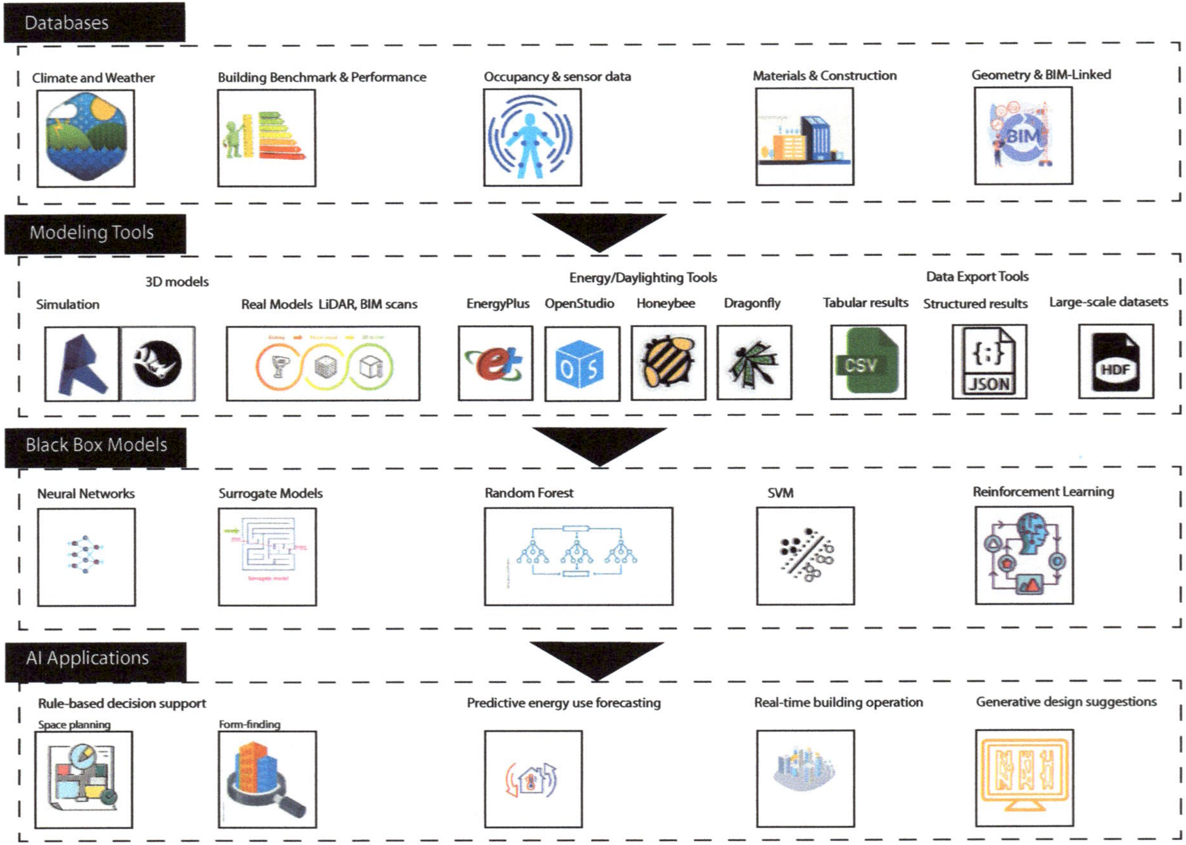

Fig. 3 Workflow of AI-enhanced sustainable building design

While some studies use actual data-driven methods for optimizing occupant behavior, others use simulation and physical results to create prediction models. For example, to determine the ideal cooling power values for maximum energy efficiency and indoor thermal comfort, Yang et al. [31] used a hybrid optimization approach in conjunction with two adaptive artificial neural networks (ANN)-based models for predicting air temperatures and predicted mean votes (PMVs). Similarly, Bunning et al. [32] proposed a random forest-based data predictive control strategy to find the best cooling settings for reducing cooling energy consumption while maintaining room temperatures within comfort limits. A three-layer feed-forward ANN model was created by Asadi et al. [33] to predict the energy consumption of space heating and cooling, sanitary hot water, and the overall percentage of discomfort hours. The model was then combined with a multi-objective genetic algorithm to determine the best energy consumption, retrofit cost, and total percentage of discomfort hours for windows, solar collector installation, external wall insulation, roof insulation, and various HVAC system variables. These initiatives made significant advances in the field of building energy optimization.

5.1 Practical Applications

AI can help with design modeling, optimize energy efficiency, and enable predictive maintenance. The aims of SDGs 7: Affordable and Clean Energy and 11: Sustainable Cities and Communities can be directly supported by the use of AI in designing and operating sustainable buildings. AI-driven strategies that support resource conservation, energy efficiency, and the general sustainability of urban environments include decentralized energy management, real-time optimization, and predictive modeling. AI plays an essential role in optimizing energy production, distribution, and consumption inside buildings and throughout the smart grid.

By dynamically modifying HVAC, lighting, and other building systems in response to real-time data and occupancy patterns, case studies of AI-driven energy management systems show how they can improve energy efficiency. AI-driven energy management systems guarantee optimal efficiency while lowering dependency on fossil fuels by combining renewable energy sources like solar and wind. Predictive maintenance solutions driven by AI, for example, save energy waste by anticipating equipment breakdowns and facilitating prompt interventions, thus prolonging the life of vital building infrastructure. SDG 7 emphasizes enhancing energy efficiency and the proportion of renewable energy sources in urban settings that align with these developments. Real-time data, smart building automation, and AI-driven digital twin support resilient, inclusive, and sustainable urban development, which advances SDG 11. Digital twins offer real-time insight for decision-making.

Meanwhile, sophisticated machine learning algorithms support data-driven analysis. Utilizing AI across the building's lifecycle, encompassing design, construction, operation, and maintenance, presents a distinctive chance to alleviate systemic inefficiencies and foster significant transformation. Moreover, with the advancement of AI technology, there is an increasing emphasis on utilizing decentralized and autonomous systems to enhance building operations and resource management. AI-driven decentralized energy systems can optimize energy generation, storage, and distribution within buildings and across smart grids, thereby boosting resilience and sustainability while diminishing reliance on centralized energy sources.

Several real-world examples of AI implementation in sustainable design exist. For example, AI-powered digital twins offer virtual simulations and predictive modeling [34], allowing stakeholders to anticipate performance outcomes and optimize building designs for sustainability and resilience [35].

AI technologies provide features that improve efficiency during the construction phase, particularly in waste reduction. Construction sites may be tracked in real-time to spot inefficiencies and reduce risks using computer vision and AI-driven analytics. AI-enabled material and equipment tracking, for instance, has been demonstrated to cut waste by more than 40% [28], which results in significant economic savings and environmental advantages [36]. Additionally, material performance, durability, and embodied carbon emissions can be predicted by AI integration [37], allowing for better material selection and building techniques.

After buildings are up and running, AI is still essential for optimizing performance and efficiency in the operation stage. Machine learning algorithms can optimize the control of several building systems, including HVAC, lighting, and security, using real-time data and occupancy patterns. AI-enabled automation can dramatically lower energy usage and operating costs while preserving occupant comfort and productivity by dynamically modifying system settings in response to shifting conditions [38]. According to studies, AI-driven building automation might save commercial buildings 20–30% on energy [39], while residential buildings could save 8.48% on energy and 7.52% on cost [40]. AI reduces downtime and enhances building performance by predicting and diagnosing maintenance and repair needs. Predictive maintenance systems driven by AI can, for instance, evaluate data from building sensors to forecast when equipment is likely to break, enabling proactive maintenance and reducing on equipment downtime. At this point, AI can assist in integrating renewable energy sources, such as wind and solar, into building systems to optimize energy use and output. AI-powered energy management systems, for example, can evaluate real-time data from building loads and renewable energy sources, appropriately modifying energy output and consumption to optimize efficiency and lower energy costs [41].

Additionally, Pieter de Wield [42] classified AI applications into intelligent or smart buildings, emphasizing responsiveness to organizational and human demands through integrating the Internet of Things (IoT), which establishes a network of connected objects, allowing wireless sensing and data collection. Digital twins also function as real-time equivalents, facilitating better interventions, cost savings, enhanced performance, and societal advantages. He demonstrated how different technologies have interrelated functions in the context of sustainable construction lifecycles. BPS models predict physical building performance, while data-driven intelligence with machine learning models optimizes energy use and forecasts maintenance requirements. Building information models (BIMs) offer thorough digital representations of buildings, while digital twins serve as real-time counterparts, providing dynamic representations of buildings.

To reduce urban heat island impacts and improve indoor environmental quality, case studies demonstrate how AI-powered simulations and predictive analytics empower stakeholders to design climate-responsive, energy-efficient buildings. Computer vision and AI-driven monitoring systems support sustainable urban growth, improving materials efficiency and reducing building waste. Reducing carbon emissions and improving urban air quality are benefits of AI-integrated mobility and transportation solutions.

5.2 Challenges and Considerations

Even if AI technology has advanced significantly, the construction sector still has a way to go before embracing new developments. This delay can be due to the intricacy of building projects, the sector's conventional character, and the need for a deeper understanding of AI's potential benefits.

There exists substantial, unexploited potential in utilizing AI for various building components, including generative design, construction automation, and lifetime optimization, techniques that would diminish the environmental impact of structures [42]. Nonetheless, it is imperative to address the issues intrinsic to technology, such as data availability, validation, and industrial acceptance. Because of worries about disruption, risk aversion, and dependence on traditional methods, the construction industry has been hesitant compared to other industries that have embraced AI more readily, such as finance or healthcare. These difficulties include initial implementation costs, privacy and data security issues, a lack of standardization, a skills gap, interoperability problems, ethical issues, and regulatory compliance.

The ideal scenario will incorporate the computational design process of integrating AI into construction as a file-to-factory process. In this context, the "Construction 5.0" concept emerges from this paradigm. "Construction 4.0" and "Construction 5.0" refer to the progression of the construction sector with the integration of new technologies and innovative methodologies. They draw inspiration from the principles of "Industry 4.0" and "Industry 5.0," embodying digital transformation and human-centered innovations in the sector [43, 44]. Construction 4.0 emphasizes the digital age and automation, whereas "Construction 5.0" incorporates human values, sustainability, and collaboration into technological evolution. In this sense, "Construction 5.0" aims to harmonize technological advancements with human requirements, promoting a sustainable and resilient construction ecosystem that aligns with global sustainability goals, especially the United Nations' SDGs 7 and 11. It also advocates incorporating renewable energy sources in construction initiatives, including solar and wind energy [45]. In this context, smart networks and energy management systems enhance the efficiency of renewable energy use, guaranteeing a sustainable power supply for both construction sites and finished structures.

6 Future Perspectives and Implications

We expect the developments in real-time adaptive control systems, generative design, and deep learning will propel the evolution of AI in sustainable architecture. Continuous monitoring and adaptive reactions to shifting environmental circumstances will be possible by combining AI, digital twins, and IoT. Developing explainable AI may also reduce concern about the dependability and transparency of sustainable building practices by boosting confidence in AI-driven decision-making. The growing usage of AI-powered generative design tools is another noteworthy development. Early in the design phase, these tools can optimize energy efficiency, daylighting, and thermal comfort by automatically generating and refining design solutions based on sustainability criteria.

AI-driven material innovation will make creating high-performance, environmentally friendly materials for energy-efficient buildings easier. The lasting incorporation of AI in architectural design and building processes will fundamentally transform building development, construction, and maintenance. Automation powered by AI in performance optimization and predictive analytics will diminish human errors, improve energy efficiency, and promote circular economic principles by reducing material waste. AI's predictive maintenance function would prolong the longevity of buildings, hence decreasing operating expenses and environmental repercussions. The adoption of AI significantly transforms design approaches. Architects and engineers will interact with AI as co-creators rather than depending primarily on human intuition and conventional optimization methods. This transition will need novel skills and compelling design professionals to cultivate proficiency in AI to enhance design-making and data-informed strategies.

7 Conclusion

Incorporating AI in sustainable building design indicates a significant change in methodologies employed by architects, engineers, and urban planners regarding energy efficiency, environmental effects, and occupant comfort. This chapter examines the role of AI-driven approaches, such as machine learning, generative design, and predictive analytics, to enhance building performance throughout the design, construction, and operation phases. Utilizing AI in the AEC sector may transition from traditional trial-and-error methods to data-driven, evidence-based decision-making procedures that improve sustainability results. AI-driven tools allow architects to investigate intricate design scenarios swiftly, recording several goals such as daylight optimization, thermal comfort, and energy efficiency. Moreover, AI-driven real-time monitoring and predictive maintenance solutions enhance long-term building efficiency and resilience. These achievements correspond with goal sustainability objectives, including SDGs 7 and 11, underscoring AI's capacity to promote ecologically sustainable urban development.

Nonetheless, although AI presents significant opportunities, problems persist regarding data accessibility, computing requirements, and ethical implications. Ensuring openness and explainability in AI-driven decisions and promoting interdisciplinary cooperation will be essential for optimizing AI's advantages in sustainable design. Subsequent research should focus on improving AI models to augment their interpretability, scalability, and adaptability to various climatic and geographic circumstances. By utilizing AI ethically and inclusively, the AEC industry can spearhead the shift towards a future in which buildings are functional, aesthetically appealing, energy-efficient, and sustainable.

References

1. Evins R (2013) A review of computational optimisation methods applied to sustainable building design. Renew Sustain Energy Rev 22:230–245. https://doi.org/10.1016/j.rser.2013.02.004
2. Nguyen A-T, Reiter S, Rigo P (2014) A review on simulation-based optimization methods applied to building performance analysis. Appl Energy 113:1043–1058. https://doi.org/10.1016/j.apenergy.2013.08.061
3. Attia S, Hamdy M, O'Brien W, Carlucci S (2013) Assessing gaps and needs for integrating building performance optimization tools in net zero energy buildings design. Energy Build 60:110–124. https://doi.org/10.1016/j.enbuild.2013.01.016
4. Shi X (2010) Performance-based and performance-driven architectural design and optimization. Front Architect Civ Eng China 4:512–518
5. Gupta P, Anand S, Gupta H (2017) Developing a roadmap to overcome barriers to energy efficiency in buildings using best worst method. Sustain Cities Soc 31:244–259. https://doi.org/10.1016/j.scs.2017.02.005
6. di Stefano AG, Ruta M, Masera G, Hoque S (2025) Integrated workflow development for data-driven neighborhood-scale building performance simulation. ASME J Eng Sustain Build Cities 6(1):1. https://doi.org/10.1115/1.4066565
7. Carlucci S, Cattarin G, Causone F, Pagliano L (2015) Multi-objective optimization of a nearly zero-energy building based on thermal and visual discomfort minimization using a non-dominated sorting genetic algorithm (NSGA-II). Energy Build 104:378–394. https://doi.org/10.1016/j.enbuild.2015.06.064
8. Shi X, Tian Z, Chen W, Si B, Jin X (2016) A review on building energy efficient design optimization from the perspective of architects. Renew Sustain Energy Rev 65:872–884. https://doi.org/10.1016/j.rser.2016.07.050
9. Beckman WA, Broman L, Fiksel A, Klein SA, Lindberg E, Schuler M, Thornton J (1994) TRNSYS The most complete solar energy system modeling and simulation software. Renew Energy 5(1–4):486–488. https://doi.org/10.1016/0960-1481(94)90420-0
10. Van Gelder L, Das P, Janssen H, Roels S (2014) Comparative study of metamodelling techniques in building energy simulation: Guidelines for practitioners. Simul Model Pract Theory 49:245–257. https://doi.org/10.1016/j.simpat.2014.10.004
11. Dogan T, Reinhart C, Michalatos P (2012) Urban daylight simulation calculating the daylit area of urban designs. SimBuild Conference 2012(5):613–620
12. Geebelen B, Van Der Voorden M, Neuckermans H (2005) Fast and accurate simulation of long-term daylight availability using the radiosity method. Light Res Technol 37(4):295–310
13. Zuo W (2011) Acceleration of Radiance for lighting simulation by using parallel computing with OpenCL
14. Westermann P, Evins R (2019) Surrogate modelling for sustainable building design—a review. In: Energy and buildings, vol 198. Elsevier Ltd., pp 170–186. https://doi.org/10.1016/j.enbuild.2019.05.057

15. Østergård T, Jensen RL, Maagaard SE (2017) Early Building Design: Informed decision-making by exploring multidimensional design space using sensitivity analysis. Energy Build 142:8–22. https://doi.org/10.1016/J.ENBUILD.2017.02.059
16. Košir M, Iglič N, Kunič R (2018) Optimisation of heating, cooling and lighting energy performance of modular buildings in respect to location's climatic specifics. Renew Energy 129:527–539. https://doi.org/10.1016/j.renene.2018.06.026
17. Geyer P, Singaravel S (2018) Component-based machine learning for performance prediction in building design. Appl Energy 228:1439–1453. https://doi.org/10.1016/j.apenergy.2018.07.011
18. Ayoub M (2020) A review on machine learning algorithms to predict daylighting inside buildings. Sol Energy 202:249–275. https://doi.org/10.1016/j.solener.2020.03.104
19. Westermann P, Evins R (2021) Using Bayesian deep learning approaches for uncertainty-aware building energy surrogate models. Energy and A I:3. https://doi.org/10.1016/j.egyai.2020.100039
20. Esteghamati MZ, Flint MM (2021) Developing data-driven surrogate models for holistic performance-based assessment of mid-rise RC frame buildings at early design. Eng Struct 245:112971. https://doi.org/10.1016/j.engstruct.2021.112971
21. Pinto G, Deltetto D, Capozzoli A (2021) Data-driven district energy management with surrogate models and deep reinforcement learning. Appl Energy 304:117642. https://doi.org/10.1016/j.apenergy.2021.117642
22. Prosser P (1993) Hybrid algorithms for the constraint satisfaction problem. Comput Intell 9(3):268–299. https://doi.org/10.1111/j.1467-8640.1993.tb00310.x
23. Ting TO, Yang X-S, Cheng S, Huang K (2015) Hybrid metaheuristic algorithms: past, present, and future. Recent Adv Swarm Intell Evol Comput. https://doi.org/10.1007/978/3-319-13826-8-4
24. Abioye SO, Oyedele LO, Akanbi L, Ajayi A, Delgado JMD, Bilal M, Akinade OO, Ahmed A (2021) Artificial intelligence in the construction industry: A review of present status, opportunities and future challenges. J Build Eng 44:103299. https://doi.org/10.1016/j.jobe.2021.103299
25. Li F, Laili Y, Chen X, Lou Y, Wang C, Yang H, Gao X, Han H (2023) Towards big data driven construction industry. J Ind Inf Integr. https://doi.org/10.1016/j.jii.2023.100483
26. Ong, Y. S., Nair, P. B., Keane, A. J., & Wong, K. W. (2005). Surrogate-assisted evolutionary optimization frameworks for high-fidelity engineering design problems. In: Knowledge incorporation in evolutionary computation, pp 307–331
27. Bigham GF, Adamtey S, Onsarigo L, Jha N (2019) Artificial intelligence for construction safety: mitigation of the risk of fall. In: Intelligent systems and applications: Proceedings of the 2018 intelligent systems conference (IntelliSys), vol 2, pp 1024–1037
28. Aste N, Manfren M, Marenzi G (2017) Building automation and control systems and performance optimization: a framework for analysis. Renew Sustain Energy Rev 75:313–330. https://doi.org/10.1016/j.rser.2016.10.072
29. Chari A, Christodoulou S (2017) Building energy performance prediction using neural networks. Energ Effi 10(5):1315–1327
30. Reynolds J, Rezgui Y, Kwan A, Piriou S (2018) A zone-level, building energy optimisation combining an artificial neural network, a genetic algorithm, and model predictive control. Energy 151:729–739. https://doi.org/10.1016/j.energy.2018.03.113
31. Yang S, Wan MP, Chen W, Ng BF, Dubey S (2020) Model predictive control with adaptive machine-learning-based model for building energy efficiency and comfort optimization. Appl Energy 271:115147
32. Bünning F, Huber B, Heer P, Aboudonia A, Lygeros J (2020) Experimental demonstration of data predictive control for energy optimization and thermal comfort in buildings. Energy Build 211:109792. https://doi.org/10.1016/j.enbuild.2020.109792
33. Asadi E, Da Silva MG, Antunes CH, Dias L, Glicksman L (2014) Multi-objective optimization for building retrofit: a model using genetic algorithm and artificial neural network and an application. Energy Build 81:444–456. https://doi.org/10.1016/j.enbuild.2014.06.009

34. Adewale BA, Ene VO, Ogunbayo BF, Aigbavboa CO (2024) A systematic review of the applications of AI in a sustainable building's lifecycle. In: Buildings, vol 14, Iss 7. Multidisciplinary Digital Publishing Institute (MDPI). https://doi.org/10.3390/buildings14072137
35. Rane N (2023) Integrating leading-edge artificial intelligence (AI), internet of things (IOT), and big data technologies for smart and sustainable architecture, engineering and construction (AEC) industry: Challenges and future directions. In: Engineering and Construction (AEC) Industry: Challenges and Future Directions (September 24, 2023)
36. Alahi MEE, Sukkuea A, Tina FW, Nag A, Kurdthongmee W, Suwannarat K, Mukhopadhyay SC (2023) Integration of IoT-enabled technologies and artificial intelligence (AI) for smart city scenario: recent advancements and future trends. Sensors 23(11):5206. https://doi.org/10.3390/s23115206
37. Gaur L, Afaq A, Arora GK, Khan N (2023) Artificial intelligence for carbon emissions using system of systems theory. Eco Inform 76:102165
38. Yan H, Yan K, Ji G (2022) Optimization and prediction in the early design stage of office buildings using genetic and XGBoost algorithms. Build Environ 218:109081. https://doi.org/10.1016/j.buildenv.2022.109081
39. Miller C, Meggers F (2017) The Building Data Genome Project: An open, public data set from non-residential building electrical meters. Energy Procedia 122:439–444. https://doi.org/10.1016/j.egypro.2017.07.400
40. Long LD (2023) An AI-driven model for predicting and optimizing energy-efficient building envelopes. Alex Eng J 79:480–501. https://doi.org/10.1016/j.aej.2023.08.041
41. de Wilde P (2023) Building performance simulation in the brave new world of artificial intelligence and digital twins: a systematic review. Energy Build 292:113171. https://doi.org/10.1016/j.enbuild.2023.113171
42. Regona M, Yigitcanlar T, Xia B, Li RYM (2022) Opportunities and adoption challenges of AI in the construction industry: A PRISMA review. J Open Innov: Technol Market Complex 8(1):45. https://doi.org/10.3390/joitmc8010045
43. Forcael E, Ferrari I, Opazo-Vega A, Pulido-Arcas JA (2020) Construction 4.0: a literature review. Sustainability 12(22):9755. https://doi.org/10.3390/su12229755
44. Perrier N, Bled A, Bourgault M, Cousin N, Danjou C, Pellerin R, Roland T (2020) Construction 4.0: A survey of research trends. J Inf Technol Constr 25:416–437. https://doi.org/10.3680/j.itcon.2020.024
45. Liu H, Ju Q, Zhao N, Li H, Skibniewski MJ (2022) [Retracted] value proposition for enabling construction project innovation by applying building information modeling. Comput Intell Neurosci 2022(1):2586307. https://doi.org/10.1155/2022/2586307

Image-Based Machine Learning for Preliminary Condition Categorization of Traditional Houses: A Case Study in Menteşe, Muğla

Veli Mustafa Yönder

Abstract In the context of sustainability, monitoring and maintenance of traditional buildings is crucial. The concept of artificial intelligence can be an important instrument for the preservation and sustainability of tangible cultural heritage assets and can be considered to be of great importance for buildings such as historic houses. This study investigates the implications of applying machine learning methods in the preservation-oriented maintenance of selected local masonry houses in Menteşe, Muğla (Türkiye). In this study, the physical condition of traditional houses is analyzed through facade photographs to test whether they can provide preliminary architectural information to the users before preparing maintenance or repair planning. The tags used in this study were categorized into two groups: those in need of maintenance and those in need of simple repair. The machine learning model is trained at different parameter values, and then confusion matrix tables, validation and loss plots are obtained. By categorizing vernacular houses according to their physical condition, this study attempts to contribute to the growing body of knowledge and prioritize a new approach for sustainable design interventions. Thus, it is thought that AI-supported analyses will have the potential to provide information about the current state of traditional structures.

Keywords Architectural conservation · Sustainability · Teachable machine · Categorization · Menteşe · Muğla

1 Introduction

Conservation, maintenance, repair, refunctioning, and rehabilitation of vernacular architectural heritage are important in sustainability design principles. In this context, it is now important to study the concept of vernacular architecture and demystify its

V. M. Yönder (✉)
Kütahya Dumlupınar University, Kütahya, Türkiye
e-mail: velimustafa.yonder@dpu.edu.tr

M. Sedrez and G. Pitts (eds.), *Artificial Intelligence-Aided Design for Sustainability*, Urban Sustainability, https://doi.org/10.1007/978-981-95-1349-9_12

relationship with sustainability principles. According to Pardo [1, p. 2], vernacular architecture is defined by: (a) the utilization of materials gathered locally; (b) "a planimetric design adapted to the climatic and topographical conditions"; (c) using construction methods and aesthetic features that develop through historical adaptation; and (d) an active role of users and regional artisans in the creation of the building. Thus, vernacular building construction is multi-layered, and both the designs and materials have distinctive characteristics. Additionally, traditional/vernacular dwellings' "structural, geometrical, aesthetic as well as decorative qualities were shaped through centuries" [2, p. 57]. Accordingly, conventional construction techniques and knowledge are among the attributes worthy of preservation. In addition to preserving tangible assets such as vernacular buildings and structural elements, tacit knowledge, and other intangible elements such as craftsmanship experience and knowledge of workmanship also need to be protected. For this reason, the conservation field of rural houses is multidimensional and can be divided into subheadings. This study focuses on tangible cultural heritage elements rather than intangible architectural heritage elements.

The first step to be taken for the conservation of buildings or groups of structures is to determine and analyze in detail the status of the building or group of buildings. In this context, determining intervention type and achieving the physical conditions of traditional structures can be considered as essential elements. The types of interventions can be classified as maintenance, simple repairs, major repairs/restorations, and reconstruction process [3]. Major repair/restoration and conservation operations are carried out under the responsibility of the conservation architect. The conservation architect who "is the generalist in the whole building conservation process" must have extensive knowledge of architectural history and modern construction methods in order to "preserve the artistic and historical value" of heritage buildings while also responding to contemporary needs [4, p. 13]. Local authorities analyze and approve every aspect of restoration work or reconstruction, like conservation boards (These boards are under the Ministry of Culture and Tourism), the General Directorate of Pious Foundations, municipalities, and relevant official offices in Türkiye. For this reason, conservation architects carefully investigate the building and use the conservation data to make their design decisions. In this context, comprehensive data on the building's current condition (e.g., outdoor-indoor photos and field notes), and detailed survey drawings are required. Moreover, computational techniques (e.g., tacheometry data, photogrammetry, terrestrial laser scanning, point cloud surveying, heritage building information modelling, digital twin, geographic information system, game engine, and parametric modeling) and artificial intelligence algorithms (e.g., scene classification, instance segmentation, object recognition, clustering, and generative models) can be utilized in preparing and analyzing conservation data and also heritage digitization stages by conservation architects. A conservation architect familiar with computational approaches must possess expertise in algorithmic methodologies, problem-solving abilities, and methodological approaches, alongside a comprehensive understanding of architectural history and traditional construction principles. For instance, Historic building information modeling (HBIM) [5], a 3D digital technique, can effectively facilitate the creation of 3D models and 2D drawings of structures and their elements. The HBIM system enables the generation of detailed 3D models, semantic characteristics, visualizations of architectural features, cost analysis, 4D analysis (at both micro and macro levels), and shop drawings. The development of HBIM systems requires point cloud data from laser scanners or data from photogrammetry techniques.

Computational methods [6, 7] and algorithmic design methods [8] have great capability to be applied to architecture. Furthermore, computational designers and architects use artificial intelligence (AI) algorithms and techniques. In this context, it is possible to evaluate facades using architectural facade databases and algorithms, as well as to generate new designs within the framework of sustainable design principles. Comparable techniques utilizing machine learning (ML) approaches have been implemented in prior research, including the application of deep learning (Pix2Pix GAN) for facade segmentation and detection as part of a scalable and automated reconstruction workflow [9], the use of improved generative adversarial networks (i.e., pix2pix) for facade generation and recognition through two facade datasets: "Center for Machine Perception (CMP) Facade dataset" and self-created traditional Chinese facade datasets [10], and the employment of Stable Diffusion model, LoRA (Low-Rank Adaptation) models, and ControlNet in urban renewal initiatives via historic arcade facade image dataset [11]. Therefore, akin methods, including deep learning algorithms (e.g., Pix2Pix, diffusion models, large language models, small language models, and generative adversarial neural nets), digital tools (e.g., ComfyUI), and image classification models (e.g., convolutional neural nets and vision transformers), have potential for use in adjacent fields, such as historic site preservation and vernacular building conservation.

Machine learning-based processes and methods can also find application in the field of building maintenance, repair, restoration, or adaptive reuse, especially in the context of a sustainable design process. In this context, this study employs an artificial intelligence-based methodology in lieu of traditional techniques (i.e., human expert-based observations and assessments for monitoring a building's physical condition). Particularly, using a machine learning-based approach through a dataset of non-orthogonal photographs, this study categorized the current needs of vernacular houses into "requiring maintenance" and "requiring simple repair" using facade images. In the study of Bahrami and Albadvi [12], deep learning methods applied to Iran's cultural heritage buildings, demonstrated the efficiency of using transfer learning (TL) with pre-trained ResNet, MobileNet, and Inception models for binary categorization (i.e., label one is including physical defects in buildings and label zero not including), showing an accuracy of 90%. A similar binary classification was made in their study, but in this study, an attempt was made to formulate the classes based on the cultural heritage conservation regulations. Additionally, in this study, a machine learning model was created using a web-based tool (i.e., Google's Teachable Machine) [13] without the need for deep code knowledge.

Traditional Menteşe houses (Muğla) can be considered as one of the valuable elements of Anatolian folk architecture (specifically for the Aegean zone), considering their structures, compositions, aesthetic values, and architectural features. Therefore, within the scope of this study, some traditional Menteşe houses were selected, and a facade-based dataset was created. An attempt was made to compile the curated dataset into two categories. Detailed creation of the training dataset and increasing data quality were also considered as important elements within the scope of this study. It can be considered that the creation of the facade dataset in terms of quality and within specific criteria is important for training machine learning (ML) models. During the training phase, various hyperparameters were tried, and experiments were performed with different values. By analyzing different values, it was possible to compare the results. Therefore, this study can be considered proof of concept with the abovementioned criteria. In addition, this study can indirectly contribute to the digital steps to be taken in the maintenance of traditional houses. This study is expected to contribute to the field by testing a digital approach that can be used by conservation boards, municipalities, museums, or civil society organizations based on local tangible heritage data.

1.1 Research Questions and Scope of this Study

Sustainability and conservation share values like responsible resource use, limited intervention to protect cultural and physical integrity, and "constructive evolution" over destructive change [14, p. 60]. In this context, maintaining and repairing existing structures is crucial for energy-efficient and sustainable urban environments. This study aims to employ an ML model on photographs of the facades of single- or two-story traditional buildings in the Menteşe district of Muğla province (Aegean region, Türkiye) to assess their existing status and aid users and the conservation architect in determining maintenance or repair options. This study's initial research issue investigates the feasibility of developing an ML-assisted system to assess the status of vernacular residences, serving as a preliminary step in the preservation of traditional dwellings through an image classification model. The second question is to examine the conditions that might enable this occurrence, should such a system be constructed, specifically if high accuracy scores are attained on validation and test datasets. If the required scores are not attained, this study emphasizes identifying areas for development and refinement in the process (e.g., model training, dataset quality, selection of misinterpreted photos). Ultimately, if this system is established, it is crucial to comprehend how conservation architects and non-expert users would get advantages within the framework of sustainable design.

2 Sustainable Design Through Traditional Buildings

In the context of sustainability, building maintenance, repair, or restoration can contribute to many economic, cultural, and social benefits as it helps keep buildings in use. Conservation of buildings with an emphasis on sustainability and providing eco-friendly interventions is exceedingly beneficial. Besides, the adaptive reuse of existing structures markedly decreases waste, conserves energy by retaining the embodied energy of construction and materials, and can preserve cultural heritage by keeping historic buildings in current use [15]. Therefore, such conservation-oriented approaches play a critical role in achieving sustainability goals. Refurbishment scenarios that emphasize preservation have the potential to enable the reuse of original materials, allow for efficient intervention planning, reduce costs, leverage embodied energy, identify areas most vulnerable to climate-induced degradation, minimize waste, and ultimately extend the building's lifetime [16]. Such approaches not only contribute to sustainability goals but are also in line with broader heritage conservation strategies. Furthermore, "adaptive reuse of historic buildings has increasingly emerged in urban conservation, in particular, in the developing countries" [17, p. 360]. This can be considered not only as a contribution to cultural heritage and urban memory, but also as a step towards sustainable development. However, preserving and developing the historic and cultural identity of the building or group of buildings should always be a priority and privilege. In some cases, it may not be possible to re-functionalize traditional houses (except for museum and touristic uses), so it may be considered to continue their regular use by applying processes such as maintenance, repair and reconstruction. Thus, the continuation of the building with the function with which it was first built can contribute to urban memory and collective memory.

Vernacular architecture functions as a valuable repository of knowledge used to optimize natural resources and the environment in relation to dwelling areas designed to respond to social needs; its forms and typological features embody accumulated social practices, preserved and adapted with novel experiences [18]. However, it is thought that this accumulation is not sufficiently reflected in today's architectural practices. Local heritage represents a tremendous resource with considerable potential for describing the norms of sustainable design and modern architecture; yet its strategies and techniques are basically ignored and rarely used in current Western building methods [19]. One of the factors affecting the sustainability of traditional houses is the use of local materials. Nearby quarries or forests are thought to influence the material properties and construction technology of the houses. In addition, the employment of local craftsmen and workers who are familiar with the conventional construction technology can provide opportunities to reduce the cost of construction.

Adaptation of vernacular buildings is also significant for energy efficiency and passive design strategies. Energy-efficient design concepts may be advanced by using recyclable materials or using materials with minimal carbon emissions. Since local techniques have been tried, proven, and tested over the course of time, energy efficiency in architecture is a factor that can be achieved through the implementation of "indigenous solutions" [20, p. 230]. Life Cycle Assessment (LCA) is a key technique used to assess the energy consumption and environmental impact of sustainable buildings and their components; from raw material extraction through construction, use and end of life, embodied energy plays a critical role [21]. Therefore, energy analysis and assessment of historic, rural, or vernacular buildings can be considered valuable in the context of sustainable conservation.

2.1 A Brief Overview of Traditional Houses in Menteşe, Muğla (Türkiye)

Muğla, located in the Aegean region of Türkiye, is renowned for its main tourist destinations, including Bodrum and Datça. Bodrum's traditional architecture also has distinctive features, and the traditional texture has been preserved. Within the scope of this study, some of the traditional houses in the center of Muğla were examined. The traditional Muğla house possesses specific characteristics [22]. Houses built from natural materials suitable for the climatic conditions of the region were used intensively by the local people. However, with the formation of new settlement areas today, a significant population lives in the region, although it is not intensively used. One of the main reasons for this is that it is close to the city center. Muğla city center is characterized by the presence of traditional Menteşe houses, which are important examples of vernacular architecture. These houses are mostly built using stone masonry methods and timber framed systems and the exterior walls (i.e., those facing the street and those adjacent to neighboring plots) are stone masonry, while the inner facade (i.e., facing the courtyard) is timber-framed character [23]. Typically two-storey (there are other variants) and composite, most of these dwellings have a courtyard (mainly used in spring and summer) and continue to be used as living space [23].

When some houses are analyzed in the planimetric layout, there are "*sofas*" (i.e., specific interior or exterior hall). In this setting, these houses conform to the traditional Turkish house typology in terms of their plan type. Most conventional dwellings feature a "*Hayat*" (i.e., exterior hall), a semi-open area that is extensively utilized during the summer and spring months. Kuban [24] asserts that the "*Hayat*" section is a crucial component of a traditional Turkish residence. These semi-open spaces, which are an important element in traditional houses in terms of creating spatial diversity and responding to the different needs of users, can be considered to have the potential to offer different experiences, especially when integrated with the use of the courtyard. Notably, the chimney designs of these houses are both architecturally unique and functionally efficient. Besides, traditional Menteşe houses are situated in a designated protected area, with their condition overseen by official authorities (e.g., conservation boards).

The selected traditional Menteşe house was analyzed on the plan layout using computational and spatial methods. In this context, the Space Syntax method serves as a computational approach for the mathematical study and visualization of spatial configurations. The space syntax analysis functions akin to the "grammar" employed to structure and examine spatial components, with the primary objective of this analysis being to comprehend users [25, p. 290]. Space syntax method is employed for both contemporary structures and the design of modern buildings, as well as for historic buildings [26] and the historic environment [27]. In this regard, a house with a courtyard in the Menteşe district was selected for this study, and a spatial configuration analysis was performed on the first-floor plan using the digital tool depthmapX [28]. The first floor comprises a sofa, rooms, and a staircase. Visibility graph analysis (VGA) and agent analysis were conducted (Fig. 1). The initial research indicates that the "*sofa*", a distinctive feature of the Turkish house, serves as a significant element for visual communication and consistency in this house. As seen in the outputs of other types of analysis, the unique situation of the *sofa* within the rooms is revealed. It can be stated that the *sofa* space stands out compared to other spaces, particularly in terms of visual permeability, accessibility, and ease of wayfinding. When the site plan of the local houses is analyzed, the streets and avenues reflect the historic texture (i.e., narrow and organic streets) and harmony with the topography. For this reason, the houses selected from the study area were chosen for their traditional texture. Besides, field observations indicate that the traditional Menteşe houses have a multi-layered character when read in both site and plan scale. Thus, these houses' conservation, maintenance, and management are of great importance.

3 Methodology

ML methodologies encompass a diverse array of algorithms and frameworks, enabling the identification of patterns within data [29]. The basic concept of artificial intelligence can be deconstructed into three components: input, hidden layers, and output. This three-layer structure is called a "black-box". In this study, in order to make the process more transparent, different parameters were applied in the training process of the model and the results were compared.

There are several studies on the application of ML and deep learning (DL) methodologies for monitoring [30], "semantic segmentation of heritage point clouds" [31, p. 81], analyzing, and documenting cultural assets. Besides, ML algorithms can be utilized for structural health monitoring (SHM) of historic buildings [32] and for facilitating decision-making processes (predicting maintenance request category via LSTM neural networks) [33]. The methodology formulated for this study adopts an ML-based approach aligned with sustainable design principles and prioritizes heritage data-based strategies alongside algorithmic parameters and optimizations (Fig. 2). Data collection and data quality are crucial and rank among the priorities of the data-driven artificial intelligence methodology. The initial phase in this context involves the collection and pre-processing of conservation data, which forms the basis of the applied ML technique. In this study, facade photographs obtained from online sources and personal archives were used. However, it is critical that the obtained facade photographs have a certain quality. Houses located in the urban protected area were specifically included in the dataset. As many buildings have courtyards, some houses do not have photos of the entrance facades. Some houses have more than one facade included in the dataset (especially houses located on corner plots). Facade photographs contain noises such as electrical cables, antennas, streetscapes, and vehicles that block the image, and cropping is required to remove them.

Automatic adjustments for tonality, contrast, and color were implemented to enhance image quality. Two categories of labels have been identified: those requiring

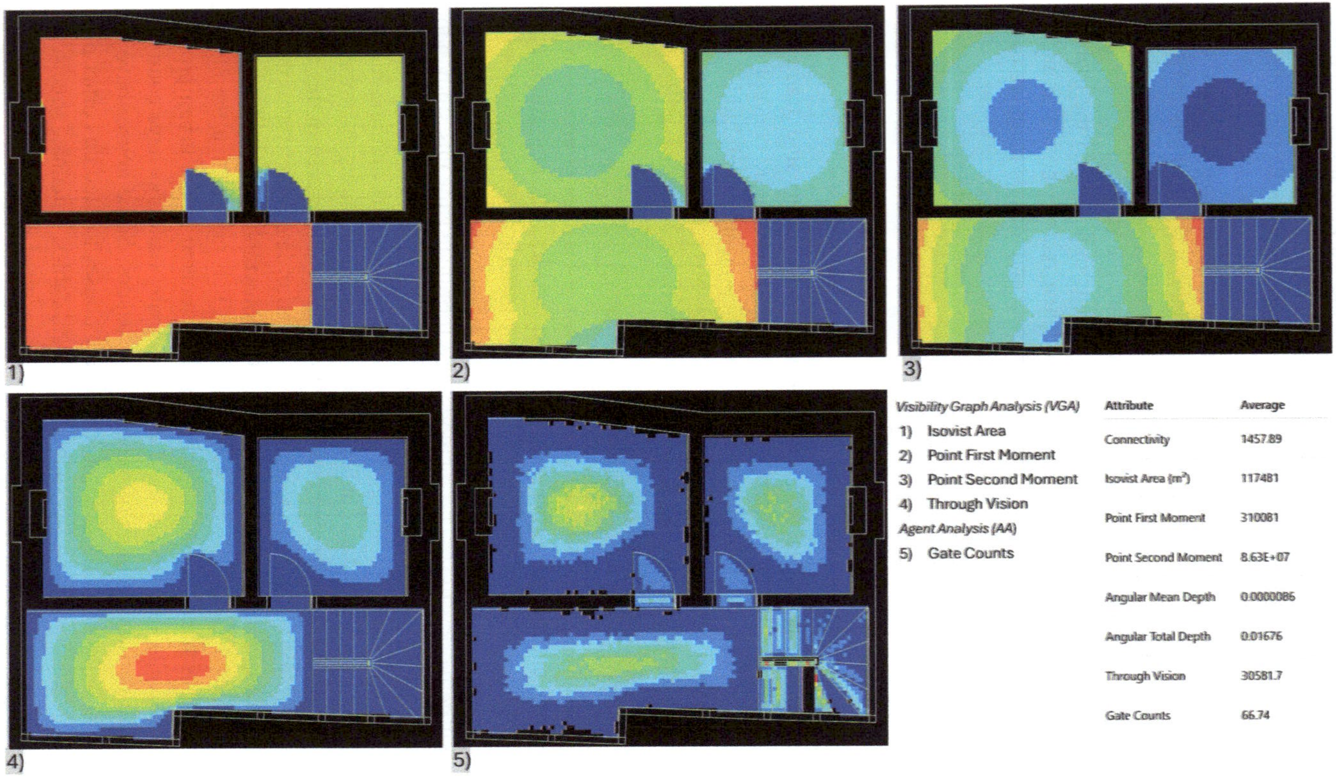

Fig. 1 Visibility graph analysis and agent analysis results (base plan was redrawn from [23])

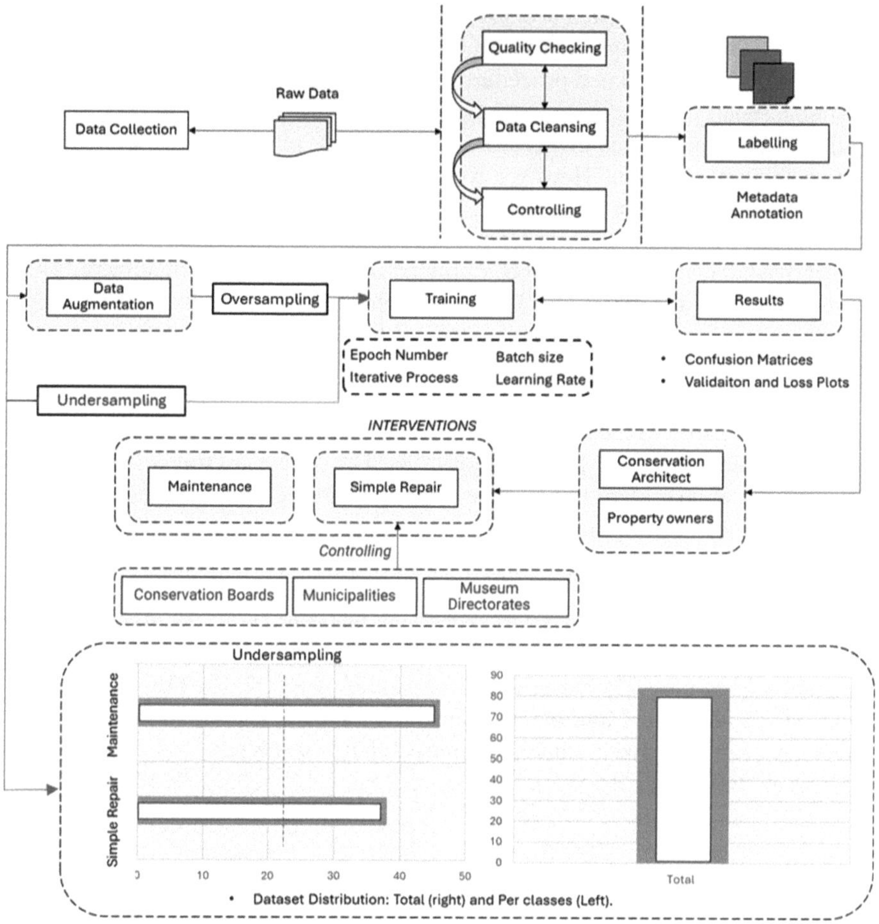

Fig. 2 Image-based Machine Learning Workflow for this Study

simple repairs and those necessitating maintenance relatively. In the regulations for the conservation of cultural heritage, classifications of the state of repair of buildings are defined in various categories within certain standards. In this study, an attempt was made to comply with these definitions, but limited data were obtained, and an imbalance occurred between classes. Specifically, the dataset was unbalanced due to an insufficient number of facade photographs necessitating simple repair. Nevertheless, there exist oversampling and undersampling techniques to address the imbalanced dataset. Initially, the oversampling technique was utilized in this study. In this context, data augmentation methods were implemented to one class (i.e., simple repair class). In addition to being important for machine learning processes, data augmentation techniques are an effective method in cases where there is very little data. It is possible to augment the image data by various methods, but only some methods have been tried because the dataset is sensitive (i.e., cultural heritage data). The dataset was augmented with GIMP (GNU Image Manipulation Program) tool. After data augmentation procedures, the dataset has thirty-eight facade images of "simple repair". Furthermore, undersampling was applied to stabilize the dataset and training was performed in this case as well. Facade photos were deleted from the class containing more data. However, a strategy was adopted to minimize the loss of quality data during this process.

The dataset was trained using Google's Teachable Machine [13], and the results were derived. This web tool has been utilized in numerous investigations [34–37]. In the study of Malahina et al. [34] study highlights the effectiveness of the grid search method in tuning hyperparameters within Teachable Machine, along with the use of data augmentation techniques such as expression variation, rotation and scaling during preprocessing to enhance data diversity and support model generalization. Teachable Machine uses MobileNet architecture as the base model and applies transfer learning for classification tasks. The concept of transfer learning can be considered as an effective method in artificial intelligence processes, as well as a technique that can be used in situations where data is relatively limited. Upon categorization of the dataset, it is systematically partitioned into two categories: training data and test data. Model configurations encompass the number of epochs, learning rate, and batch size parameters. Training was conducted at 25, 50, 75, and 100 epochs, respectively. Subsequently, learning rate values of 0.1, 0.01, 0.001, and 0.0001 were evaluated sequentially. Likewise, batch size values of 16, 32, 64, 128, and 256 were evaluated. Conducting experiments with different parameters and comparing the results has been effective in finding the most suitable settings.

4 Experiment Results

In this study, the hyperparameter settings, such as number of epochs, batch size, and learning rate values, were entered with different values. Besides, datasets are automatically divided into two groups: 85% training, 15% testing. 224 × 224 pixel color images were entered. Training was performed at several epoch values, and testing persisted until higher accuracy levels were attained. Accuracy metrics and the confusion matrix (Fig. 3) were derived for each configuration of settings. Loss graphs were also presented, with the accuracy plots of the testing and training conducted using Google's Teachable Machine (Fig. 4). Values for (A) in Fig. 3; Epochs: 50- Batch Size: 16- Learning Rate: 0.01, and values for (B) Epochs: 50- Batch Size: 16- Learning Rate: 0.001 were recorded.

Due to working with limited data (especially the "Simple Repair" dataset), various tests and experiments (with data augmentation and without) were conducted during model training. As a result of this situation, only the results of two different learning rate values were significant (Figs. 3 and 4). The results of the confusion matrix indicate higher prediction rates in each class. Based on the parameter values shown in Fig. 4, training (blue) and test (orange) accuracies (left column), as well as training

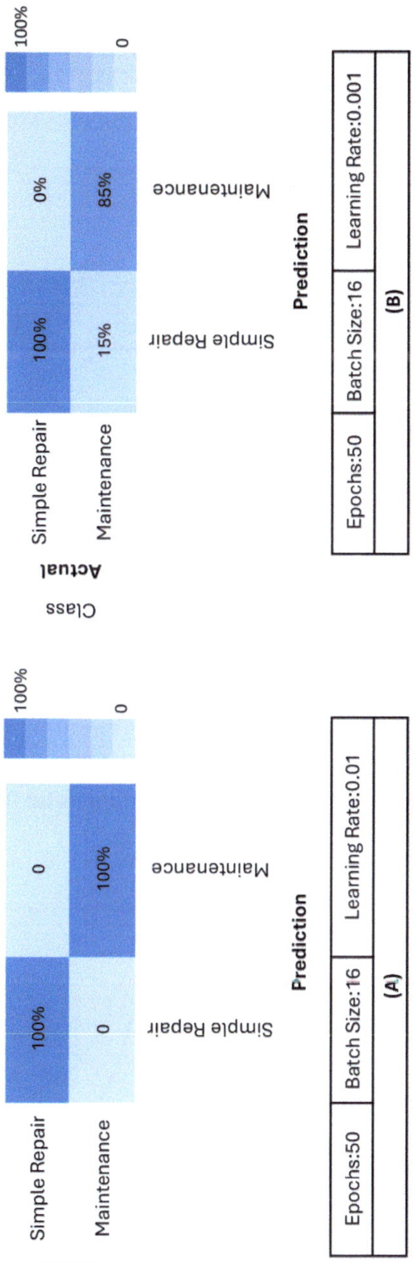

Fig. 3 Normalized outputs of confusion matrices for different experiments (upper) and settings (lower) (redrawn by author)

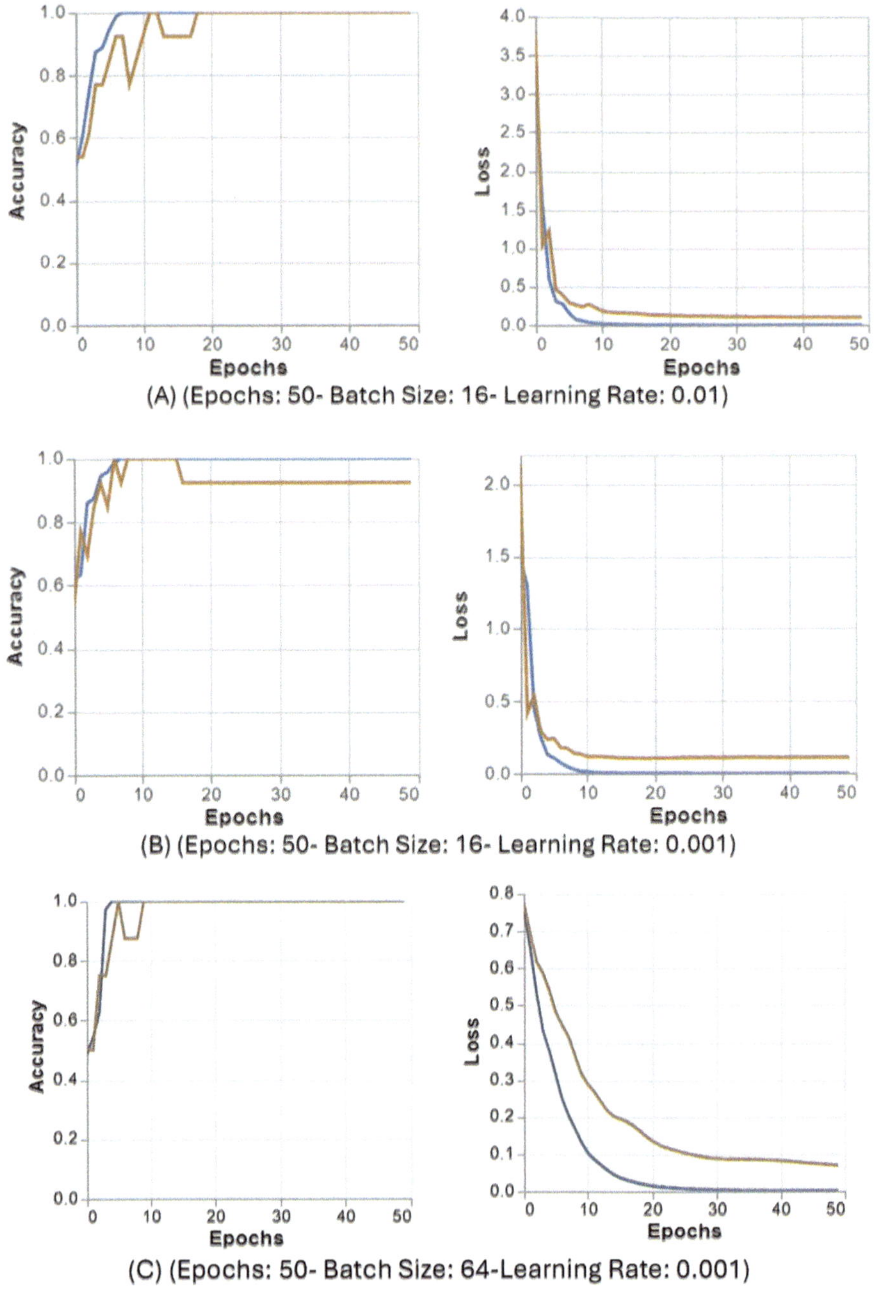

Fig. 4 Accuracy and loss plots for different experiments [training (blue) and test (orange)]

(blue) and test (orange) loss values (right column), were derived via the web tool. The accuracy values for both training and testing are elevated, with a minimal disparity between them (Fig. 4). After 50 epochs of training, the loss graphs for the training and test datasets exhibit slight variation. Efforts were made to avoid problems such as overfitting or underfitting during the training, but lack of qualified data made the process difficult.

Furthermore, to reduce the potential risk of data leakage, the model was trained solely on a limited set of original images without applying any data augmentation (Fig. 4, (C)). A downsampling method was applied to address the issue of an unbalanced dataset. Despite the small dataset, both training and testing metrics showed stable and consistent trends, demonstrating the model's ability to generalize in this limited setting. However, the limited sample size may not fully represent real-world variability.

5 Discussion

The maintenance, repair, rehabilitation, refurbishing, reconstruction, and repurposing of traditional buildings facilitate their reuse by individuals in daily activities, thereby reintegrating the building or group of buildings into the urban or rural environment. Enhancing the physical state of vacant or dilapidated rural buildings is highly beneficial to both the built environment and the populace, while also reducing the expenses associated with new construction. From this perspective, they (e.g., maintenance, repair, restoration, and reconstruction, etc.) are an exceedingly successful approach for sustainable urban development and architecture. This study identifies the first measures for the preservation and maintenance of historic dwellings within the framework of sustainable architecture, facilitated by the web-based machine learning tool. This research is crucial for the complex process of evaluating historic structures, mainly traditional residences. Traditional dwellings provide significant visual diversity in their floor plans, along with strong spatial relationships in their sectional and facade elements. This work aims to evaluate the physical state of Menteşe (Türkiye) vernacular residences using machine learning methodologies and to organize preliminary information for the conservation architect and users.

Without the involvement of professional designers and without the use of modern frameworks and construction processes, vernacular/rural architecture is characterized by anonymous forms and extensive use of local materials and techniques. Rural/traditional buildings hold extensive intricate information and tacit knowledge that can only be uncovered by architects specializing in conservation. The machine learning-based approach described within this chapter can help the conservation architect and users. Modern machine learning processes are both data-driven (i.e., data management, data quality, data curation, and data cleaning, etc.) and model-driven (i.e., algorithm development, designing advanced codes, programming abilities, hyperparameters, and increasing hardware capacity etc.). In this context, various methods can be tried to improve and enhance model performance and data standards. For the training of machine learning algorithms, it is also important that the dataset of traditional houses is organized and of desired quality. Since facade photographs are used as the data set, it is obvious that it is a complex and layered training set.

The research was conducted to prioritize scalable, ML-driven solutions for the classification of the physical condition of the exterior appearance of these culturally significant local forms. The findings of this study tested the potential of machine learning in the assessment of conservation status to provide scalable solutions for the preservation of local architectural heritage in line with sustainability principles. Automation of conservation status classification can increase the efficiency and scalability of conservation activities, minimize human errors, and align with technical mobile infrastructures and practical design practices. Consequently, the conservation architect may comprehend detailed facts regarding the structure and subsequently build, alter, and eventually finalize decisions based on this insight.

6 Conclusion

For complete automation or semi-automation of cultural heritage assessment processes, a theoretical connection needs to be established between ML approaches and conservation architects. Architects working in the field of vernacular heritage conservation should have extensive expertise in local construction methods, building analysis and assessment (e.g., physical condition, material analysis, and surveying etc.), and local craftsmanship. ML techniques accelerate the process of assessment and minimize human mistakes, thus becoming a crucial tool for large-scale conservation initiatives. Machine learning algorithms offer a new field for the preservation and documentation of cultural heritage with an approach that can be an effective paradigm for conservation architects during the monitoring, pre-design, design, and construction stages. In order to increase research and development related to local heritage, it is essential that modern algorithms and computer-aided systems are included in the heritage process. There is a need for the creation of more complex systems, computational architects, restoration specialists, and digital repositories of local architectural structures to support sustainable cities and the built environment. In subsequent research, it is intended to create a database that is both extensive and complete, as well as a system that is founded on the concept of ML.

Within the limitations of this chapter, this study involved training completed with a limited amount of data in the ML process. Owing to the labor schedule, it was unfeasible to gather and evaluate data from all residences in the Menteşe district. Instead, certain streets and routes were designated for data collection. Although data augmentation increases the robustness of the model, this study is limited by the potential risk of data leakage if the test set contains rotated, translated or cropped images. Therefore, although a limited number of data augmentation techniques were applied to some of the images of the single class (to obtain less similar images and increase the number of samples), future work will focus on organically expanding the dataset to reduce this risk.

References

1. Pardo JMF (2023) Challenges and current research trends for vernacular architecture in a global world: a literature review. Buildings 13(1):162
2. Samalavičius AL, Traškinaitė D (2021) Traditional vernacular buildings, architectural heritage and sustainability
3. Ministry of Culture, Republic of Türkiye, and Supreme Council for the Conservation of Immovable Cultural Assets (1999) Principle Decision No. 660: Classification, Maintenance, and Repair of Immovable Cultural Assets, 19:1–5
4. Feilden B (2003) Conservation of historic buildings. Architectural Press, Elsevier
5. Murphy M, McGovern E, Pavia S (2009) Historic building information modelling (HBIM). Struct Surv 27(4):311–327
6. Sorguç AG, Selçuk SA (2013) Computational models in architecture: Understanding multi-dimensionality and mapping. Nexus Netw J 15:349–362
7. Caetano I, Leitão A (2020) Architecture meets computation: an overview of the evolution of computational design approaches in architecture. Archit Sci Rev 63(2):165–174
8. Terzidis K (2004, September) Algorithmic design: a paradigm shift in architecture. In: Architecture in the network society [22nd eCAADe conference proceedings/ISBN 0-9541183-2-4] Copenhagen (Denmark), pp 201–207
9. Bacharidis K, Sarri F, Ragia L (2020) 3D Building Façade Reconstruction Using Deep Learning. ISPRS Int J Geo Inf 9(5):322. https://doi.org/10.3390/ijgi9050322
10. Yu Q, Malaeb J, Ma W (2020, October) Architectural facade recognition and generation through generative adversarial networks. In: 2020 International conference on big data & artificial intelligence & software engineering (ICBASE). IEEE, pp 310–316. https://doi.org/10.1109/ICBASE51474.2020.00072
11. Kuang Z, Zhang J, Huang Y, Li Y (2023) Advancing urban renewal: an automated approach to generating historical arcade facades with stable diffusion models. ArXiv, 2311(11590). https://doi.org/10.48550/arXiv.2311.11590
12. Bahrami M, Albadvi A (2024) Deep learning for identifying Iran's cultural heritage buildings in need of conservation using image classification and grad-CAM. ACM J Comput Cult Herit 17(1):1–20. https://doi.org/10.1145/3631130
13. Carney M, Webster B, Alvarado I, Phillips K, Howell N, Griffith J, Chen A (2020, April) Teachable machine: approachable web-based tool for exploring machine learning classification. In: Extended abstracts of the 2020 CHI conference on human factors in computing systems, pp 1–8
14. Rodwell D (2003) Sustainability and the holistic approach to the conservation of historic cities. J Archit Conserv 9(1):58–73
15. Akadiri PO, Chinyio EA, Olomolaiye PO (2012) Design of a sustainable building: a conceptual framework for implementing sustainability in the building sector. Buildings 2(2):126–152
16. Loli A, Bertolin C (2018) Towards zero-emission refurbishment of historic buildings: a literature review. Buildings 8(2):22
17. Yung EH, Chan EH (2012) Implementation challenges to the adaptive reuse of heritage buildings: towards the goals of sustainable, low carbon cities. Habitat Int 36(3):352–361
18. Guerrero Baca LF, Soria López FJ (2018) Traditional architecture and sustainable conservation. J Cult Herit Manag Sustain Dev 8(2):194–206
19. Correia M, Carlos G, Mecca S, Dipasquale L, Guillaud H, Vegas F, Castro A (2014) Lessons learned and VerSus outcomes. In: VerSus: heritage for the future. Vernacular knowledge for sustainable architecture. FUP, pp 16–20
20. Nasir O, Kamal MA (2021) Vernacular architecture as a design paradigm for sustainability and identity: the case of Ladakh, India. Am J Civ Eng Archit 9(6):219–231
21. Chwieduk D (2003) Towards sustainable-energy buildings. Appl Energy 76(1–3):211–217
22. Aladağ E (1991) Muğla evi: Muğla'daki geleneksel konutların sıhhilestirilmesi ve yeniden değerlendirilmesi konusunda bir arastırma. Hamle Matbaacılık

23. Yönder VM, Hamamcıoğlu Turan M, Timur BA, Nalça C, Keke DGB, Özel E, Yönetken EE, Özdemir E, Eken E, Durmuşlar F, Güneş H, Şener İE, Ergin M, Birgin Ö, Ünal Y (2016) "Survey, Analysis and Evaluation of Some Portions of Muğla Historical Urban Site"—Architectural Restoration Department. Fall Semester. Izmir Institute of Technology (Iztech). Retrieved May 14, 2025, from Iyte.edu.tr website: https://restoration.iyte.edu.tr/mugla-tarihi-kentsel-alaninin-bazi-bolumleri-icin-tespit-analiz-ve-degerlendirme-calismasi/
24. Kuban D (1995) The Turkish hayat house. İstanbul
25. Dettlaff W (2014) Space syntax analysis–methodology of understanding the space. PhD Interdisc J 1:283–291
26. Chen Y, Chen J, Zheng L (2022) Analysis and research on space syntax in urban heritage building: the conservatório de Macau as an example. Int J Arts Hum Stud 2(1):66–74
27. Lyu Y, Abd Malek MI, Jaafar NH, Sima Y, Han Z, Liu Z (2023) Unveiling the potential of space syntax approach for revitalizing historic urban areas: A case study of Yushan Historic District, China. Front Archit Res 12(6):1144–1156
28. Varoudis T (2012) DepthmapX multi-platform spatial network analysis software. Version 0.30 OpenSource, Available online: http://varoudis.github.io/depthmapX. Accessed on 10 Jan 2025
29. Bishop CM, Nasrabadi NM (2006) Pattern recognition and machine learning, vol 4, no. 4. Springer, New York, p. 738
30. Fiorini L, Conti A, Pellis E, Bonora V, Masiero A, Tucci G (2024) Machine learning-based monitoring for planning climate-resilient conservation of built heritage. Drones 8(6):249
31. Matrone F, Martini M (2021) Transfer learning and performance enhancement techniques for deep semantic segmentation of built heritage point clouds. Virt Archaeol Rev 12(25):73–84
32. Mishra M (2021) Machine learning techniques for structural health monitoring of heritage buildings: a state-of-the-art review and case studies. J Cult Herit 47:227–245
33. D'Orazio M, Bernardini G, Di Giuseppe E (2023) Improving Cultural Heritage conservation: LSTM neural networks to effectively processing end-user's maintenance requests. VITRUVIO-Int J Archit Technol Sustain 8:18–29
34. Malahina EAU, Iriane GR, Belutowe YS, Katemba P, Asmara J (2024) A grid-search method approach for hyperparameter evaluation and optimization on teachable machine accuracy: a case study of sample size variation. J Appl Data Sci 5(3):1008–1025
35. Wong JJN, Fadzly N (2022) Development of species recognition models using Google teachable machine on shorebirds and waterbirds. J Taibah Univ Sci 16(1):1096–1111
36. Forchhammer S, Abu-Ghazaleh A, Metzler G, Garbe C, Eigentler T (2022) Development of an image analysis-based prognosis score using Google's teachable machine in melanoma. Cancers 14(9):2243
37. Jeong H (2020) Feasibility study of Google's Teachable Machine in diagnosis of tooth-marked tongue. J Dent Hyg Sci 20(4):206–212

Integrating Generative AI into Climate-Aided Design: Methods, Tools, and Professional Implications

Angelo Figliola and Maurizio Barberio

Abstract The study discusses incorporating Generative AI (GenAI) tools into the Climate-Aided Design (CADE) model, emphasising their ability to solve sustainable design problems. Using a custom-developed chatbot with expertise in climate data analysis and architecture principles, the research explores ways that GenAI can assist architects in designing sustainable buildings right from the design onset. The research approach integrates text-to-image models with Large Language Models to assess architectural responses in varying climatic conditions through two case studies: context-aware Athene and context-less Bari. By comparing the approaches, the research discusses both the potential and limitations inherent in current GenAI tools in handling urban contextual data in conjunction with climatic information. The study detects a generational shift in the utilisation of digital instruments, in which the democratisation of GenAI technologies presents a challenge as well as an opportunity for both architectural practice and education. Though the tools enhance creativity and accessibility, analytic rigour should still be ensured. The study concludes that integrating CADE methodology into text-to-image pipelines can bridge the gap between instant visual appeal and performance-driven design, particularly for small to medium-sized offices and educational contexts.

Keywords Generative AI · Sustainable design · Text-to-images · Data-driven · Environmental design

A. Figliola (✉)
Sapienza Università di Roma, Rome, Italy
e-mail: angelo.figliola@uniroma1.it

M. Barberio
Politecnico di Bari, Bari, Italy
e-mail: maurizio.barberio@poliba.it

M. Sedrez and G. Pitts (eds.), *Artificial Intelligence-Aided Design for Sustainability*, Urban Sustainability, https://doi.org/10.1007/978-981-95-1349-9_13

1 Introduction

The Architecture, Engineering and Construction sector (AEC), in its entire supply chain, from design to construction, is responsible for a significant share of global carbon dioxide emissions, contributing substantially to anthropogenic climate change [1]. Building construction and operation are responsible for around 40 to 50 per cent of the total energy consumption worldwide. According to the Intergovernmental Panel on Climate Change (IPCC) in their fourth Assessment Report (AR4), the increase in the atmospheric concentration of carbon dioxide (CO2), methane (CH4), and nitrous oxide (N2O) has mainly occurred during the last two centuries because of human activities exceeding any precedent pre-industrial numbers [2]. Within this scenario, new community policies for decarbonisation have been defined, such as the Paris Agreement at an International scale and the recent approval of the Energy Performance of Buildings Directive (EPBD) in the EU, which aims to reduce greenhouse gas emissions by 55% by 2030 and to keep the temperature rise within the 1,5 degrees. Environmental Design (ED) plays a crucial role in decarbonising the building's stocks, and one of the main approaches consists of reducing environmental impact through a performance-based approach from the early stages of the design process thanks to the interoperability of software that allows for appropriate analyses at different design phases ensuring optimisation of specific key performance indicators (KPI's). In the last decade, a series of tools were developed to support architects' and technicians' choices, particularly in the early-stage design phase from concept development onwards, as these have a significant influence on building performance. Alongside a well-established performance-based computational design workflow, recent technological advancements have significantly accelerated the development of various Artificial Intelligence (AI) tools, profoundly transforming a wide range of disciplines, including architecture [3]. Conversing with AI, generating images from textual prompts using neural networks, and training predictive models: this technological abundance marks the new frontier of creative professions, creating a further step beyond the data-driven and performance-based computational processes that now represent established realities in the field of ED. On the one hand, this technological abundance fosters creativity by promoting a transdisciplinary and holistic approach, while on the other, it could provide effective solutions to thorny issues related to digital and ecological transition. It is, therefore, crucial to investigate the possibility of employing such tools in the early-stage design phase and evaluate their effectiveness in the context of Climate-Aided Design (CADE). In this regard, the essay examines the potential and challenges of using recent GenAI tools, such as chatbots and text-to-image, in the context of CADE to provide a methodology and a critical evaluation of their applicability in addressing concrete issues and to provide solid solutions considering the UN Sustainable Development Goal(s) (SDGs). Specifically, the proposed methodology and workflow should support architects and technicians in developing "inclusive, safe, resilient and sustainable" cities (Goal 11) and provide valuable solutions to fight climate change (Goal 13).

2 Ai for Sustainability

AI is an emerging technology based on algorithms and automated instructions that simulate the learning and decision-making processes of the human brain. Methods and tools related to AI have become an integral part of the digital technologies used for designing sustainable and resilient architectures [4]. The general term AI encompasses a wide range of technologies with increasing complexity, including Machine Learning (ML), algorithms with self-learning capabilities without specific programming, Deep Learning (DL), Reinforcement Learning (RL), and Convolutional Neural Networks (CNN), which use artificial neural networks to adapt and learn from vast amounts of data without explicit instructions [5]. ML algorithms are currently being used to integrate architectural design at various scales and to analyse environmental performance, leveraging their ability to analyse large amounts of data and use it to predict performance, thus avoiding complex and time-consuming simulation processes [6], effectively democratising data-driven and performance-based processes. There are currently numerous solutions on the market which allow the exploration of different design variants through integration with major 3D modelling software and comparison based on specific KPIs defined by the designer. ML processes are used to quickly evaluate design benchmarks such as daylighting [7], solar radiation, energy consumption, and outdoor comfort, varying the parameters that describe a specific design solution [8]. Underpinning these processes is a parametric digital model through which design variants are generated, providing the designer control over functional layout, geometry, and materials. The integration of DL and RL in architectural design offers significant advantages in optimising the energy efficiency of buildings. Through the automatic generation and simulation of various design layouts combined with performance evaluation [9], it is possible to explore different solutions to select the one that optimises energy efficiency without increasing construction costs [10].

3 GenAI as a Form of Augmented Intelligence

The extensive use of DL algorithms on vast amounts of multimodal data underpins the development of Large Language Models (LLMs), widely used in various professional and research fields [11] and GenAI. LLMs are fundamental to the development of commercial software like ChatGPT and Gemini, powering multiple natural language processing applications. Numerous studies employ these models to perform a range of operations aimed at automating processes and supporting data-driven decision-making, such as querying and exploring datasets and external databases, structuring unstructured data, and generating parameters through specific functions. Unlike LLMs, GenAI uses DL models capable of autonomously creating original content on demand, such as text, images, videos, audio, or software code, in response to a user's prompt or request. This technology is at the core of software development like Midjourney, DALL-E, Adobe Firefly, Flux (Krea.ai), and ComfyUI, which allow the generation of images from textual prompts. The integration of GenAI into architectural design processes has a largely unexplored potential to impact architectural practice significantly. Its use as a design assistance tool transforms it into a form of "augmented intelligence", enhancing architects' abilities to investigate new design concepts [12]. Moreover, AI's capacity to semi-autonomously generate architectural meta-projects creates new opportunities to innovate the creative process, stimulating critical debate within the discipline [13]. Before the advent of GenAI, designers used AI techniques to navigate variable requirements and different potential solutions, fostering creativity and innovation. With the introduction of GenAI techniques, such as text-to-image and text-to-3D, a paradigm shift has occurred in architectural design. These technologies allow for a highly exploratory approach, in some ways surpassing traditional generative computational design.

3.1 GenAI vs Sustainability: A Possible Alliance

GenAI introduces a greater element of randomness in responding to the designer's inputs, which can lead to unexpected and innovative results, adding a level of unpredictability to the creative process. Combining creativity and algorithmic efficiency, AI can provide the designer with many useful tools for the design process: starting from a text-to-image process, it is subsequently possible to create three-dimensional models through traditional techniques, manually modelling shapes that reproduce the concept developed by AI. On the other hand, when text-to-image techniques are integrated into a design process based on parametric modelling, GenAI can be used to facilitate architectural visualisation. This allows for a quicker and more efficient evaluation of the architectural impact of different design solutions within the reference context. However, it is important to emphasise that the advent of GenAI presents both opportunities and challenges for sustainable design. While GenAI allows for faster and less effortful generation of design concepts, it is important to note that currently, no processes or tools have been developed capable of generating inherently sustainable projects using GenAI from simple textual instructions. Consequently, the use of GenAI could complicate the integration of lean sustainable design practices into current design practices, such as those proposed by the CADE methodology.

4 Climate-Aided Design (Cade)

CADE is an environmentally oriented design methodology focused on "protection" to reduce, from the early-stage phase, the failure rate of a design proposal through the efficient use of energy and material resources and increasing the likelihood of the project's integration into the specific context. The well-established framework on which the CADE methodology is based comprises several key phases.

Initially, the process begins with the analysis of design requirements and the definition of KPIs. Then, a preliminary climate analysis is conducted, accompanied by the creation of future climate scenarios useful for inspiring and guiding early design choices. Another fundamental step is the analysis of vernacular architecture, aimed at extracting the principles that characterise it. After collecting the preliminary data, a critical analysis of the results is carried out, leading to the identification of KPIs related to the building's function. In this phase, meta-design hypotheses are also developed through the creation of simplified 2D and 3D concepts, considering the psychrometric chart and the main passive strategies. The design concepts are then evaluated using parametric software, with a particular focus on the project's environmental sustainability about current and future climate scenarios, through a data-driven and performance-based approach. Finally, the selected meta-project is further developed, qualifying the technological and material aspects. If necessary, the design, environmental, and technological aspects are also refined.

4.1 Performance-Based Design and Climate-Aided Design (Cade)

Although the concepts of climate-aided design and performance-based architecture emerge from different starting points, they often overlap in contemporary practice [14]. On the one hand, a design proposal that emphasises energy efficiency, low carbon emissions, and resilience to extreme weather events may appeal to a performance-based framework that includes explicit metrics for various climate scenarios [15]. In this context, climate-oriented techniques, from massing orientation and shading devices to strategic placement of thermal mass, serve as tools to reach or optimise specific energy and comfort thresholds [16]. In contrast, even when climate-related factors are not the sole focus, performance-based processes may incorporate climate adaptation elements along with other benchmarks that can be related to structural or economic factors. Thus, climate-aided design strategies help projects remain comfortable and energy efficient, while performance-based controls confirm compliance with increasingly stringent regulations. This combination emphasises how climate influences architectural form and is one of several factors that need to be carefully measured and verified. The key motivations of the two approaches reveal the philosophical differences between them. A bioclimatic ethic, which sees architecture as embedded in the local environment and aims to combine human shelter with more extensive biological processes, is reflected in climate-aided design. Alternatively, evidence-based verification is the foundation of performance-based architecture, which assigns explicit benchmarks, like keeping a building's annual energy use below a certain threshold or ensuring structural integrity under specific load conditions, and then systematically verifies that the building fulfils those promises. However, the two paradigms are not mutually exclusive; climate factors frequently influence the shape of performance-based goals, while climate-oriented solutions often thrive best in an environment where systematic testing and validation are the norm.

5 A Possible Methodology to Combine Climate-Aided Design and GenAI

Design based on regenerative principles lays a solid foundation in the development of the concept and the critical reading of the present and future environmental and sociocultural context. Therefore, it is essential to adopt performance analysis tools integrated into three-dimensional modelling software to have full control over the process and effectively use the analyses as a guide in concept development to enhance both quantitative and qualitative performance. In this regard, disruptive technological innovations such as GenAI and LLMs require the study of new methodologies to ensure that design proposals generated through them adhere to the principles of CADE.

5.1 Climate-Aided Design: A Methodological Framework

Considering the above, the research aims to investigate and propose a methodology within the CADE framework that employs tools based on GenAI and LLMs, such as chatbots and text-to-image technologies. Specifically, software like Midjourney v.6 has been used to generate images from textual prompts. At the same time, ChatGPT 4o, as an LLM, has been adopted to create a specialised chatbot able to analyse external files and directly generate images using DALL-E technology. The combination of GenAI and LLM aims to create a correlation between site-specific climate analysis and the generation of concepts from textual prompts, acquiring the specific instructions needed to inform various design solutions. In this way, it is possible to merge qualitative data with quantitative parameters, focusing particularly on passive strategies for which choices made in the ultra-early design phase are crucial. To train the AI on specific climate data, a custom chatbot, based on ChatGPT 4o, has been configured to read and analyse environmental data from EPW files and technical reports (e.g. IPCC report on climate change) necessary to provide guidance on improving climate resilience and optimise performance, create site-specific images and design concepts from prompts, and analyse the AI-generated images for further contextual insights. In this regard, it is fundamental to categorise specific questions/instructions that a designer might ask when establishing a relationship between the project and the site's climate conditions during the early design phase. The textual instructions provided by chatbots can be transformed into prompts to be used in the text-to-image process, elaborating a design concept based on the images. Using traditional 3D modelling and advanced Text-to-3D technologies, volumetric conceptual

3D models can be created to match the textual suggestions and the images generated by the software.

5.2 *From Concepts Generation to Performance Evaluation*

The methodological process, described in Fig. 1, summarises the steps from concept generation to the evaluation of concepts created using GenAI and LLM. The first step involves configuring and training the chatbot, followed by the analysis of climate files (EnergyPlus EPW) for pilot sites, both current and future scenarios. This step is essential to identify potential climate challenges and assess their impact on building performance. At this stage, sustainable design criteria are defined about the climate data and urban context, considering factors such as optimal orientation, glazing-to-wall ratio (WWR), building shape, shading devices, nature-based solutions, and the integration of renewable energy sources. Subsequently, geometric features and passive strategies are integrated into the design. The collected data and analyses are used to generate textual prompts to be employed in creating images in Midjourney and DALL-E. Each image generated from prompts is then evaluated based on the previously defined sustainable design criteria, assigning a score on a scale of 1 to 10 for each criterion. The best images, selected based on the score obtained, are transformed into 3D models using conventional modelling techniques and advanced methods such as image-to-3D. Finally, environmental analysis is conducted using intuitive benchmarks, such as beneficial/harmful solar radiation, to further refine the selection of the concepts. In the second stage of the experimentation, a further improvement has been made: because architecture is context-sensitive and site-specific, the analysis of the building site and then the 3D model integrated into it has been integrated. The sites and urban contexts are evaluated through image analysis, allowing for the extraction of site-specific information crucial for concept development. Thus, the selected 3D models are placed in context and further evaluated by the AI chatbot to select the two best options, which will be analysed through a data-driven process.

5.3 *Cade Assistant: A Specialised Chatbot for Data-Driven Text-To-Image Processes*

The aim of training a specific design assistant called Climate-Aided Environmental Design is to bridge the gap between environmental analysis and architectural design as well as urban planning in the GenAI framework by leveraging a multifaceted approach rooted in data analytics and visual assessment in the ultra-early stage of the design processes, focusing on climate resilience. The training process involved several stages of machine learning, natural language processing, and domain-specific fine-tuning and has been designed to enable the chatbot to function as a specialised

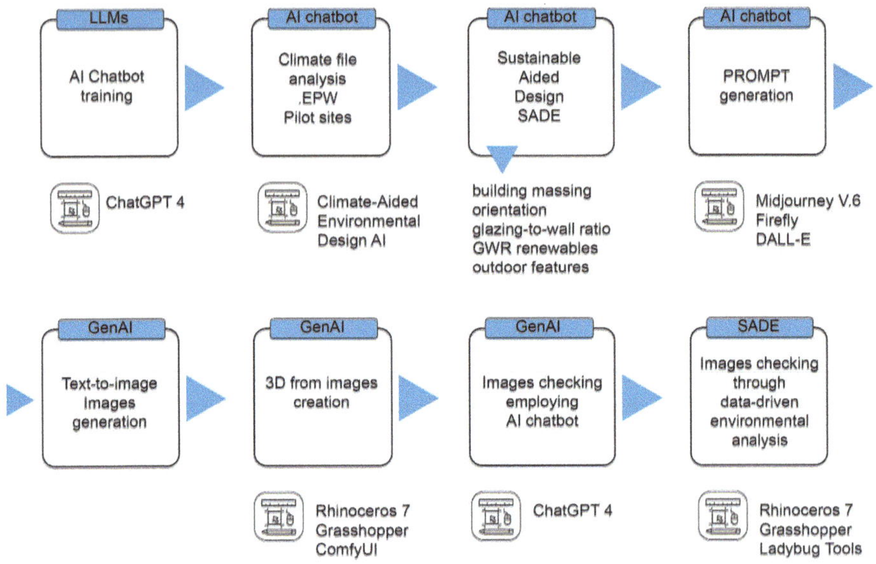

Fig. 1 Methodology framework

assistant focusing on the following areas: climate data analysis, integration of historical architectural insights, sustainable design recommendations, images and prompts generation and visual assessment of concept images through a score-based approach tailored for a specific building site. Specifically, the General Language Model is trained through unsupervised learning by using a vast dataset of publicly available text, including books, research papers, technical documents, and web content. Fine-tuning for environmental design is done using a specific dataset specifically related to climate science, architecture and urban planning, historical architectural models, climate data analysis, case studies and research papers. The process is implemented by reinforcement learning from human feedback (RLHF) done by human experts such as environmental scientists, architects, and urban planners.

5.4 Interoperability and Risk Assessment

Thanks to the capability of reading and analysing EPW files (.epw) and CSV data (.csv) from current and future scenarios, the chatbot has been trained to calculate detailed risk indexes to offer a critical understanding of the potential impact of various climate hazards, such as heavy rainfall, flooding, drought, and extreme heat. The training process has included the calculation of sophisticated metrics like R20mm, days with precipitation $\geq$ 20 mm, R95pTOT, annual total precipitation on very wet days, Rx5 day, maximum 5-day cumulative precipitation, CWD, consecutive wet days, CDD, consecutive dry days, CDD5, events of consecutive dry days, Tx90p, days with temperature exceeding the 90th percentile, su30, days with maximum temperature > 35 °C, HWI, heat wave intensity, Tx99p, days with temperature exceeding the 99th percentile, and TR, tropical nights with minimum temperature > 20 °C, that is important for understanding the climate change patterns. Through the analysis of climate files and specialised reports (such as the IPCC report), it is possible to create a scientific overview that gives insight into the most important environmental variables that can have direct consequences on the design processes.

5.5 Place Identity and Cultural Context

To explore the integration of Artificial Intelligence (AI) in architectural design through the lens of place identity and cultural context, the training process proposes a framework where place identity is not only preserved but also critically examined to enhance sustainable design outcomes. Thus, the chatbot can analyse historical architectural styles and traditional technological systems adapted to different climate zones and propose architectural models and systems that are not only historically informed but also aligned with contemporary sustainability practices. The process includes reviewing ancient and vernacular architectural models and their climate adaptations, suggesting traditional technological systems for heating, cooling, and ventilation and integrating these insights with modern sustainable design principles. Starting from that, the chatbot can generate concept images using DALL-E to prepare structured prompts to be used in Midjourney. Beyond numerical data analysis, the chatbot has been trained to analyse images of architectural projects or urban spaces, providing insights that blend visual features with climatic data for a holistic understanding of environmental impacts. These analytical capabilities, based on a rule-based analytical approach combined with computer vision techniques and expert knowledge in environmental design, allow it to recommend actionable strategies for enhancing resilience and sustainability within the context of climate change, focusing on adaptive strategies that meet the unique climatic challenges of various regions. In conclusion, the analysis spans detailed climate metrics and indices (quantitative data), visual and contextual evaluations (qualitative insights) and tailored strategies for climate-resilient design. At the end of the research development (developing and testing), the chatbot will be available to architects and technicians who work in the related field to be used as environmental expert assistants in the ultra-early stage of the design processes that involve GenAI.

5.6 Evaluating the Compatibility of Ai-Generated Images with Environmental Sustainability Features

Among the main objectives of the pre-trained chatbot is the capability to assess the compatibility of AI-generated images with environmental sustainability features that come from the data analysis. Among the main objectives of the pre-trained chatbot is the capability to assess the compatibility of AI-generated images with environmental sustainability features that come from the data analysis. This compatibility is assessed through a score-based approach using specific ranges defined during the fine-tuning process, choosing between a series of ranges proposed by the chatbot and considering the clarity and efficiency of the evaluation scale. In particular, the score-based process emphasises the evaluation of environmental aspects related to specific building sites and climate zones that come from the climate file analysis. Each site and climate zone presents unique challenges and opportunities for sustainable design, and the analysis incorporates these local specifics to ensure tailored, effective evaluation and design suggestions for improving the evaluated concept. When analysing images to score environmental sustainability, key visible features are the primary focus. These include: Windows-to-Wall Ratio as the percentage of windows to the overall wall surface; building shape through the assessment of the overall geometry and form of the building; materiality (qualitative) by identifying and evaluating the materials used in the building's façade; the presence and effectiveness of shading devices like overhangs, louvres, and shades; green features such as visible green roofs, walls, and landscaping that contribute to sustainability; orientation to take advantage of natural light and ventilation; visible features that support natural ventilation and orientation according to prevailing wind direction in the specific season.

5.7 Score-Based Approach

The score ranges from 1 to 10 and is based on the presence and accurate representation of sustainability features following this values scale: Score 1, low compatibility, no sustainability features are present or visible; Score 2 to 3, poor compatibility, few sustainability features are present, and they are not prominent or correctly represented; Score 4 to 5, moderate compatibility, some sustainability features are present and are somewhat aligned with sustainable design practices; Score 6 to 7, good compatibility, many sustainability features are present, and they appear to be well-integrated into the design; Score 8 to 9, high compatibility, most of the sustainability features mentioned in the prompt are present and accurately represented; Score 10, excellent compatibility, the image clearly illustrates the sustainability concept stated in the prompt, showing all the sustainable design solutions implemented.

To further refine the score-based assessment, the evaluation framework was enhanced by extending the analysis to the urban context, with emphasis on key main aspects such as shadows cast by surrounding buildings (based on building density), solar access, the amount of sunlight that the building gets during the day and throughout the year, and natural ventilation considering the urban morphology.

6 Case Studies: Context-Aware and Context-Less Approaches

The research was developed in two different steps, with different levels of complexity, and it presents two distinct approaches to climate analysis. The first case study was developed considering only the climate zone, while the second integrated a specific urban context. The study focuses on Bari for the context-less approach and on Athene for the context-aware one. Both share similar climatic characteristics, representing the Mediterranean climate (classified as "Csa" in the Köppen-Geiger system).

6.1 From Ai Chatbot to Text-To-Image Processes

In the following paragraph, the shared part of the methodological framework adopted for the two experimentations is presented (Fig. 1). Based on the information discussed in Bari (Italy, 41.117143, 16.871871) and Athens (Greece, 37.986693, 23.735078), the following common steps encapsulate the process from EPW data analysis to sustainable design concept generation and evaluation.

The methodology (Fig. 1) began with the analysis of EnergyPlus Weather (.epw) files for the pilot sites, considering the current weather data along with projections for 2030, 2050, and 2080. This climate data analysis aimed to identify potential climate challenges and assess their impact on building performance and sustainability.

The building site analysis followed two distinct approaches. In Athens, a context-aware analysis was conducted, assessing the urban context including existing green spaces, surrounding building heights, orientations, and shadow patterns. In contrast, in Bari, the analysis was context-less and focused solely on climatic data without accounting for the urban context. In both cases, the EPW files were analysed to extract climate patterns, including temperature ranges, precipitation, wind speeds, and solar radiation. The seasonal breakdown of this data provided insights into temperature extremes, prevailing wind directions, and solar radiation patterns.

Sustainable design criteria were then defined by considering how local climate trends and urban conditions affect building design. The focus was placed on resilience and sustainability, evaluating aspects such as Window-to-Wall Ratio (WWR), building shape, materiality, shading devices, nature-based solutions, orientation, and natural ventilation.

Building orientation was addressed specifically for Mediterranean climates. Designs aimed to orient the building's longer axis along the north–south direction to maximise natural daylight and support passive cooling via prevailing winds, while minimising east and west facade exposure to limit thermal gain from the rising and setting sun.

For the optimisation of the Window-to-Wall Ratio (WWR), a range between 30 and 40% was considered optimal, balancing energy efficiency with natural lighting needs. Solar shading devices were recommended, particularly on south-facing facades, to regulate solar gain across seasons.

Passive strategies and site-specific geometric features were also integrated into the design. Building elements were conceived to reduce heat absorption and enhance natural ventilation, with green roofs, terraces, and other features improving insulation, managing stormwater, and supporting biodiversity. Overhangs, louvres, and external blinds were proposed for south and east–west facing windows to help regulate solar gains.

The generation of images played a central role in the early design stage. Using the CAED chatbot (DALL·E) and prompts derived from the climate and site analysis, a series of images were produced to support architectural decision-making. Structured prompts (eight in total) were developed for Midjourney to generate images of sustainable buildings tailored to the conditions in Bari and Athens. A total of 64 images were generated.

These images were then evaluated by the authors using the CAED chatbot and a score-based approach. Each AI-generated image was rated on a scale from 1 to 10 according to the previously defined sustainable design criteria. This allowed for the selection of the best-performing images, based on the highest scores.

Human supervision and expert knowledge were applied to ensure the consistency and reliability of the evaluation process. The best-scoring design concepts—four for Athens and two for Bari—were then transformed into 3D models using either traditional modelling techniques or an image-to-3D transformation process.

The final step of the methodology varied depending on whether the urban context was considered. In Bari, where the analysis was context-less, the 3D models were evaluated by the AI chatbot to determine the best candidate for further data-driven validation. A comparison was then conducted between models generated by Midjourney (using image-to-3D technology) and those produced by the CAED chatbot (DALL·E) using conventional 3D modelling. In Athens, where the approach was context-aware, the 3D models were placed within the site modelled using a GIS-to-BIM workflow. The AI images were further evaluated by the chatbot to select the most suitable design for validation. The adopted methodology involved project proposal evaluation (combining 3D models with the urban context), a contextual compatibility analysis, and the final selection of the best design proposal.

6.2 Samples Validation Through Image Analysis

To ensure consistency and evaluate the initial sustainable features of the images produced by the process, two strategies were adopted: chatbot analysis and data-driven environmental analysis using 3D models created from selected images. The images generated through text-to-image technology were uploaded into the chatbot to analyse the design's efficiency in sustainable principles for a specific building site. The process for scoring images (each criterion is scored on a scale from 1 to 10) analysis involves evaluating each image of a building design based on predefined sustainability criteria: WWR, assessing the proportion of windows to the overall wall surface; building shape, evaluate the overall geometry and form, promoting energy efficiency and natural ventilation; materiality, identify and assess the sustainable, high-performance materials used in the building facade; shading devices, check for the presence and effectiveness of shading elements; green features, look for visible green roofs, walls, and landscaping that contribute to sustainability; orientation, assess how the building is positioned to take advantage of natural light and ventilation; ventilation, evaluate features that support natural ventilation, such as operable windows and vents.

6.3 Samples Validation Through Data-Driven Design

The design options selected (context-aware and context-less) by the score-based approach and transformed into 3D massing were further analysed using Ladybug, an environmental design plug-in integrated into Grasshopper, Rhino's add-on. The chosen benchmark to evaluate the environmental performance in the early stages of the design process was the BENEFIT / HARMFUL RADIATION. The choice was made considering the suggestions of the CAIED chatbot and to provide a clear benchmark with minimal computation time. Specifically, BENEFIT / HARMFUL RADIATION gives positive and negative values: negative values represent harmful summertime sun energy that can increase cooling loads during hot temperatures, and positive values represent helpful wintertime sun energy that can offset heating loads during cold temperatures.

In the case of **Athens**, the sample testing (Fig. 2) involved 3D models realised using traditional 3D modelling techniques. The prompt was generated by a custom AI chatbot (ChatGPT 4o, DALL-E text-to-image) and was based on the analysis of climate files. The results showed an area of 2661 m^2 of the volume exposed to harmful radiation, and 11,543.40 m^2 exposed to beneficial radiation.

For **Bari**, different methods were tested. One sample was generated using 3D-from-image processing with ComfyUI. The prompt, also generated by the custom AI chatbot and based on climate file analysis, was: "Design a visionary high-rise building for Bari, integrating cutting-edge sustainable technologies. The rectangular footprint is optimised for wind flow, with aerodynamically shaped corners to reduce wind load and enhance natural ventilation. East and west facades feature dynamic solar shading systems that adjust in real-time to the sun's position, combined with high-performance, electrochromic glazing to control solar heat gain. The roof and selected terraces are equipped with rainwater harvesting systems and solar water heating panels, contributing to the building's water and energy efficiency goals for LEED Gold certification.“ This concept was generated using Midjourney V.6, and

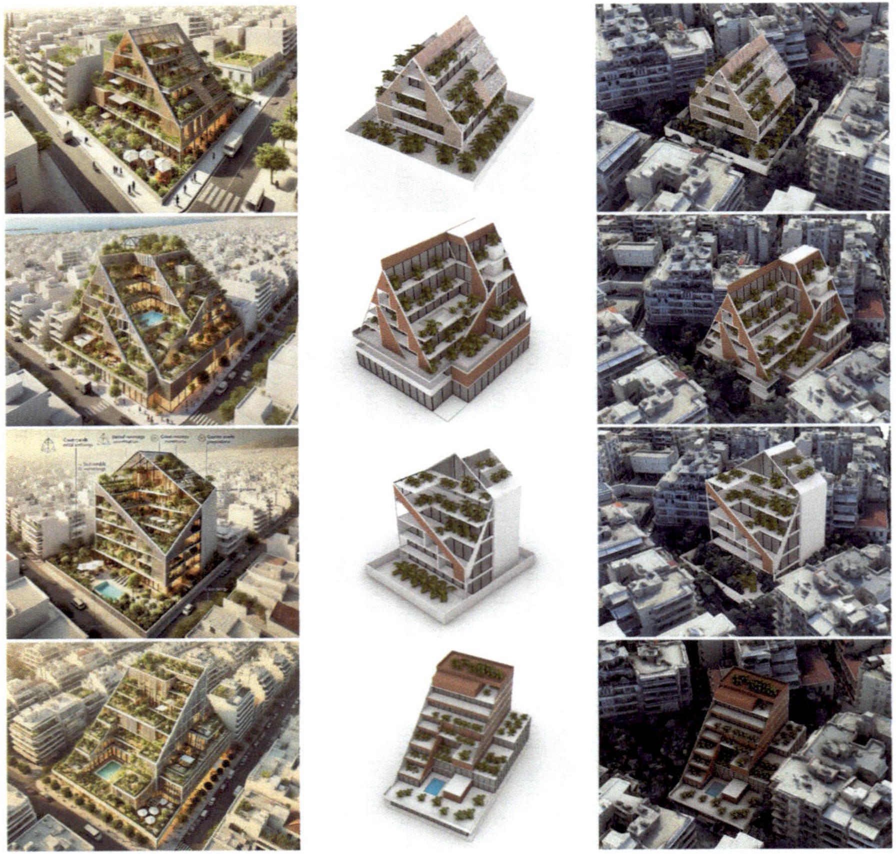

Design Proposal 1:
A pyramidal shape optimises sunlight access for upper levels, reducing artificial lighting dependency. Integrated greenery acts as a microclimate buffer, mitigating heat and enhancing air quality. While upper terraces benefit from sunlight and passive cooling, lower levels may suffer from limited light and ventilation due to surrounding buildings. **CAED Auto Score: 8/10**. Recommendation: Use reflective surfaces and light wells to improve lighting in lower levels.

Design Proposal 2:
An angular design provides shading for lower levels while incorporating green terraces as communal spaces. It integrates well with the dense urban fabric but may result in uneven sunlight distribution. **CAED Auto Score: 7/10**. Recommendation: Use light-coloured materials to reflect heat and enhance thermal comfort.

Design Proposal 3:
A stepped structure maximises natural light and ventilation across all levels. Green terraces improve thermal insulation and air quality, addressing the lack of surrounding greenery. Lower levels, however, might still struggle with reduced airflow. **CAED Auto Score: 8/10**. Recommendation: Implement vertical gardens and façade greenery for better thermal performance.

Design Proposal 4:
Staggered terraces create extensive green spaces and communal areas, enhancing urban livability. The design effectively manages solar gain and ventilation but may result in higher construction and maintenance costs. **CAED Auto Score: 9/10**. Recommendation: Ensure efficient irrigation systems and integrate renewable energy solutions, such as solar panels.

Fig. 2 Athene, sample testing

PROMPT

Design a visionary high-rise building for Bari, integrating cutting-edge sustainable technologies. The rectangular footprint is optimized for wind flow, with aerodynamically shaped corners to reduce wind load and enhance natural ventilation. East and west facades feature dynamic solar shading systems that adjust in real-time to the sun's position, combined with high-performance, electrochromic glazing to control solar heat gain. The roof and selected terraces are equipped with rainwater harvesting systems and solar water heating panels, contributing to the building's water and energy efficiency goals for LEED Gold certification.

Fig. 3 Bari, sample testing

the analysis (Fig. 3) revealed 738.70 m^2 of harmful radiation and 330.40 m^2 of beneficial radiation, with a sustainability score of 5.5/10.Bari, sample testing. 3D Models realized traditional 3D modelling techniques. Prompt, generated by custom AI chatbot ChatGPT 4 and based on DALL-E text-to-image. Area (m^2) of the volume with harmful radiation: 17,231.50. Area (m^2) of the volume with benefits radiation: 4,350.00. Sustainability score 9/10. A second Bari sample was tested using traditional 3D modelling techniques, with a prompt generated by the same AI chatbot (ChatGPT 4) and based on DALL-E text-to-image output. In this case, the area of harmful radiation reached 17,231.50 m^2, while the beneficial radiation accounted for 4,350.00 m^2, resulting in a sustainability score of 9/10.

7 Results and Discussion

Analysing the results of the research, a series of shared considerations can be made for both approaches. The images that achieved the overall best score (from 63/70 to 65/70) from the AI images analysis were created directly by the CAIED chatbot, specifically with the DALL-E engine. That demonstrates a strong coherence between the climate data analysis, the specific building sites, the sustainable design criteria and the generated images (design concept). The building massing and orientation are functional for exploiting passive strategies suggested by the AI chatbot. The information provided by the chatbot enables customisation based on specific building orientations, window sizing, and architectural aesthetics. Some features of the building system, such as the exact depth, should be optimised through detailed design considerations, which include solar angle calculations and potentially employing dynamic shading solutions for greater flexibility across seasons. The images created by CAIED demonstrate attention to resilient outdoor spaces (nature-based solutions) proposed in response to the hot and cold climates. On the contrary, the design concepts generated by Midjourney are more focused on the aesthetic values of the buildings. The AI-generated images show only the south/ east-facing facades; the others need to be reconstructed following the CAIED insights. Further analysis of the images allows the chatbot to provide more information about optimal orientation and recommended Window-to-Wall Ratio (WWR).

7.1 Data-Driven Analysis

The results of the data-driven analysis on the selected model generated by DALL-E for Athens and by Midjourney for Bari can be summarized as follows: harmful solar radiation is almost always related to opaque surfaces, such as roofs, the concept images provide strategies to mitigate the harmful effects such as green roofs or PV panels; the building massing and orientation are functional for exploiting passive strategies suggested by the AI chatbot, such as natural ventilation in the Mediterranean climate.

7.2 Context-Less and Context-Aware Analysis

The comparative analysis of both approaches yielded the following insights. In the context-less analysis (Bari, Fig. 4), the interference between the design proposal and the urban context is not considered, and it is difficult to define the orientations of the buildings; some information needs to be deducted from knowledge-based experiences.

Regarding the context-aware analysis (Athens, Fig. 5), the text-to-image process does not allow full control and manipulation of the building context that plays a crucial role in environmental design, especially in the early stages of the design process. Although the chatbot was trained to analyse the project site starting from images (plan and axonometry), taking into consideration some fundamental aspects, the images proposed do not seem to consider specific geometric characteristics of the project sites, unlike the textual suggestions, which demonstrate a strong coherence with the analyses.

A comprehensive description of the outputs created is available here: https://miro.com/app/board/uXjVLJ4rZcw=/.

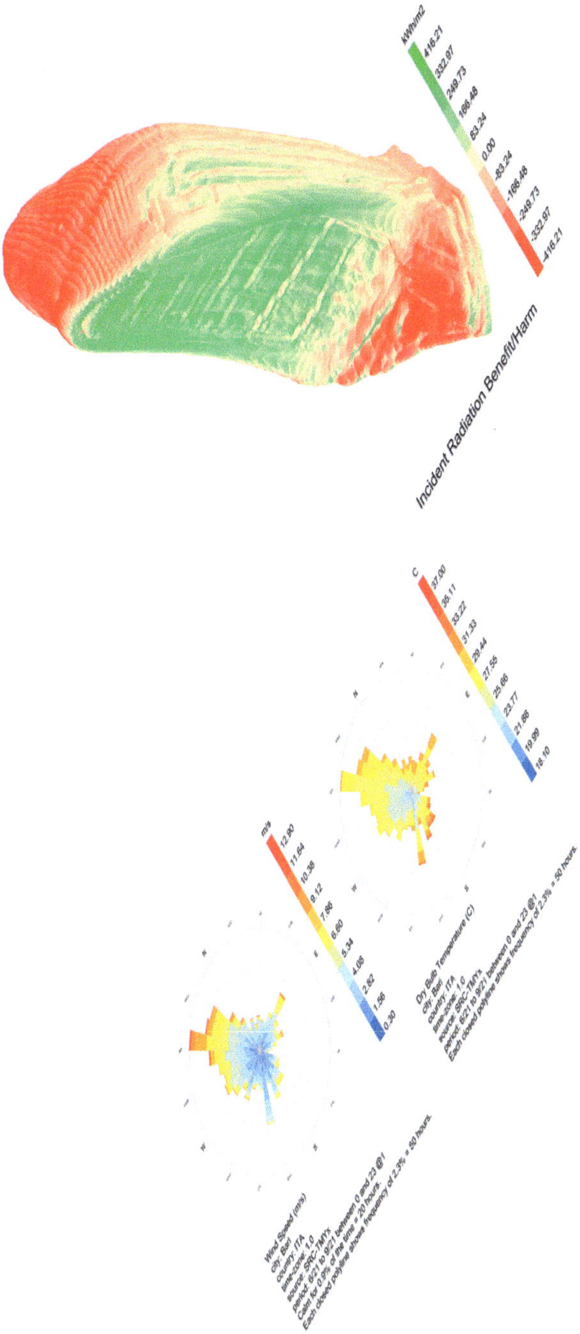

Fig. 4 Bari, chosen output analysis

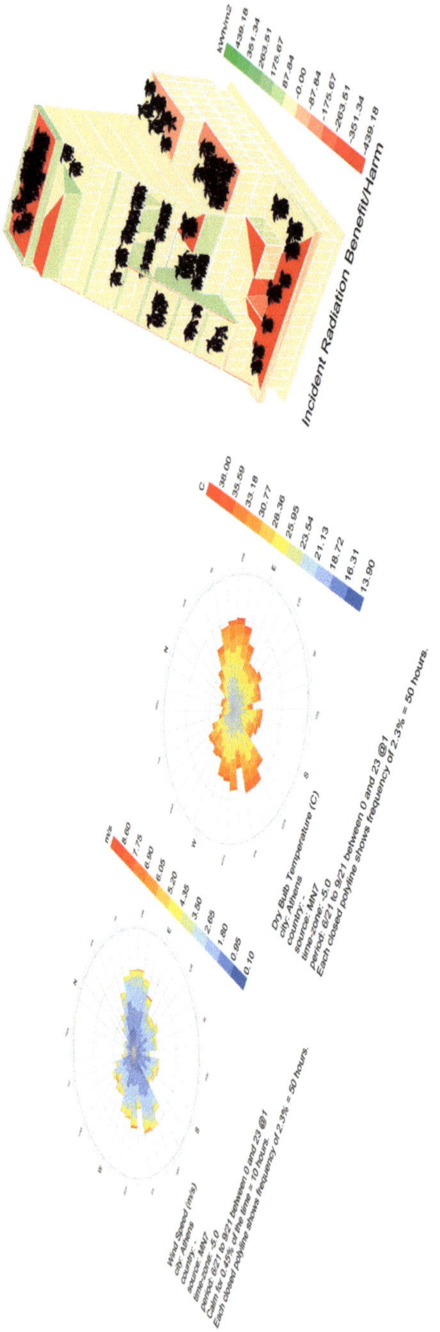

Fig. 5 Athene, chosen output analysis

8 Conclusions

The comparison between the context-aware and context-less approaches offers a valuable understanding of the influence of urban fabric on sustainable design. Although both approaches exhibit the potential of AI-augmented design workflows, the context-aware approach offers more integrated and site-specific solutions. The context-less approach, however, cannot represent the intricate dynamics of urban interaction that significantly contribute to building performance despite its merits for a purely climatic response. Therefore, performing detailed solar and thermal analyses is crucial to refine these general estimates to the specific context of each project, thus maximising comfort and energy efficiency.

In addition to the technological paradigm shift, we would like to draw attention to the different ways in which digital tools for architectural design are used among different generations of designers. The generation of "digital" designers that went through the transition from digital to analogue technology had a very strong understanding of computer programs, tending to create custom solutions through both visual and text programming to realise their design intent. The epistemological construct of this generation was founded on a deep understanding of tools, such as disassembling and rebuilding, to create innovative solutions.

Conversely, contemporary (and future) generations, raised in a software-dominated ecosystem, have a different milieu in which tools are offered by software companies with fewer options for individualisation or customisation. The widespread use of text-to-image technologies illustrates this trend: its usability tends to eclipse its rational consistency or usefulness of output. This tendency is an expression of general social media trends in which the desire for visual appeal tends to overshadow deep analytical examination.

8.1 Professional Implications

This generation gap has two important implications for the future of architectural practice. The erosion of 'making' culture: With fewer practitioners familiar with the underlying mechanics of their tools, those who preserve this knowledge have a competitive advantage. Being able to create bespoke tools, devise unorthodox processes, or 'hack' the constraint of standardised technology is an ever scarcer and, thus, valuable skill. The primacy of form over content: In a marketplace that rewards instant visual impact, the people who can balance depth and instantaneity will succeed. The problem is not only one of preserving analytical rigour but also of conveying difficult concepts in terms that engage with contemporary attention spans.

However, it's known that architectural sustainability is more than visual design and is closely related to the performance of the building concerning energy use, material use, water efficiency, indoor environmental quality, and connection to the natural environment. Over the past few decades, performance-driven and computational design has facilitated the tailored development of data-informed processes to measure sustainability right from the initial design phase, owing to software interoperability.

As GenAI technologies reduce the barrier to entry for "design content" creation, the differentiator will lie in being able to integrate these tools with structured thinking and long-term vision. The study illustrates that though AI tools can successfully analyse climatic data and produce design solutions, the inclusion of urban context data greatly increases the validity and applicability of the outcome. The context-aware approach is more comprehensive, considering climate and intricate interactions with buildings and the urban context.

8.2 Future Outlook

The proliferation of text-to-image technologies calls for research into new methods of verifying that the design proposals developed using this technology are feasible to develop further. Being able to develop and set up our chatbot, which has been trained using climatic information for a site, can potentially see the incorporation of the SADE methodology into text-to-image workflows, blending creativity with a scientific methodology, like how computational methods did previously.

Further development needs to concentrate on the capacity of AI to process and synthesise urban context information, perhaps by developing self-learning algorithms capable of analysing climatic and urban parameters together. Such integration would close the gap between context-aware and context-less methodologies, resulting in more integrated and applicable design outcomes. This development needs to be tempered by an awareness of how generations approach and use these tools.

The democratisation of these tools can result in a better understanding of how to incorporate environmental information at the earliest point in the design process and can benefit small to medium-sized offices. In addition, didactics can be gained from the application of this methodology in design processes, particularly in situations where there is a rigid separation between disciplines that inhibits the synthesis of various types of knowledge into an integrated framework. This pedagogical dimension turns out to be especially important in closing the divide between those familiar with the inner mechanics of digital tools and those who interact mainly with their interfaces.

Acknowledgements The authors are designated as first and co-first authors in the preparation of the article: Conceptualisation, A.F. and M.B.; methodology, A.F. and M.B.; investigation, A.F. and M.B.; writing—review and editing, A.F. and M.B. The research has adopted ChatGPT 4o for developing the specialised chatbot, and the training process was entirely written and guided by humans. The chapter could contain extracted parts from ChatGPT 4o since the use of LLMs is an essential part of it.

References

1. Yahaya HL, Vivek SM, Shehu UM, Auwal AM (2022) Carbon footprint management: A review of construction industry. Clean Eng Technol 9:100531. https://doi.org/10.1016/j.clet.2022.100531
2. Altomonte S (2008) Climate change and architecture: Mitigation and adaptation strategies for a sustainable development. Sustain Dev 1(1):97–110
3. Cengizoglu FP (2024) Bibliometric analysis on artificial intelligence aided architectural design. J Des Stud 6:231–245. https://doi.org/10.46474/jds.1525949
4. Ferrante T, Romagnoli F (2023) Support or automation in decision-making: the role of artificial intelligence for the project. TECHNE - J Technol Archit Environ 25:68–77. https://doi.org/10.36253/techne-13713
5. Vissers-Similon E, Dounas T, De Walsche J (2024) Classification of artificial intelligence techniques for early architectural design stages. Int J Archit Comput. https://doi.org/10.1177/14780771241260857
6. Galanos T (2019) Machine-Learned Regenerative Design. In: Naboni E, Havinga L (eds) Regenerative Design in Digital Practice. Eurac Research, Bolzano, pp 95–99. https://issuu.com/kadk/docs/regenerative_design_in_digital_practice_lowres/104. Accessed 14 Feb 2025
7. Dieguez AP, Pacheco L, Karataş H, Drożdż D, Chronis A, Rossi G (2025) Daylight factor prediction using machine learning: a two-way study using numerical encoding and regression models, versus image encoding and pix2pix. Build Environ 274:112743. https://doi.org/10.1016/j.buildenv.2025.112743
8. Płoszaj-Mazurek M, Rynska E (2024) Artificial intelligence and digital tools for assisting low-carbon architectural design: merging the use of machine learning, large language models, and building information modelling for life cycle assessment tool development. Energies 17:2997. https://doi.org/10.3390/en17122997
9. Ji Y, Wang W, He Y, Li L, Zhang H, Zhang T (2023) Performance in generation: an automatic generalizable generative-design-based performance optimization framework for sustainable building design. Energy Build 298:113512. https://doi.org/10.1016/j.enbuild.2023.113512
10. Karimi H, Adibhesami MA, Hoseinzadeh S, Salehi A, Groppi D, Astiaso Garcia D (2024) Harnessing deep learning and reinforcement learning synergy as a form of strategic energy optimization in architectural design: a case study in Famagusta, North Cyprus. Buildings 14:1342. https://doi.org/10.3390/buildings14051342
11. Rane N (2023) ChatGPT and similar generative artificial intelligence (AI) for building and construction industry: contribution, opportunities and challenges of large language models for Industry 4.0, Industry 5.0, and Society 5.0. SSRN. https://doi.org/10.2139/ssrn.4603221. Accessed 14 Feb 2025
12. Vissers-Similon E, Dounas T (2025) Textual and visual interpretation in a text-to-image accelerated architectural design process. In: Han J, Lombardi D, Cece A (eds) Advances in the integration of technology and the built environment. AAB 2024. Lecture Notes in Civil Engineering. Springer, Singapore. https://doi.org/10.1007/978-981-96-4749-1_18
13. del Campo M, Leach N (2022) Machine hallucinations: architecture and artificial intelligence. Wiley. https://onlinelibrary.wiley.com/toc/15542769/2022/92/3. Accessed 14 Feb 2025
14. Canuto R, Koenig R, Chronis A, Galanos T, Celani G (2024) From performative to predictive-performative design: a review of current trends in performance-based design and their impact on urbanism. Int J Archit Comput 23:307–330. https://doi.org/10.1177/14780771241281883
15. Shi X (2010) Performance-based and performance-driven architectural design and optimization. Front. Archit. Civ. Eng. China 4:512–518. https://doi.org/10.1007/s11709-010-0090-6
16. Luo S-L, Shi X, Yang F (2024) A review of data-driven methods in building retrofit and performance optimization: from the perspective of carbon emission reductions. Energies 17:4641. https://doi.org/10.3390/en17184641

AI-Driven Circular Architectural Design

Burak Pak and Tuba Kocaturk

Abstract Circular architectural design reimagines built environments by embedding regenerative resource flows into early design decisions. This chapter investigates how artificial intelligence can transform early-phase architectural design (RIBA Stages 0–2), where decisions most impact circularity. Moving beyond AI as merely a performance optimiser, we position it as a *futuring* tool enhancing strategic definition and concept design through Tony Fry's *design futuring* framework. We demonstrate that when guided by principles of *design intelligence, redirective practice, temporal awareness,* and *ethical design,* AI becomes a catalyst for *"sustain-able"* transformation. The chapter presents applications showing how AI reveals hidden material flows, generates reuse-based alternatives, provides real-time feedback, and visualizes adaptation scenarios. Through examples, we discuss how AI augments creativity while embedding circular principles at the core of spatial projects. We explore how AI-driven circular design advances sustainability objectives, linking circularity to broader environmental goals while balancing technical optimization with social equity. In conclusion, this research advances design discourse by elucidating how AI can transform early phases of circular architectural design. The five identified directions are: extending human perception to reveal patterns in complex datasets and translate them into actionable intelligence; expanding solution spaces with sustainable alternatives that position circularity as the logical choice; embedding long-term foresight into design decisions via sophisticated predictive modeling; foregrounding ethical considerations through transparent sustainability metrics; and enhancing creative potential through collaborative human-AI partnerships that transcend individual expertise boundaries.

Keywords Circular architectural design · Artificial intelligence · Early design phase · Design futuring

B. Pak (✉) · T. Kocaturk
The Design for Circular Cities Master's Programme, School of Architecture and Built Environment, Deakin University, Geelong, Australia
e-mail: b.pak@deakin.edu.au

T. Kocaturk
e-mail: kocaturk@deakin.edu.au

M. Sedrez and G. Pitts (eds.), *Artificial Intelligence-Aided Design for Sustainability*, Urban Sustainability, https://doi.org/10.1007/978-981-95-1349-9_14

1 Introduction

Circular architectural design introduces a paradigm shift in shaping built and natural environments by reusing, refurbishing, remanufacturing, and recycling building materials and spatial assets. Unlike the traditional "take-make-dispose" model, circular design integrates efficiency and regeneration into the planning process, ensuring design products function as a part of dynamic ecosystems rather than extractive systems [1, 2]. Strategies operate across short, medium, and long loops of reuse and repurposing, from modular building systems to waste-to-material recovery. Importantly, this view extends beyond material recovery to encompass social, economic, and ecological dimensions, seeking equitable, resilient cities [3]. In fact, the United Nations (UN) notes that reducing resource use and adopting circular models is "central to achieving the Sustainable Development Goals", highlighting that circularity must be linked deliberately to SDGs (e.g., SDG 11 on cities, SDG 12 on consumption, SDG 13 on climate) to realize its full potential [4].

Meanwhile, artificial intelligence (AI), including generative models, machine learning, and advanced visualization, is transforming design practice by processing vast datasets and recognizing complex patterns [5]. AI can augment spatial decision-making, optimize resource flows, and reveal insights far beyond human scale. Prior research has applied AI to material optimization, waste management, and resource monitoring in construction; for example, AI-driven computer vision has been used to sort demolition waste or predict reuse potential [6]. However, most work still focuses on performance metrics or single-loop recycling. There is a notable gap in exploring AI as a *redirective* agent in early-stage design, one that reimagines spatial form and program in light of future circular outcomes, rather than simply refining conventional options.

By incorporating *design futuring,* researchers and practitioners can move beyond reactive approaches and proactively prefigure emerging urban dynamics. This approach aims to offer a robust foundation for subsequent studies and practical applications, by positioning AI as a key enabler of circular value creation through novel spatial design. Drawing on the identified knowledge gap, this chapter addresses the following research questions:

- How can emerging AI technologies be harnessed as redirective agents to transform the future of early circular architectural design?
- In what novel ways can AI models enhance circular value creation in spatial systems?
- What specific affordances do AI techniques offer for enabling circular architectural practices and fostering new design agencies in the built environment?

To answer these, we review AI developments through Gartner's 2024 Hype Cycle, then apply Tony Fry's *design futuring* framework [7] to reframe AI as a tool for long-term, ethical design. We detail AI's role in each early design stage: defining strategy and briefs (Stage 0–1) and generating concepts (Stage 2). Throughout, we link theory to practice with brief examples, from tools like Autodesk Spacemaker and Archistar.ai for site analysis, to generative design platforms (Hypar, Autodesk Generative Design) and material intelligence tools (Materiom, Granta) for concept exploration. Finally, we discuss how AI-driven circular design aligns with broader sustainability objectives and suggest future research. By integrating AI with circular economy thinking, this work extends the frontier of computational design toward truly circular design futures.

2 Background and Context

Circular architectural design proposes a regenerative approach to cities. It moves away from the linear "take-make-dispose" model towards one that continuously recirculates resources. For example, circular design might prioritize adaptive reuse (short loops) or building retrofitting (medium loops) to extend lifetimes, while long loops involve systematic deconstruction and material banks [8]. These strategies reduce waste and enhance resilience by emphasizing efficiency, adaptability, and regeneration. They also interlink with social and ecological goals: a circular building strategy can provide local jobs in refurbishing facilities (addressing SDG 8), reduce embodied carbon (SDG 13), and improve community well-being (SDG 11) [9]. However, achieving this requires systemic intelligence that can coordinate spatial, social, and ecological systems.

Operating across multiple intervention scales, circular architectural design envisions short, medium, and long loops to extend the lifecycle of spatial assets. Short loops prioritize adaptive reuse, reduction, and efficient building use, minimizing resource extraction through modular construction and resource conscious planning. Medium loops emphasise extending the lifespan of architectural assets through repair, refurbishment, and repurposing; integrating strategies such as building retrofitting, flexible space programming and circular construction systems. Long loops promote creative material recovery and systemic regeneration, encouraging recycling and repurposing of materials from demolished buildings while challenging sprawl and disposable architecture in favour of regenerative spatial strategies [8].

Beyond material recovery, circular architectural design seeks to integrate social, economic, immaterial, and ecological dimensions, with a vision of development that prioritises sustainability, equity, cultural continuity and ecosystem regeneration.

2.1 Emergent AI Technologies and Their Relevance to Circular Architectural Design

AI has rapidly matured and now offers potent capabilities for circular design. Gartner's 2024 Hype Cycle for Artificial Intelligence tracks AI innovations from nascent to mainstream. Technologies such as generative AI (e.g., large vision or language models), multi-agent systems, and knowledge graphs are reaching maturity [10]. The cycle highlights that combining AI methods (composite AI) can enable richer, context-aware solutions. In the built environment, such advances mean AI can go beyond single-use optimizations to learn complex site patterns, stakeholder needs, and long-term outcomes.

Several studies have already begun linking AI with circular architectural design. Oluleye et al. [11] surveyed AI use in construction and found applications in systemic circularity, such as algorithms for material reuse and waste reduction. Özsoy [12] analysed AI models in circular economy transitions, showing how predictive analytics can restructure supply chains for efficiency. A bibliometric review by Noman et al. [13] identified that most AI-CE research centers on waste management and reverse logistics, with fewer studies on spatial or social aspects. Angelopoulos et al. [14] introduced an *IDEAL-CITIES* framework where AI supports resource management in smart cities. More recently, Zocco et al. [15] proposed using deep learning and robotics to automate circular material flows. These works underscore AI's promise but also the need for broader, built-environment perspectives (Table 1).

Table 1 Affordances of AI Technologies and their location in Gartner's Hype Cycle [11] and former studies [11, 12]

AI Model	Strength	Gartner's Hype Cycle
ANN (Artificial Neural Network)	Models complex, non-linear relationships	Plateau of Productivity
DT (Decision Tree)	Handles categorical data effectively	Plateau of Productivity
GP (Gaussian Process)	Provides flexible regression and uncertainty quantification	Slope of Enlightenment
GA (Genetic Algorithms)	Excels at global optimisation	Slope of Enlightenment
SVM (Support Vector Machine)	Provides robust classification on diverse data	Plateau of Productivity
RF (Random Forest)	Uses ensemble learning to reduce overfitting	Plateau of Productivity
GB (Gradient Boosting)	Offers high accuracy and robustness in classification	Plateau of Productivity
CNN (Convolutional Neural Network)	Extracts rich semantic features from images	Plateau of Productivity
Generative Adversarial Networks (GANs)	Synthesises high-fidelity, realistic visual data	Peak of Inflated Expectations
Advanced Natural Language Processing (NLP)	Demonstrates deep contextual and creative comprehension	Peak of Inflated Expectations

Computer vision is not in the table because often conceptualised as a broader field rather than a singular algorithmic method. It comprises a spectrum of techniques, most prominently, convolutional neural networks (CNNs) that extract, interpret, and act on information from visual inputs

Taken together, this background suggests AI is positioned to enhance circular strategies at multiple scales. AI tools exist today for rapid site analysis (e.g., GIS-based planning with Ladybug Tools), design generation (Hypar, TestFit), and materials selection (Materiom, Granta). However, their integration into an architectural mindset, one that values social equity and future adaptability, is just emerging. In the next sections, we use the design futuring lens to examine how AI can be woven into the early-phase of design, ensuring that technological innovation advances sustainability in lockstep with SDGs.

2.2 Design Futuring: Key Themes

Tony Fry's [7] Design Futuring framework is a call to reimagine the role of design in securing sustainable futures. According to Fry, design must be redirected toward what he terms *the Sustainment*: a continuous process of sustaining life and environments into the future. Four interrelated themes stand out in his approach: design intelligence, redirective practice, temporal awareness, and ethical design. Together, these form the foundations of futuring in design.

Fry introduces *design intelligence* as a broadened capacity to understand and leverage design's impact on the world. This is not "intelligent design" in a narrow sense, but rather an expansive literacy about how designed things shape our lives and futures. In Fry's view, true design intelligence would mean *everyone* (not just professional designers) can "read the qualities of the form and content of the designed environment" around them. Such literacy enables society to judge which actions or designs will "increase or decrease futuring potential" [7].

A cornerstone of Fry's thesis is that design must become a redirective practice. By this, he means design should consciously redirect existing trajectories of development (which are often unsustainable) toward more viable futures. Instead of designing for short-term market or stylistic goals, a *redirective* designer asks how each project can shift society away from "defuturing" outcomes and toward sustain-able futures [7].

Futuring inherently requires a keen temporal awareness, an understanding that design operates on, and must answer to, long-term horizons. Fry stresses that designers must learn to "design in time," not just in space [7]. In his view, time is "a medium, not just a measure", and designing with a futures perspective means considering how the effects of design unfold over years, decades, or even centuries.

Redirective practice is inherently ethical, it asks designers to do what is "good" for humanity's future, not just what is immediately profitable or convenient. This might manifest as an architect refusing to design a wasteful luxury development and choosing instead to work on affordable, eco-efficient housing. Fry's futuring aligns design with deep ethical inquiry, reviving the idea of design as a form of *praxis* (ethical action) rather than just production [7].

2.3 Early Design Phase According to RIBA's Plan of Work

In order to establish a common ground for "early design phases", this study builds on the RIBA Plan of Work overlay for Design for Manufacturing and Assembly (DfMA) [16] (Table 2). This document reframes the traditional stages (0–4) to prioritize construction strategies that integrate off-site fabrication and efficient on-site assembly. In the early phases, Stages 0 (Strategic Definition), 1 (Preparation and Brief), and 2 (Concept Design), project teams identify opportunities for DfMA approaches, set strategic goals, and formalize feasibility analyses [16].

During Stage 0, the best means of achieving the client's requirements is clarified, including deciding whether to embrace DfMA categories such as 2D or 3D structural systems. The outcome of this stage might be a refurbishment, new build, or hybrid option, with DfMA recognized early as a potential way to reduce waste and accelerate construction. In Stage 1, the client-approved brief and project objectives are finalized. Here, teams initiate DfMA thinking by considering off-site manufacturing possibilities for repetitive building components and analysing supply chain capacity. They also refine the sustainability and cost goals set in Stage 0. By Stage 2, the

Table 2 DfMA Project Early Stages (0–2), summarised from RIBA's Plan of Work [16]

Stage	Name	Outcome	Core DfMA Tasks
0	Strategic Definition	Client Requirements confirmed	Consider MMC opportunities across portfolios; evaluate DfMA impact on Business Case including end-of-life considerations; assess MMC impact on project team structure
1	Preparation and Briefing	Project Brief approved and site accommodation confirmed	Initiate DfMA thinking with focus on repeatability; incorporate MMC categories into Project Brief and Programme; conduct R&D with manufacturers; consider DfMA in Feasibility Studies; address team setup including contracts and IP issues
2	Concept Design	Architectural Concept approved and aligned with Project Brief	Embed Modern Methods of Construction categories into Architectural Concept; identify DfMA solutions for Sustainability; ensure Cost Plan and strategies consider DfMA; address Strategic Engineering aspects; initiate planning authority discussions

architectural concept is developed in more detail, embedding DfMA-related solutions into layouts and structural choices. Early identification of modular elements or standard component sizes improves downstream coordination. Feasibility studies further validate DfMA strategies for cost, sustainability, and time [16].

3 Design Futuring AI-Driven Circular Architectural Design in Early Phases

In architecture, early design stages offer maximum leverage for circular outcomes because decisions made at this stage influence 50–80% of a project's lifecycle environmental impacts [3, 17]. Integrating circular design principles from the outset, design for reuse, adaptability, and regenerative systems, minimizes waste and emissions, emphasizes designing for durability and reuse to maintain materials in closed loops [18]. Early design choices thus fundamentally shape circularity and sustainability in the built environment [19].

Decisions made in Stage 0 (Strategic Definition) and Stage 1 (Preparation and Brief) influence whether a project will follow a conventional linear path or a circular, sustainable one. By Stage 2 (Concept Design), the project's basic configuration and character are established. Here we discuss each stage in turn, highlighting how AI tools can be harnessed in line with Tony Fry's principles to future-proof spatial design.

3.1 Strategic Definition with Futuring Intelligence (Stage 0)

Strategic Definition is about defining the project's fundamental purpose and viability before design proper begins. The RIBA Plan of Work notes that an *open mind* is required at Stage 0: the best solution might not even be a new building at all [16]. This aligns with redirective practice which challenges conventional development trajectories in favour of more sustainable and future-oriented alternatives. A circular economy mindset at Stage 0 involves considering options like renovating an existing structure, building nothing, or pursuing non-architectural solutions to meet the client's needs. Fry's ethical and temporal advocacy comes into play here, deciding in the client's Business Case to pursue a path that has long-term community and environmental benefits, rather than prioritizing short-term economic gains. At Stage 0, AI plays a key role in supporting intelligence gathering, scenario analysis, stakeholder engagement and strategic optimization.

3.1.1 Data-Driven Strategic Insights (Stage 0)

At the outset, the design team defines the project's purpose and viability. AI can amplify *strategic intelligence* by analyzing data about existing assets, site context, and future scenarios. For instance, a machine-learning model can ingest cadastral data, environmental maps, and supply-chain information to evaluate whether to reuse an existing building or build anew. Computer vision frameworks (e.g. TensorFlow Object Detection or Detectron2) could scan an existing structure to inventory reusable components – beams, panels or fixtures – quantifying how much embodied material can be salvaged. Such analysis might reveal that refurbishment yields large carbon and cost savings, whereas a new build might require extensive new resources.

AI can synthesize large-scale data to inform whether a project should proceed and how. For instance, if considering reuse of an existing building, AI vision models (like CNNs) can rapidly audit the building's condition and components. A deep learning model can identify facade materials and even structural features from imagery [20]. Combined with public data (e.g., age of building, construction type), this provides a quick estimate of what reusable materials are available on-site or in the local building stock. Such analysis feeds into the circular Business Case, e.g., if AI reveals that a site has reclaimable steel and brick worth reusing, the Stage 0 team might conclude that refurbishment is the best strategy (saving cost and carbon). Conversely, if no suitable structure exists, AI could help identify nearby sources of reclaimed materials or predict material flows in the region (perhaps using machine learning on demolition permit databases [20] or other publicly available datasets). This data-driven approach expands strategic possibilities, enabling design decisions that extend beyond an isolated project, embodying *design intelligence* at a broad scale, to contribute to a regional circular ecosystem where materials are continuously repurposed rather than discarded.

3.1.2 Scenario Planning and "Futuring" Simulations (Stage 0)

Decisions at Stage 0 often involve uncertainty about the future context, demographic trends, climate impacts, evolving user needs. AI can help mitigate this uncertainty by enhancing temporal awareness through scenario modeling and predictive simulations. Tools like Gaussian process models or system-dynamics simulations (augmented by machine learning) can project how different factors, such as population growth, transportation patterns, or climate trends, might affect the long-term viability of a development. For instance, AI could model how heat stress, flooding risks, or shifting material availability would impact a site over the next 30 years, helping designers make data-informed choices about building resilience and adaptability.

More qualitatively, natural language AI (like chatbots) can help generate scenario storylines for the future of a site or community, which, when combined with visualization AI (like DALL·E or GANs), allow stakeholders to literally *see* future scenarios

[21]. In this AI-augmented process design teams draft future scenarios (e.g., population growth, climate stress) and then use AI (language models for narrative, generative image models for visuals) to flesh out and compare these scenarios. A language model generates detailed scenario storylines, and a generative image model creates scenario visualizations. Finally, the implications of each scenario are derived to inform design decisions [21].

Visual and textual AI outputs can help clients and communities grasp circular concepts early. For example, an AI generative model (or AR application like ARki) might produce a simple animation showing how a building's modules could be taken apart. Text generators can draft contrasting narratives ("Option 1: Conventional design—typical outcome" vs. "Option 2: Circular retrofit—future benefits and reuse"), making the implications of circular choices tangible. By educating stakeholders, AI supports the *ethical design* goal of inclusive decision-making.

A recent EU project "Reincarnate" illustrated this by using AI-generated visualizations of adaptive reuse options to bring diverse stakeholder visions to the table [22]. By visualizing possibilities (e.g., an AI shows one future where an old factory is refurbished into housing vs. another where it's demolished for new build), stakeholders and clients can better grasp the long-term implications at hand. This use of AI fosters a more inclusive, informed dialogue, aligning with the *ethical design* principle. It helps ensure that the chosen strategic direction is not only technically sound but also socially and environmentally considered.

3.1.3 Strategic Optimization (Stage 0)

Although detailed optimization is usually a Stage 2 activity, even at Stage 0 AI can assist in weighing high-level options. For example, a practical approach would be using decision-tree models that integrate Circular Economy principles (similar to frameworks proposed by Lozano and Lozano [23]). These models can evaluate early-stage project criteria, helping stakeholders weigh options based on sustainability, cost, material efficiency, and long-term viability. Inputting project criteria into such a system could yield a recommended strategic route (e.g., "Option A: renovate existing shell, Option B: new build with modular design; Option A has 40% lower lifecycle emissions"), along with the key factors influencing that choice. This use of AI supports *redirective practice* by systematically highlighting the most sustainable path, countering biases that might default to "business as usual." It also externalizes the ethical criteria, making sure that factors like carbon, community impact, end-of-life plan are part of the strategic equation, not overshadowed by just cost or program needs.

AI-based planning methods help achieve consensus on an adaptive reuse strategy, effectively accelerating the early design process and aligning stakeholder needs through forward-looking co-creation. An example is Autodesk's Spacemaker tool being applied in real-world development projects: multi-disciplinary teams report that they can weigh dozens of design variations (balancing daylight, noise reduction, density, and more) in hours rather than weeks, leading to more sustainable masterplans and a more collaborative decision process among architects and developers [24].

In summary, AI at Stage 0 can act as an intelligence amplifier and foresight tool. This would then strengthen the team's capacity to make a well-informed, future-conscious Go/No-go decision and to define a project brief that is aligned with circular economy principles from the very start. By embracing these tools, the Stage 0 team can ground the project in *design futuring*. The outcome of Stage 0, per RIBA [16], is confirming "the best means of achieving the Client Requirements", with AI and Fry's framework, "best" can be understood not just in immediate functional terms, but as the option that best preserves future value and reduces harm.

3.2 *Informing the Brief with Redirective Intelligence (Stage 1)*

This stage involves development of the project brief and evaluating feasibility. This is where the high-level vision from Stage 0 is translated into specific requirements and constraints. Key tasks include site analysis, refining project objectives, initial sustainability and cost appraisals, and assembling the project team. For circular architectural design, Stage 1 is critical to embed specific circular goals into the brief (e.g., "at least 50% of materials by mass to be reclaimed or recycled content," or "design for disassembly principles shall guide the structure and envelope"). It's also the stage to identify opportunities and challenges: what circular strategies are viable given the context and budget? AI can enhance Stage 1 by providing knowledge and decision support that make the brief both well-informed and ambitiously sustainable.

3.2.1 Site and Context Analysis (Stage 1)

AI-driven remote sensing, GIS, and machine learning can provide comprehensive site analysis by identifying key factors relevant to circular design. For instance, in a data-rich future scenario, AI can map out the material flows around the site, are there industries that produce useful by-products nearby? Is there an active marketplace for salvaged components in the region? While these insights might not be immediately visible, AI can process and analyse large datasets (e.g., construction waste records, demolition permits, local circular economy networks) to highlight potential resources that could be integrated into the project brief.

If Stage 0 decided on reuse of an existing structure, Stage 1 can refine this approach by using AI-powered 3D scanning and object recognition to catalogue the building's reusable components. For example, AI can systematically identify and quantify salvageable beams, façade panels, and internal fixtures, providing an accurate inventory of materials that can be reintegrated into the new design. By Stage 1's end, the project brief can outline concrete reuse strategies rather than vague sustainability goals. For instance, it could specify: "Reuse 80% of on-site structural steel; any shortfall to be met via local reclaimed steel suppliers". This level of precision and accountability enabled by AI-driven audits ensures that *redirective practice* moves beyond intention to implementation, embedding circular principles directly into the project's DNA.

3.2.2 Knowledge Synthesis and Precedent Mining (Stage 1)

Preparing a brief often involves researching precedents and best practices. AI can function as an intelligent research assistant, quickly pulling insights from vast technical literature and past project data. For example, a language model could summarize lessons from similar circular building projects worldwide (what design strategies they used, what outcomes achieved). Large architectural firms are even training AI on their own project archives [25], an archival Artificial Neural Network that can retrieve past designs and outcomes. By querying such systems, a team might ask: "Show us previous office building projects that achieved cradle-to-cradle certification or high material reuse" and get relevant case studies. This *design intelligence* gathering, powered by AI search and classification, equips the team to write a brief that leverages collective knowledge. It aligns with Fry's [7] idea of mobilizing "sediments of knowledge" to build design intelligence. Rather than each project starting from scratch or limited experience, AI makes broad knowledge accessible at Stage 1, thus raising the baseline of what can be imagined and demanded.

3.2.3 Stakeholder Alignment Through AI Visualization (Stage 1)

Similar to Stage 0, early stakeholder engagement is valuable in Stage 1, especially to set sustainability goals. AI-generated visuals (Stable Diffusion, Midjourney) and even text (chatbot LLM-like tools) can help communicate circular design concepts to clients and users who may not be familiar with them. For instance, if designing for disassembly is a key strategy, an AI could generate a simple graphic or animation showing how the building's components could be taken apart and reused. This makes abstract principles concrete, facilitating buy-in. It's part of the *ethical design* stance, educating and bringing the client and community on board with futuring principles. AI could also generate multiple concept *scenarios* in narrative form for the brief: "Option 1: Conventional Design, describes likely outcomes; Option 2: Circular Design, describes how the building would function and benefit community in 20 years." These narratives, enhanced by AI, illustrate future use and end-of-life, reinforcing *temporal awareness* into the process of drafting the brief.

3.2.4 Multicriteria Decision Support (Stage 1)

By Stage 1, key constraints, such as budget, spatial requirements, regulatory conditions, become clearer. However, trade-offs often arise, such as: upfront cost vs. long-term savings, or architectural flexibility vs. reuse of an existing structure. AI can assist in multicriteria decision-making, systematically evaluating different design approaches based on a range of financial, environmental, social criteria. Advanced AI techniques, such as Gaussian process classifiers or decision-tree ensembles, can categorize project options (as feasible or infeasible) and rank them based on circular economy principles. For example, if multiple sites are under consideration, an AI model could assess which site offers the best potential for circular development (perhaps scoring factors like proximity to public transport, availability of recycled materials, community needs for regeneration, etc.). This data-driven evaluation ensures that the most practical option is prioritized in the project brief. It also reinforces *redirective practice* by steering decision-making toward long-term environmental and economic benefits rather than cost efficiency. For example, if an AI model determines that a slightly more expensive site enables significantly higher material reuse and lower life-cycle emissions the data can justify its selection as a strategic investment in circularity rather than a financial burden.

3.2.5 Setting Measurable Targets (Stage 1)

AI can also help in formulating the targets themselves. Using predictive models, the project team can simulate whether certain targets are realistic. For instance, if the brief aims for "90% construction waste diversion," AI can analyse current data to see if that is attainable or if the target should be higher/lower. This makes the brief's sustainability goals both ambitious and evidence backed. Fry's [7] ethical design principle would argue for pushing boundaries in response to the urgency of sustainment, and AI can provide data-driven support needed to justify these choices. For example, if a project aims for X% material reuse, AI can determine that achieving this goal would require integrating modular Y and Z systems or sourcing reclaimed materials from specific regional suppliers. By the end of Stage 1, the brief and feasibility analysis should clearly reflect a circular, futuring approach, essentially a blueprint for redirective design in Stage 2. Thanks to AI, the brief can be richer in information and forward-looking. It can anticipate future regulatory trends, for instance, predicting that by the time the building is completed, sustainability standards might become more stringent, allowing the brief to pre-emptively incorporate compliance measures rather than requiring costly adjustments later. It can also document the rationale for its requirements, improving accountability. This stage, with AI support, bridges the gap between lofty Stage 0 ideals and the concrete design work to follow, ensuring the design team enters Stage 2 with both inspiration and clarity.

Multi-criteria optimization algorithms (Gaussian processes or ensemble models) can be used to rank feasible options. If several sites are considered, an AI model might score each for combined circular criteria: material reuse potential, proximity to transit, and projected emissions reduction. The AI could then recommend the site that, while slightly more expensive, offers far higher life-cycle sustainability. Thus, the brief can justify ambitious goals (and budget) with data-backed evidence.

3.3 Concept Design: Generative and Data-Driven Circular Solutions (Stage 2)

This stage is where the stakeholders develop the foundational design concept defining the project's overall layout, form, and spatial organization while making preliminary selection of systems and materials. This is an inherently creative yet profoundly consequential phase as decisions made at this point, such as the choice of structural system, fundamental geometry, and spatial orientation, have far-reaching implications for the project's long-term performance, adaptability and circularity. With a futuring-oriented brief, the design team can leverage AI in multiple ways during this stage as described in the following sections.

3.3.1 Generative Design Exploration (Stage 2)

Using GANs, variational autoencoders, or evolutionary algorithms, designers can generate a multitude of concept options that align with the brief's constraints. AI accelerates speed and variety, whereas traditional methods might take weeks to develop a handful of sketches, an AI trained on circular design principles can quickly produce dozens of design schemes within days. This breadth prevents tunnel vision and fosters redirective thinking. AI-generated configurations may challenge conventional approaches, suggesting, for instance, an unexpected reuse of an existing building frame or a circular courtyard scheme that maximizes daylight and material efficiency. Rather than dismissing these outliers, the design team can refine and adapt them.

In this respect, AI serves as a catalyst for design intelligence, while the iterative dialogue between human intuition and machine computation forms a hybrid intelligence [26]. For instance, an AI might propose a structural grid that precisely matches standard lengths of reclaimed steel, an efficiency insight a designer might not identify so early. Architects can then incorporate such optimizations. Chaillou's [26] *ArchiGAN* project exemplifies this synergy, where nested GAN models generated full apartment designs while allowing human input at each step. The result is not AI or human, but a richer combination of both.

3.3.2 Real-Time Performance Feedback (Stage 2)

As the concept evolves, AI models, such as artificial neural networks (ANNs) or genetic programming (GP)-based predictors, offer instant feedback on various performance metrics. This is akin to having a sustainability consultant and engineer "in the room" at the sketch stage. For example, as an architect adjusts massing or layout, an AI plugin might flag, "This option uses 20% more concrete than the previous, consider longer spans using steel beams from the old building to reduce new concrete." Such real-time prompts keep the design aligned with circular goals continuously, rather than being an afterthought during final reviews.

By embedding life-cycle awareness into design workflows, AI operationalizes *temporal awareness* by constantly linking design choices to long-term outcomes like material recovery or energy efficiency. Gaussian process emulators, for example, can predict daylight performance [27], but could also assess deconstruction time or reuse potential. By integrating these into the design software, the team effectively has a futuring lens on every decision, such as AI warning that a proposed material assembly would hinder future separation later, prompting a more circular alternative instead.

3.3.3 Optimization and Decision Support (Stage 2)

Beyond passive feedback, AI can actively optimize design concepts. Multi-objective optimization algorithms, leveraging genetic algorithms or Bayesian optimization, allow design team to set parameters like "minimize embodied carbon, maximize daylight, maintain area = X ± 5%." AI then explores design variations that best achieve these constraints adjusting footprint, orientation, or structural grid to identify solutions that significantly reduce material use while meeting functional requirements. This process can reveal satisficing design configurations [28] that manual trial-and-error might overlook.

These exemplify *design intelligence* at work: synthesizing vast datasets and competing requirements into actionable insights and decisions. Rather than replacing the designer's intuition, AI enhances decision-making by offering high-quality alternatives. The decision which option to adopt still involves value judgments, where the ethical stance comes in. AI might give one scheme that has ultra-low carbon but a higher cost and another that is moderate on both; the team, guided by project values, determine the best path forward. Importantly, because these AI-generated embed sustainability principles from the outset, any selection is inherently more circular conventional approaches. In this way, AI *redirect practice*, embedding sustainment principles into the design process itself, making circularity the default rather than an afterthought.

3.3.4 Detailing Circular Features in Concept (Stage 2)

Stage 2 also defines key systems, structure, facade, mechanical, electrical, and plumbing (MEP) strategy, where AI can assist in designing these systems for circularity. For example, an AI might be used to configure an optimal structural grid and demountable connection scheme that balances structural efficiency with ease of disassembly. Decision-tree algorithms act as automated watchdogs, ensuring no concept includes "red-list" toxic materials or hard-to-recycle composites by cross-referencing material selections with health databases. Generative design can be applied to optimize facade panel layouts for salvaged panels which come in different dimensions. AI can rapidly test countless panel arrangements to maximize reuse, making it feasible to incorporate non-standard, irregular components, elements often dismissed in linear design processes. By embedding such detailed optimization into early design, AI strengthens value retention and recovery, ensuring buildings are conceived from materials that have been used before or can easily be repurposed.

3.3.5 Visualization and Communication of the Concept (Stage 2)

Once a concept design is formulated, AI aids in effectively communicating its benefits. High-quality visualizations, renderings and diagrams, can be produced with AI assistance to illustrate circular features such as a future scenario where the building is disassembled or adapted. Diffusion models or GANs trained architectural imagery generate compelling storytelling visuals that highlight the design's future qualities. This not only persuades clients but also to broader audiences that AI-driven circular design is both viable and aesthetically compelling, countering the misconception that sustainability is purely utilitarian.

Throughout Stage 2, human designers remain in control, with AI is a collaborative partner, sometimes an impartial analyst, sometimes a creative artist, and sometimes an optimization engine. Together, they enact what Fry [7] describes as a new mode of designing: informed, attuned to the future, and ethically grounded. The resulting concept design is likely to be more resilient (able to adapt over time), more resource-effective, and more *sustain-able* (in Fry's sense of enabling ongoing sustainment) far beyond what conventional design approaches typically achieve.

3.4 Synthesis: AI-Driven Early-Phase Circular Design

Our exploration of AI applications across the early design stages (0–2) demonstrates how computational intelligence can systematically address the challenges of circular design at multiple scales. Tony Fry's design futuring framework provides the conceptual underpinning for these applications, ensuring that AI serves not merely as a technological enabler, but as a redirective agent for sustainable transformation.

At Stage 0 (Strategic Definition), AI serves as an intelligence amplifier and foresight tool, aggregating complex datasets to reveal hidden material flows, model future scenarios, and evaluate strategic options. Tools like CNN-based vision models can rapidly audit existing structures, while predictive simulations help stakeholders visualize long-term implications of different development paths. Examples like Autodesk's *Spacemaker* tool show how AI can accelerate stakeholder alignment by presenting sustainability metrics in easily digestible formats, helping teams make well-informed, future-conscious Go/No-go decisions aligned with circular economy principles.

During Stage 1 (Brief Development), AI enriches the brief with data-driven insights, precedent mining, and evidence-based target setting. AI-powered material audits create precise inventories of reusable components, while machine learning algorithms identify regional opportunities for material flows. This precision enables briefs to contain concrete, measurable circularity targets rather than vague aspirations. AI also facilitates knowledge transfer across projects, allowing teams to learn from global precedents and visualize complex circular concepts for client buy-in.

By Stage 2 (Concept Design), AI becomes a creative partner in exploring design alternatives. Generative algorithms produce multiple design options aligned with circular principles, while real-time feedback systems continuously evaluate performance against sustainability metrics. Multi-objective optimization allows designers to balance competing priorities while maintaining circularity as a core value. Tools like *ArchiGAN* demonstrate how human designers and AI can work collaboratively, each complementing the other's strengths to produce innovative solutions that might not emerge through conventional methods.

Throughout these early design phases, AI consistently serves Fry's four key principles:

- It enhances *design intelligence* by making invisible relationships visible and complex data actionable
- It supports *redirective practice* by systematically prioritizing sustainable alternatives
- It reinforces *temporal awareness* by linking present decisions to future outcomes
- It embeds *ethical considerations* by making sustainability metrics explicit and transparent

The integration of AI across these stages creates a continuous thread of circular thinking, where decisions at each phase build upon and reinforce those made earlier. This coherence is essential for truly circular outcomes, as it prevents sustainability considerations from being marginalized or eliminated as the design progresses.

In contrast to conventional approaches where circular principles might be applied superficially or too late in the process, AI-augmented circular design embeds these values from the very beginning. This early integration is critical, as decisions made in the early phases influence 50–80% of a project's lifecycle environmental impacts [3, 17]. By harnessing AI's computational power to address these high-leverage decisions, designers can dramatically enhance the circularity and sustainability of architectural outcomes.

The specific tools and approaches outlined in this section represent just the beginning of AI's potential in early-phase circular design. As these technologies mature and become more integrated into standard architectural workflows, we anticipate even more powerful applications emerging. The key challenge going forward will be ensuring that these technical capabilities remain firmly grounded in the ethical and future-oriented framework that Fry's approach provides.

4 Discussion and Future Directions

Tony Fry's futuring approach provides a compass to ensure technological advancements redirect design toward sustainment rather than unsustainable practices. In this context, AI emerges as a critical enabler, augmenting design intelligence, optimizing spatial and material flows, and supporting real-time decision-making within circular design frameworks [31]. More than a tool, AI fosters new modes of co-creation, resilience, and adaptive design, embedding circularity from the conceptual phase through machine learning, generative adversarial networks (GANs), reinforcement learning (RL), and neural networks. These technologies not only enhance performance but also reframe the design process itself, making sustainability an intrinsic rather than optional outcome.

This chapter has demonstrated that AI-driven circular design aligns with and extends prior research on computational design, digital fabrication, and sustainability-oriented decision-making [29, 32, 33]. While previous studies have explored AI's potential for optimizing energy performance or material efficiency, this work highlights AI's broader role in *futuring*, embedding long-term adaptability, ethical considerations, and systemic resilience into spatial practice. By synthesizing AI's predictive, generative, and analytical capabilities within a circular economy framework, this chapter builds on existing literature while proposing an expanded, ethically grounded approach to computational design (Fig. 1).

An AI-augmented, futuring-led approach to early-stage design yields concepts radically different from conventional methods: buildings repurposed rather than demolished, materials sourced from multiple sites, and components designed for reintegration into the circular economy decades into the future. This approach redefines circular architectural practice as one that does not foreclose future options but instead lays a foundation for continuous adaptation and sustainment.

Ultimately, AI serves as both an accelerator and an amplifier of our collective design intelligence in several key ways. First, AI extends human perception by identifying patterns in vast datasets, translating complex insights into actionable design intelligence. It quantifies future sustainability impacts, providing evidence-based recommendations, such as reducing future waste to support informed decision-making. By making futuring potential measurable and visible, AI democratizes design literacy, engaging both professionals and the wider public in long-term spatial thinking.

Fig. 1 Affordances of Artificial Intelligence (AI) technologies and applications across early design, futuring principles and SDG goals, showing how AI acts as a redirective agent, circular catalyst, and enabler of design capacities

Second, AI broadens the solution space, generating viable, sustainable alternatives that reduce the risk of departing from business-as-usual. Generative AI can rapidly develop reuse-based concepts, enabling architects to confidently propose retrofits over demolitions with concrete evidence. By optimizing for circularity without compromising functional goals, AI ensures that *sustain-ability* becomes the most rational, cost-effective, and viable design choice, facilitating the shift Fry [7] calls for.

Third, AI operationalizes Fry's concept of *designing-in-time* by embedding long-term foresight into design decisions. From predicting end-of-life material recovery to simulating building performance over decades, AI ensures that projects are designed not only for present needs but for future adaptability. Digital twins and machine learning allow buildings to evolve dynamically, tracking performance in use and informing future redesigns, treating buildings as part of a continuous lifecycle rather than static entities.

Fourth, AI's impact is shaped by the values programmed into it. When *sustainability* metrics are embedded into AI-driven design tools, they inherently prioritize ethical considerations such as resource stewardship, material health, and transparency. AI can test whether sustainment-driven designs still meet functional needs, easing the perceived tension between ethics and economics. Furthermore, explainable AI ensures accountability in design decisions, making sustainability efforts more transparent and defensible.

Fifth AI is not a replacement for human designers but an extension of their *agency*. Generative design platforms facilitate an iterative dialogue where AI proposes solutions that human designers critique, refine, and reframe, creating a hybrid *design intelligence* that merges computational power with human judgment. AI also expands the creative palette by drawing on precedents, biomimetic principles, and broader datasets, infusing the design process with distributed creativity. In this collaborative paradigm, authorship is shared, and design intelligence is amplified beyond individual expertise.

Aligning AI-driven circular design with global SDGs remains a priority. As highlighted, circular strategies intersect multiple goals, from clean energy and climate action (SDGs 7 & 13) to responsible production (SDG 12) and sustainable cities (SDG 11), but the synergy is not automatic [30]. The findings also indicate that circularity should be integrated with social equity and climate agendas. In practice, this means coupling AI-optimized material reuse (SDG 12) with inclusive stakeholder processes (SDG 10) and resilience planning (SDG 13).

In conclusion, by weaving AI into the early design of architecture, we lay the foundation for continuous adaptation and sustainment. AI has the potential to fundamentally reshape early-stage design by expanding our capacity to "*future*" and embed circular principles at the conceptual core of spatial projects. Future work should explore these ideas in real-world projects, test emerging tools, and develop new AI applications tailored for circularity. By continuing to align technological innovation with ethical design values, we can guide AI to drive truly *sustain-able* design futures.

References

1. Achterberg E, Hinfelaar J, Bocken N (2016) Master circular business with the value hill. Circle Economy. https://www.circle-economy.com/resources/master-circular-business-with-the-value-hill
2. Ellen MacArthur Foundation (2024) Introduction to an adaptive strategy for circular design. https://www.ellenmacarthurfoundation.org/adaptive-strategy-for-circular-design/introduction
3. Williams J (2019) Circular cities: Strategies, challenges and knowledge gaps. Sustainability 11(6):1757. https://doi.org/10.3390/su11061757
4. UNEP (Noronha L) (2024) A circular economy for a healthy people and planet. Speech, G7 Environment Ministers. https://www.unep.org/news-and-stories/speech/circular-economy-healthy-people-and-planet

5. Ghisellini P, Cialani C, Ulgiati S (2016) A review on circular economy: the expected transition to a balanced interplay of environmental and economic systems. J Clean Prod 114:11–32. https://doi.org/10.1016/j.jclepro.2015.09.007
6. Samal CG, Biswal DR, Udgata G, Pradhan SK (2025) Estimation, classification, and prediction of construction and demolition waste using machine learning for sustainable waste management: a critical review. Constr Mater 5(1):10. https://doi.org/10.3390/constrmater5010010
7. Fry T (2009) Design futuring: sustainability, ethics and new practice. Berg, Oxford
8. Fusco Girard L, Nocca F (2019) Moving towards the circular economy/city model. Sustainability 11(22):6253. https://doi.org/10.3390/su11226253
9. Ellen MacArthur Foundation (2022) Circular economy in cities: overview. https://ellenmacarthurfoundation.org/topics/cities/overview
10. Gartner (2024) Hype cycle for artificial intelligence 2024. Gartner Inc.
11. Oluleye BI, Chan DW, Antwi-Afari P (2023) AI for systemic circularity in construction. Sustain Prod Consum 37:268–283. https://doi.org/10.1016/j.spc.2022.12.002
12. Özsoy T (2023) Yapay zekâ ve döngüsel ekonomi dönüşümü. Nişantaşı Üniv Sos Bilim Derg 11(2):369–389
13. Noman AA, Akter UH, Pranto TH, Haque AKMB (2022) Machine learning and artificial intelligence in circular economy: a bibliometric analysis. Ann Emerg Technol Comput 6(2):13–40
14. Angelopoulos CM, Katos V, Kostoulas T et al (2019) IDEAL-CITIES: a trustworthy and sustainable framework for circular smart cities. DCOSS 2019:443–450. https://doi.org/10.1109/DCOSS.2019.00089
15. Zocco F, Haddad WM, Corti A, Malvezzi M (2024) Deep-learning vision and robotic manipulation for circular economy. IEEE Access. https://doi.org/10.1109/ACCESS.2024.3350278
16. RIBA (2020) RIBA plan of work 2020: Overview. https://www.architecture.com/-/media/files/ribaplanofwork2020overviewpdf.pdf
17. European Commission (2020) Circular economy action plan: For a cleaner and more competitive Europe. https://ec.europa.eu/environment/pdf/circular-economy/new_circular_economy_action_plan.pdf
18. Stahel WR (2019) The circular economy: a user's guide. Routledge, London
19. McDonough W, Braungart M (2002) Cradle to cradle: remaking the way we make things. North Point Press, New York
20. Raghu D, Markopoulou A, De Wolf C (2022) Enabling component reuse using ML and Google Street View. In: CAADRIA 2022, vol 2, pp 577–586
21. Van Laar B, Greco A, Remøy H, Gruis V, Hamida M (2023) Leveraging AI to create circular built environments. Cutter Consortium. https://www.cutter.com/article/leveraging-ai-create-circular-built-environment
22. Reincarnate Project (2025) AI-augmented scenario planning. https://www.reincarnate-project.eu/
23. Lozano R, Lozano FJ (2024) Developing a decision-making tree for circular economy. Sustain Dev 32(3):1589–1598. https://doi.org/10.1002/sd.2736
24. Autodesk (2025) Spacemaker. https://www.autodesk.com/eu/campaigns/spacemaker
25. RIBA (2024) RIBA AI report 2024. https://www.architecture.com/knowledge-and-resources/resources-landing-page/riba-ai-report-2024
26. Chaillou S (2019) ArchiGAN: Generative design for architectural layouts. NVIDIA Developer Blog. https://developer.nvidia.com/blog/archigan-generative-stack-apartment-building-design
27. Rastogi P, Khan ME, Andersen M (2017) Gaussian-process emulators for building simulation. Proc Build Simul 2017:1700–1707
28. Simon HA (2019) The sciences of the artificial. 3rd edn. MIT Press, Cambridge (Orig publ 1969)
29. Grasser-Parger A (2024) Community-driven design: Collaborative objects and the expanded role of architects. Int J Archit Comput 22(2):257–272

30. Schroeder P, Anggraeni K, Weber U (2018) The relevance of circular economy practices to the Sustainable Development Goals. J Ind Ecol 23(1):77–95. https://doi.org/10.1111/jiec.12732
31. Rodrigues RC, Duarte RB (2022) Generating floor plans with deep learning. Int J Archit Comput 20(3):630–644
32. Leach N (2021) Architecture in the age of artificial intelligence: an introduction to AI for architects. Bloomsbury, London
33. Yang S, Dortheimer J, Sprecher A, Yang Q (2024) Design workshops and chatbots: experiential learning with generative-AI tutors. Int J Archit Comput 22(2):216–237

The Promise of AI in Sustainable Design

Maycon Sedrez and Gregory Pitts

Abstract This book examines how AI can contribute to sustainability in architecture and design. It highlights AI's potential to enhance creativity, optimize design processes, and support sustainable outcomes, while also emphasizing the need for critical thinking, new methodologies, and ethical reflection. As AI reshapes the designer's role, the field must adapt through interdisciplinary collaboration and a renewed focus on human values.

Keywords Design · AI · Architecture · Lessons

1 Design Intelligence

In the first chapter of this book, we proposed a fundamental question: "How can AI contribute to sustainability?". Throughout the subsequent chapters, we and our contributors have explored various dimensions of this inquiry, although the broad scope of design as a subject and the rapid evolution of AI technology allow us to address only a fragment of this complex challenge. We have uncovered that AI has the potential to support design studio methods, enhance creativity, reduce cognitive load, and help architects to understand and simulate complex concepts with unprecedented speed. However, this collaboration demands critical thinking at the highest levels. The effective use of AI requires not only a deep understanding of architectural principles, theories, vocabulary, and AI interpretations but also the development of entirely new methodological approaches.

AI is being integrated into our professional and personal lives at super speed, often without sufficient time for reflection on its implications. This book captures a particular moment in the history of a field undergoing profound transformation.

M. Sedrez (✉) · G. Pitts
School of Architecture and Built Environment, Deakin University, Geelong, VIC, Australia
e-mail: m.sedrez@deakin.edu.au

G. Pitts
e-mail: gregory.pitts@deakin.edu.au

M. Sedrez and G. Pitts (eds.), *Artificial Intelligence-Aided Design for Sustainability*, Urban Sustainability, https://doi.org/10.1007/978-981-95-1349-9_15

The traditional role of the designer is being redefined, the process of design is being altered, and the very notion of architectural authorship, creativity and agency is being questioned in ways that demand our full attention and thoughtful response. This sentiment is captured by Neil Leach [1, 2] who suggests that AI will have a deep impact on the processes, but less of an impact on building forms. As such, design methods need to be reinvented and hopefully integrate a higher level of reflection.

Architecture and design now stand at another critical intersection, one that differs from previous technological transitions in magnitude. Where past shifts in architectural practice offered clear alternatives to the *status quo* and timelines for adaptation, today's AI revolution presents fewer options for those wishing to remain on familiar ground. One might reject using AI in design, but its ubiquity is undeniable, and we dare to say, unstoppable. The integration of AI into architectural workflows is not merely an option but increasingly a necessity for practitioners. We predict that the professions of designer and architect will face pressures that require the re-evaluation of professional pathways, including required knowledge and attributed tasks.

Still, this technological change offers extraordinary potential. AI can support design far beyond traditional processes, enabling architects to analyse environmental impacts, optimize material usage, and visualize sustainable solutions more quickly and efficiently. AI technologies enable us to expand architectural methods and ideation in directions that were previously unimaginable. This moment calls for critical discussion, and engagement that acknowledges both the transformative potential and the inherent limitations of AI technologies. This collaboration must be guided by values that go beyond technological capability—a commitment to environmental response, social equity, and cultural sensitivity that remains fundamentally human.

We call educators, practitioners, and researchers to actively engage with the evolving paradigm of AI in architecture. This moment requires open-minded exploration and bold experimentation aimed at shaping a future where AI becomes a supportive assistant within architectural education and practice. However, adopting AI alone is not enough; architects need to develop adaptable frameworks that meaningfully integrate AI technologies while safeguarding the ethical foundations of our discipline: environmental responsiveness, social responsibility, and cultural sensitivity. Understanding the distinct but complementary roles of human and artificial intelligence is crucial. Designers must cultivate an awareness of their own creative and analytical capacities while learning how to leverage AI to augment these abilities. Equally important is rethinking what and how we teach the next generation of architects. They must be equipped with both technical literacy and, particularly, critical thinking to be able to discuss architecture generated by the machine.

The chapters in this volume offer a wide array of perspectives and methodologies that address these urgent challenges. Together, they provide a rich collection of insights into how AI can act not as a replacement for the designer, but as a powerful enabler of sustainability. These contributions showcase innovative applications of AI across different contexts, while also grappling with the boundaries and responsibilities of human agency in a machine-augmented design process.

2 Key Lessons

This book informs several aspects of knowledge formation regarding AI in Design. We identified four prominent lessons that permeate, are enabled by, or recommended by these chapters.

- Critical Thinking

Critical thinking has never been more essential to architectural practice than in this era of AI integration. Identified by the World Economic Forum [3] as a crucial soft skill for the future workforce, critical thinking requires deliberate cultivation in design education and professional development. Being truly critical demands that designers effectively synthesize knowledge from architecture history, theory, tectonics and social contexts to evaluate AI outputs rather than accepting them at face value. As AI tools become increasingly sophisticated in generating architectural forms and solutions, architects must develop a heightened capacity to interrogate these outputs—questioning their assumptions, examining their biases, and assessing their appropriateness for specific contexts. Critical thinking also involves metacognition—awareness of one's own thought processes. This self-awareness is essential for maintaining human agency in design processes increasingly mediated by artificial intelligence.

- New Workflows

The integration of AI into architectural practice necessitates the development of entirely new design methodologies. Because AI encompasses a vast array of tools with different capabilities and limitations, practitioners must create, apply, test, evaluate, and assess new frameworks. This methodological innovation has already begun to emerge in academic discourse, as evidenced by recent proceedings' themes related to AI from leading computational design conferences including CAAD Futures 2023: Interconnections, co-computing beyond boundaries, eCAADe 2024: Data-driven intelligence, and CAADRIA 2024: Accelerated design [4–6]. The development of such methods creates a rich space for discourse within the profession—an opportunity to reexamine fundamental assumptions about design thinking and to establish new paradigms that maintain architecture's humanistic core while embracing computational intelligence.

- Historical Knowledge

A sophisticated understanding of AI's historical evolution is essential for its thoughtful integration into architectural practice. Designers must familiarize themselves with the roots of artificial intelligence, its historical relationship to design computing, and the social and economic forces that have shaped its development. Such historical perspective helps practitioners distinguish between genuine innovation and cyclical trends, recognizing patterns that may indicate either promising directions or potential pitfalls. Historical knowledge also provides crucial context for understanding the needs that current AI tools address and the outputs they produce.

Educational institutions have a particular responsibility to situate AI technologies within broader narratives of architectural and technological history, helping students develop a critical perspective that extends beyond immediate technical concerns.

– Interdisciplinary Collaboration

The effective implementation of AI in sustainable architectural design demands unprecedented levels of interdisciplinary collaboration. The complexity of today's environmental and social challenges cannot be addressed by architects working in isolation, regardless of the computational tools at their disposal. Instead, meaningful progress requires deep collaboration across disciplinary boundaries. Enabling architects to co-design with computer scientists is a possibility for new interfacing professionals who can effectively combine and communicate methods from both disciplines. Thus, the designer profession needs to be redefined by the practitioners and professional bodies.

3 Conclusion: A Responsible Future for AI in Sustainable Design

While we have only begun to address the complexity of the challenge in integrating AI into the design process, the contributions in this volume reveal that AI holds significant potential to enhance architectural practice, supporting design thinking, reducing cognitive load as well as accelerating our ability to model and simulate sustainable solutions. Yet this promise comes with an industry wide responsibility. AI's integration into architectural workflows demands more than just technical proficiency; most tools are easy to access, except when it comes to coding and programming AI and ML from a designer perspective. Integration of AI in Design requires critical thinking, ethical awareness, and most importantly, the development of new design methodologies. In addition, architects need to interpret AI-generated outputs, but more so, question the inbuilt assumptions, biases, and appropriateness of AI tools within specific contexts.

AI may or may not radically reshape architectural forms, but it can already be seen as transforming the processes through which architecture is conceived and delivered. This shift redefines professional roles, challenges traditional notions of authorship, and calls for new educational models that combine technical literacy with critical thinking. The urgency of climate and social challenges underscores the need for interdisciplinary collaboration and thoughtful use of emerging technologies. AI should not be seen as a replacement for the designer but as a collaborator—a tool that, when guided by human values, can extend our ability to imagine and create a more sustainable built environment. As with any tool, is interchangeable, adaptable and capable of evolving.

This volume captures a transformational moment in the development of the architecture and design profession. The future of design will not be defined by 'whether' AI is used, but by 'how' we choose to integrate it into our everyday working processes. Because of the complex challenges involved in the profession today, academic and professional bodies need to collaborate on new ways to embrace technology in ethical, purposeful, responsible, and critical ways. Although this volume marks only a small portion of work at the beginning of this exploration, hopefully, we can reflect on this period of time and recognise that we responded with care, creativity, and a renewed commitment to the social and environmental responsibilities of our discipline.

Acknowledgements Thanks to all of the contributors to this book for their valuable time and insight. We hope this work serves as a catalyst for future publications and collaborative efforts in AI-Aided Design (AIAD), fostering deeper critical engagement within the architectural community.

References

1. Leach N (2021). Architecture in the age of artificial intelligence: an introduction to AI for architects. Bloomsbury Publishing
2. Leach N (2018) Design in the age of artificial intelligence. Lands Archit Front 6(2):8–19. https://doi.org/10.15302/J-LAF-20180202
3. World Economic Forum (2025). Future of Jobs Report 2025: Insight Report. Switzerland. https://www.weforum.org/reports/the-future-ofjobs-report-2025/. Accessed 15 Apr 2025.
4. eCAADe. https://cyprusconferences.org/ecaade2024/. Accessed 10 May 2025
5. CAADRIA. https://caadria2024.org/. Accessed 10 May 2025
6. CAAD Futures. https://www.caadfutures2023.nl/132056. Accessed 10 May 2025

FSC
www.fsc.org
MIX
Papier aus verantwortungsvollen Quellen
Paper from responsible sources
FSC® C105338

If you have any concerns about our products,
you can contact us on
ProductSafety@springernature.com

In case Publisher is established outside the EU,
the EU authorized representative is:
Springer Nature Customer Service Center GmbH
Europaplatz 3, 69115 Heidelberg, Germany

Printed by Libri Plureos GmbH
in Hamburg, Germany